The Sun Recorded Through History

Astrophysics and Space Science Library

For other titles published in the series, go to
http://www.springer.com/series/5664

The Sun Recorded Through History

J. M. Vaquero
Universidad Extremadura,
Cáceres
Spain

and

M. Vázquez
Instituto de Astrofísica de Canarias,
Tenerife
Spain

Springer

José M. Vaquero
Universidad Extremadura
Departamento de Física Aplicada
Avenida de la Universidad, s/n
10071 Cáceres
Spain
jvaquero@unex.es

M. Vázquez
Instituto de Astrofísica de Canarias
c/ Vía Láctea s/n
38200 La Laguna
mva@iac.es

ISBN 978-0-387-92789-3 e-ISBN 978-0-387-92790-9
DOI 10.1007/978-0-387-92790-9
Springer Dordrecht Heidelberg London New York

Library of Congress Control Number: 2009921141

Printed on acid-free paper

Springer is part of Springer Science+Business Media (www.springer.com)

Preface

The Sun is nowadays observed using different techniques that provide an almost instantaneous 3-D map of its structure. Of particular interest is the study of the variability in the solar output produced by the dissipation of magnetic energy on different spatial and temporal scales – the so-called magnetic activity. The 11-year cycle is the main feature describing this phenomenon. Apart from its intrinsic scientific interest, this topic is worth studying because of the interaction of such processes with the terrestrial environment. A fleet of space and ground-based observatories are currently monitoring the behaviour of our star on a daily basis.

However, solar activity varies not only on this decadal time-scale, as has been attested mainly through two methods: (a) records of the number of sunspots observed on the solar surface from 1610, and (b) the records of cosmogenic isotopes, such as ^{14}C and ^{10}Be, measured in tree-rings and ice-cores, respectively.

The study of the long-term behaviour of solar activity may be complemented by the study of historical accounts describing phenomena directly or indirectly related to solar activity. Numerous scientific and non-scientific documents have reported these events and we can make use of them as a proxy of solar activity in past times.

In this book we shall review these descriptions of solar activity in the past, providing, on the one hand, primary material for the history of astronomy and, on the other hand, verifying or rebuffing current ideas concerning the time variability of the Sun on the scale of centuries. We shall concentrate on documents that provide information on these topics before the discovery of photography around 1840. Modern drawings will also be included. The lower temporal limit of our study will be set by the archaeoastronomy of prehistoric sources.

The first chapter provides the necessary background on the Sun, with special emphasis on the observing techniques and the influences of the telescope and the Earth's atmosphere on the information obtained from solar observations. A list of books on solar physics is included at the end of this chapter.

Naked-eye observations offered the first possibility to distinguish certain structures, eventually called sunspots, on the apparently pure solar surface. In the second chapter we give an overview of these records and their adequacy to reveal long-term variations of solar activity.

The discovery of the telescope was a turning point in the history of science, with special impact on our knowledge of the Universe and, of course, of the Sun. For centuries the eye and the hand were combined by astronomers to produce excellent drawings of the observed solar structures, most of them on sunspots. This chapter summarizes the work of different solar astronomers until the invention of photography and its application to solar observations. These drawings can be used not only as a tool for informing us about the temporal variation of solar activity, but also to extract physical knowledge about the structures observed. The Wilson effect and the determination of solar rotation are two of these applications described at the end of the chapter.

Chapter 4 is dedicated to one of the most fascinating spectacles given by Nature, total solar eclipses. When the skies were clear, historical documents have always reported these phenomena. In the 18th century, the pioneering work of E. Halley made it possible to forecast solar eclipses with greater accuracy; this, together with the advances in navigation, enabled scientific expeditions to be carried out in order to observe these events.

Since the beginnings of astronomy, astronomers have tried to measure the relevant scales of our accessible vicinity, the Solar System. The development of trigonometry and the art of measuring small angles on the sky were essential tools for this purpose. In Chapter 5, we describe in some detail first the measurements of the solar diameter and then the transits of Mercury and Venus across the solar disk, a phenomenon that for centuries was essential to measuring the Earth–Sun distance. Nowadays, planetary transits in our Solar System are an excellent tool for calibrating current and future observations of exoplanets transiting the disk of other suns.

The mythology of several cultures of the people living in northern latitudes is connected with the aurorae, an event known to originate from transitory phenomena on the Sun. Step by step, the scientists brought this topic to the field of science, showing its relation with transitory events occurring on the solar atmosphere such as flares and coronal mass ejections.

The final aim of the present work is to complement previous studies on the reconstruction of solar activity in the past. The reference to the excellent work made by D.V. Hoyt and K.H. Schatten is our starting point. With this idea in mind, we summarize the available data in the last chapter, proposing tasks to be done in the future.

Many people have been involved, in different ways, in the preparation of this book. At the IAC, R. Castro elaborated and retouched a substantial number of the figures, and the Library staff (M. Gómez and L. Abellán) provided an excellent service in tracing old publications. Parts of this work were written at the CHCUL and IDL-CGUL (University of Lisbon, Portugal).

J.A. Bonet, J. Casanovas, M.C. Gallego, B. Ruiz Cobo, J. Sánchez Almeida, F. Sánchez Bajo, S. Sofia, R.M. Trigo, R. Vílchez Gómez and A. Wittmann have critically read different drafts of individual chapters of the book and gave valuable comments, advice and suggestions.

Figures, data and different suggestions have also been kindly supplied by A. Ardanuy, J.A. Bonet, P. Hingley, J.M. Pasachoff, Y.A. Nagovitsyn, P. Ribeiro, P. Rocher, F.R. Stephenson, I.G. Usoskin, E. Vázquez Dueñas, D.M. Willis, A. Wittmann and H. Wöhl. We would also like to thank the NASA ADS service, which provides a wonderfully efficient service to the scientific community.

Anna Fagan and Robert Chatwin helped to make this book readable in English. We alone, however, bear the responsibility for its content. We thank the Springer staff, especially Dr. Harry Blom, for his confidence in our work. We acknowledge the excellent work done by Lydia Shinoj and her team (Integra) during the production of the book.

Finally, our families showed great patience and gave us their full support during the lengthy process of writing this book, which we dedicate to our wives and children.

Badajoz and La Laguna
November 2008

J.M. Vaquero
M. Vázquez

Contents

1

The Sun

Our Sun is an ordinary star of spectral type G2V. However, there are several reasons why its study deserves special attention. For example, it is the only star where we can directly observe details on its surface; this allows astronomers to test theories of great relevance. Let us mention only one of them: the identification of the process of nuclear fusion in its interior laid the foundation for discovering its age and for understanding the evolution of the stars. Another reason is that the Sun clearly influences the Earth's environment at different time scales, producing events that have impressed both astronomers and laymen alike. The latter aspect is the one we will be addressing in this book.

In this chapter we will provide the necessary background on the Sun for interpreting the knowledge hidden in the historical, scientific and non-scientific documents.

1.1 The Solar Structure

During the 19th century it became evident that the age of the Earth could be estimated as hundreds of millions of years. This stimulated research about possible energy sources that were able to keep the Sun shining for such a long period of time.

Basically, we can make a distinction between the interior of the Sun and its atmosphere. The Sun's interior can be further divided into the following layers, starting from the centre (see Table 1.1):

- *Core*: this is the region where nuclear fusion takes place. Hydrogen is converted into helium and, since the Sun is mainly composed of H and He, its nuclear fuel lasts for 10^{10} years in total. The temperature is about 1.5×10^7 K.
- *Radiative zone*: here the energy is transported outwards by radiation.

J.M. Vaquero, M. Vázquez, *The Sun Recorded Through History*,
Astrophysics and Space Science Library 361, DOI 10.1007/978-0-387-92790-9_1,

Table 1.1. Basic characteristics of the main zones of the solar interior. $\rho^1_{\text{atm,SL}}$ is the density of the Earth's atmosphere at sea level. $R_\odot$ stands for the solar radius

Name	Extension in $R_\odot$	Temperature	Density [g/cm^3]
Core	0–0.25	1.5×10^7–7×10^6	150–20
Radiative zone	0.25–0.70	7×10^6–2×10^6	20–0.2
Tachocline	thin		
Convective zone	0.70–1.0	2×10^6–7×10^3	0.2–$1/10000\rho^1_{\text{atm,SL}}$

- *Tachocline*: in this thin zone, shearing motions occur between the fluid motions of the upper lying convection zone and the stable radiative zone; these motions are able to produce the magnetic fields that eventually emerge at the surface.
- *Convection zone*: because of the lower temperature, atoms become only partially ionized, which increases the opacity and gives rise to convective motions.

We will now briefly discuss the solar atmosphere from which the radiation originates.

The *photosphere* is a layer that is only about 400 km thick and where more than 90% of the solar radiation is emitted (especially in the visible). This layer is often referred to as the solar surface.

Above the photosphere the temperature rises from a minimum of about 4500 K to several 10^4 K in the *chromosphere*. In the subsequent transition zone the temperature increases very sharply to several 10^5 K, and the outermost layer of the solar atmosphere is called the *corona*. The chromosphere and the corona cannot be observed under normal conditions because these layers are very faint in comparison with the solar surface. The first observations of the corona were made during total solar eclipses. The temperature there is several 10^6 K. In Table 1.2 and Figure 1.1 we give the basic parameters of these layers.

Table 1.2. Basic characteristics of the solar atmosphere

Name	Extension	Temperature	Density [g cm^{-3}]
Photosphere	400 km	7000–4500	$\sim 10^{-7}$
Chromosphere	$\sim 10^4$ km	10^4	10^{-12}
Transition region	thin		
Corona	$R_\odot$	10^6	10^{-17}

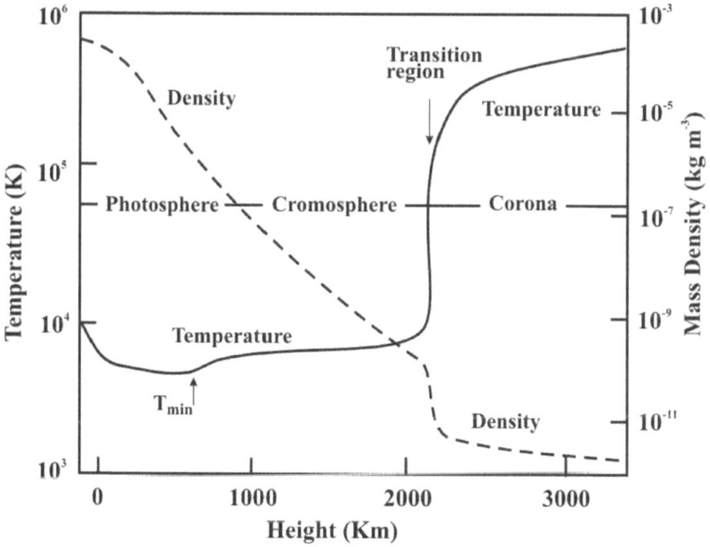

Fig. 1.1. Variation of temperature and density in the solar atmosphere. Adapted from Athay (1976).

1.2 The Photosphere

1.2.1 The Solar Spectrum

The mean temperature of the photosphere is 5770 K. According to this value, the Sun will emit most of the energy in the visible range. At a wavelength of 550 nm its flux outside the atmosphere is $1.96\,\mathrm{J\,m^{-2}\,s^{-1}}$, corresponding to a photon flux of 5.4×10^{18} photons $\mathrm{m^{-2}\,nm^{-1}\,s^{-1}}$.

One of the primary objectives of early solar astrophysics was the measurement of the spectral distribution of solar irradiance. It soon became evident that the terrestrial atmosphere filters out an important part of the solar radiation (Figure 1.2). Table 1.3 summarizes the main atmospheric components contributing to this absorption.

The solar constant, S, is defined as the integrated solar spectral irradiance over all wavelengths. It is given in $\mathrm{Wm^{-2}}$ and corrected to 1 AU.[1] The derived value from daily averages from six satellites over 1978–1998 is $S = 1365.1\,\mathrm{Wm^{-2}}$ (Cox, 2000).

The prism experiment carried out by I. Newton in 1665 opened the possibility to study solar radiation in different colours. One bright sunny day, Newton darkened his room and made a hole in his window shutter, allowing just one beam of sunlight to enter the room. He then took a glass prism and placed it in the sunbeam. The result was a spectacular multicoloured band of

[1] 1 AU = 1 astronomical unit = mean Sun–Earth distance = 149 598 500 km.

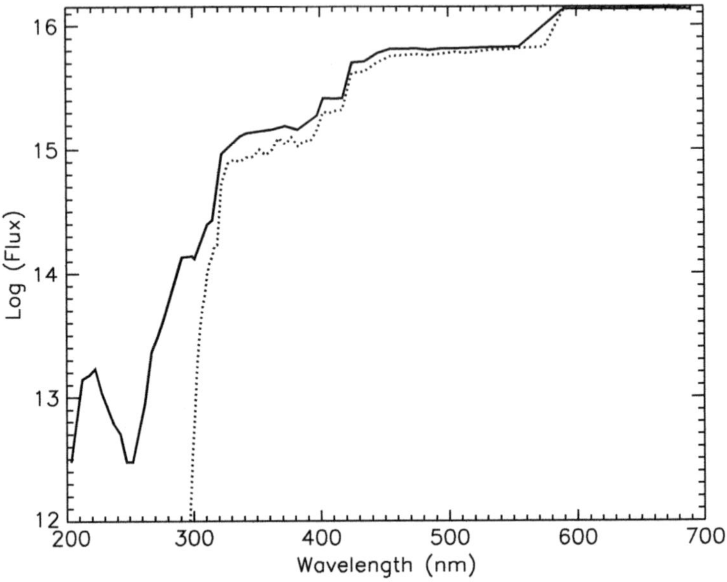

Fig. 1.2. The solar spectrum at an altitude of 40 km (solid line) compared with that recorded at the surface (dotted line).

light just like a rainbow, called a colour spectrum. In a second experiment, he placed another prism upside-down in the way of the light spectrum after passing through the first prism. The band of colours combined again into white sunlight. However, Newton thought that colour was not a physical property but a psychological phenomenon.

William Wollaston (1766–1828) in 1802 published a paper describing a solar spectrum and seven dark lines within it. The importance of these lines was not realized by Wollaston or his readers.[2] He used a slit one-twentieth of an inch wide, and viewed directly through a prism of flint glass held in front of his eye (Wollaston, 1802).

Table 1.3. Main components contributing to the absorption of radiation in the terrestrial atmosphere. Wavelengths are expressed in microns

Absorbing Agent	Absorbing Window
Atomic oxygen, nitrogen	0–0.085 (X - rays)
Molecular oxygen, nitrogen	0.085–0.2 (Far UV)
Ozone (O_3)	0.2–0.35 (Near UV)
CO_2, CH_4, H_2O, NH_3	Infrared bands

[2] Wollaston suggested that the lines were the edges of the primary colours.

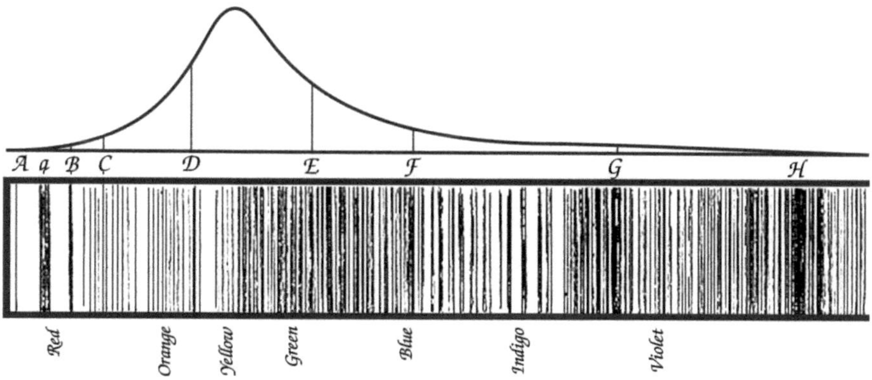

Fig. 1.3. Reproduction of Fraunhofer's original 1817 drawing of the solar spectrum. The more prominent dark lines are labelled alphabetically; some of this nomenclature has survived to this day. From Denkschriften der K. Acad. der Wissenschaften zu München 1814–15, pp. 193–226.

Joseph Fraunhofer (1787–1826) invented the spectroscope and the diffraction grating and in doing so transformed spectroscopy from a qualitative art to a quantitative science by demonstrating how one could measure the wavelength of light accurately. Examining the spectrum of solar light passing through a thin slit, he noticed a multitude (574) of dark lines (Figure 1.3).

The right interpretation of these dark features was done rapidly. John Herschel (1792–1871) demonstrated that when a substance is heated and its light passed through a spectroscope, each chemical element gave off its own set of characteristic bright lines of colour. The combined use of a prism and a narrow slit was the basic design of a spectrograph. The invention of the Bunsen burner, around 1850, and the development of the basic laws of radiation by Robert Kirchhoff (1824–1887) allowed the development of spectroscopy and the distinction between the different types of spectra (continuum, emission and absorption). Figure 1.4 shows one of the first spectroscopes built, by C.A. Steinheil (1801–1870) in Munich.

1.2.2 Limb Darkening and Optical Depth

A very well-known phenomenon on the Sun, visible with even small instruments, is limb darkening. The Sun appears brighter near the centre of its disk than near the limb. When we look at the centre of the solar disk in the visible range, near the centre we look into deep and hence hot regions (the temperature increases with depth). Towards the limb, we get radiation from higher and hence cooler levels (Figure 1.5). This is valid for the visible part of the solar spectrum.

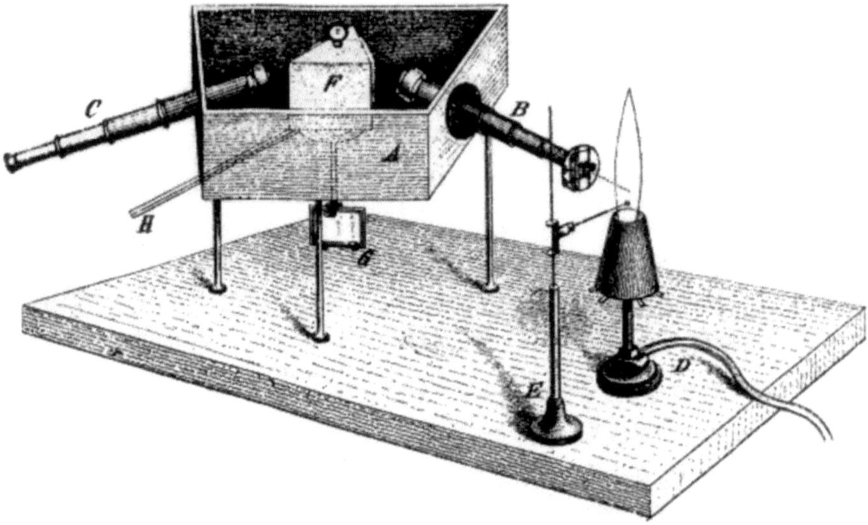

Fig. 1.4. One of the first spectroscopes. Adapted from Kirchhoff and Bunsen (1860).

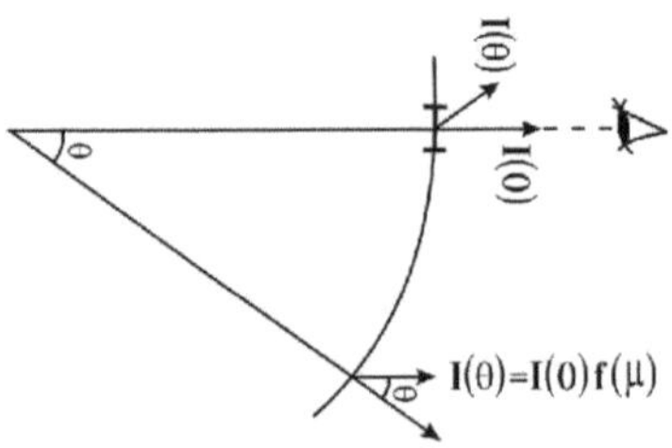

Fig. 1.5. The centre-to-limb variation of the photospheric brightness.

Elste and Gilliam (2007) describe different measurements of this parameter and the associated problems. The explanation of this effect lies in the interaction between the radiation and matter in the solar atmosphere.

The absorption coefficient, which determines how deep we see into the solar atmosphere, increases rapidly toward the blue part of the spectrum. This means that, in the UV, we see higher parts of the solar atmosphere. At observations below $\lambda = 150.0\,\mathrm{nm}$, limb darkening changes to limb brightening. This phenomenon can be interpreted as follows: At wavelengths shorter than 150 nm, we look into areas above the temperature minimum of the Sun, which occurs at a height of about 500 km above solar surface level (see Figure 1.1). In summary, limb darkening is mainly a geometrical effect, but the depth we are observing when we look at the Sun depends also on the properties of the solar material which absorb the radiation.

The optical depth, τ, measures how opaque the solar matter is to radiation passing through it. It is measured along the vertical path, dz, and in stellar atmospheres is defined so that $\tau = 0$ at sufficiently large distances from the star:[3]

$$d\tau_\lambda = -\kappa_\lambda dz = -\kappa_\lambda \cos\theta dh$$

where κ is the extinction coefficient which is wavelength dependent. The coefficient per particle has the units of a cross-section (cm^{-2}); per unit of volume is cm^{-1} and per unit of mass cm^2/g.

The radiation received from the Sun can be expressed as an integral that adds up the contribution of the different photospheric layers

$$\mathrm{I} = \int_0^\infty \mathrm{B}(\tau)\mathrm{e}^{-\tau}\mathrm{d}\tau$$

with $B(\tau)$ the emission of the layer with optical depth τ, usually approximated by the Planck function. From this expression one finds that layers with $\tau \sim 1$ are those contributing to most of the observed signal. When $\tau \gg 1$ (deep layers) then $e^{-\tau} \sim 0$ and no light emerges from these layers. From the Eddington–Barbier approximation, we have $I = B(\tau = \cos\theta)$, where θ is the heliocentric angle. Towards the limb we observe radiation from upper and cooler layers, producing the observed limb darkening.

The absorption spectral lines are formed above the continuum, at heights depending on the atomic transition involved and the physical parameters of the atmosphere.

1.2.3 Granulation

Under excellent observing conditions, the photosphere exhibits a cellular pattern, called granulation, the cells being about 1000 km in diameter and a lifetime of 5–10 minutes (Figure 1.6). Solar granulation is the visible manifestation of the convection zone that lies below the photosphere. Hot matter rises in the bright granules, cools and then descends in the intergranular lanes. Whereas the upflow is relatively smooth, the downflow is more turbulent and in the downflowing areas, turbulent motions occur that can induce shock waves that penetrate into the overlying chromosphere and contribute to its heating.

Granules show a broad range of sizes, with the small ones being more abundant than the larger ones. The contrast of the granulation is given by the standard deviation of the brightness fluctuations in a selected rectangular field.

$$\Delta \mathrm{I}_{\mathrm{rms}} = \sqrt{\frac{1}{NM}\sum_{n=1}^{N}\sum_{m=1}^{M}\left[\frac{I(n,m)}{\overline{I}} - 1\right]^2}$$

[3] Actually, the surface level is defined as $\tau_{500} = 1$ where the subscript refers to the wavelength at which τ is given (500 nm).

Fig. 1.6. Solar granulation observed in white light with the SST at the Roque de los Muchachos Observatory (La Palma). Courtesy: J.A. Bonet (IAC).

where N, M are the dimensions of the granular field and $\overline{I}$ is its mean brightness. Values are wavelength-dependent with a maximum around 13% in the green.

Spectroscopic observations and theoretical development show that granulation is the upper manifestation of the solar convection zone. For monographs and reviews on solar granulation see Bray et al. (1984) and Muller (1999).

Solar convection is also present at other spatial and temporal scales. Supergranulation was first detected as a pattern in the velocity field and the typical cell size is about 30 000 km. In the centre, the upflow is about $50\,\mathrm{m\,s^{-1}}$, the downflow is about $100\,\mathrm{ms^{-1}}$; the lifetime of the supergranular cells is in the order of a day. For a general review on solar convection see Nordlund (2003).

1.2.4 Sunspots

General Characteristics

Sunspots are the oldest known direct manifestations of solar activity. Most consist of a central dark region, known as the umbra (temperature about 4000 K) and a surrounding less dark filamentary region, the penumbra (temperature about 5000 K). Sunspots without penumbra are usually called *pores.* The sunspots are darker than their surroundings because they emit less energy per unit area.

Sunspots appear in groups, and a morphological classification of their evolution in nine classes or steps was proposed by M. Waldmeier (1912–2000) at the Zürich Observatory (Figure 1.7) in 1938. This classification scheme delineates characteristic evolutionary stages of sunspot groups, though not all

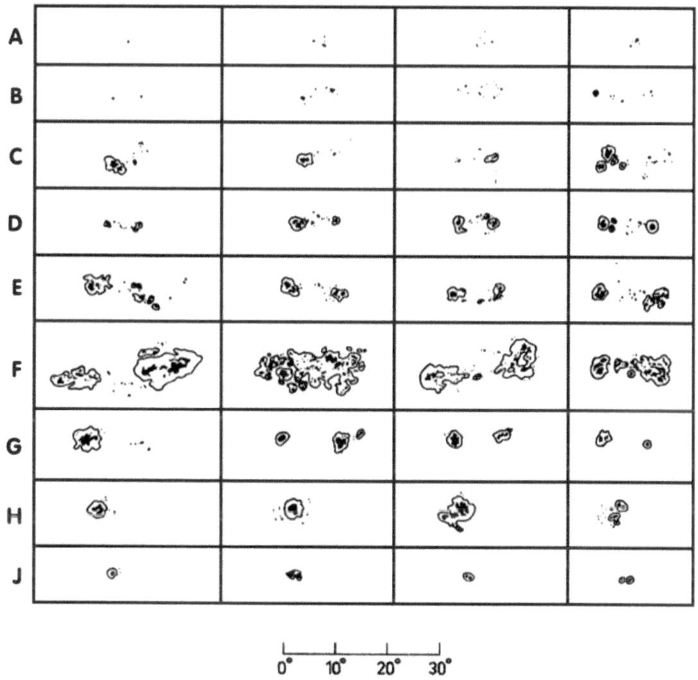

Fig. 1.7. The Zürich morphological classification.

groups go through each stage. Most groups go only part way through the steps and then either rapidly go backwards through the classes or decay to the final class.

Areas and Lifetimes

Half of all sunspot groups have lifetimes, T, of less than two days, and only 10% last for more than 11 days. Waldmeier (1955) derived an empirical formula relating both parameters

$$\mathrm{T(days)} = 0.1\mathrm{A}_{\mathrm{max}}$$

where $\mathrm{A}_{\mathrm{max}}$ is the maximum area expressed in millionths of solar hemisphere.

Fine Structure

Very often the umbrae of individual sunspots within a group are divided into different parts by a bright structure known as a light-bridge (hereafter LB). We will call these individual umbrae "umbral cores" (UCs). A schematic view of the fine structures observed in sunspot umbrae is shown in Figure 1.8.

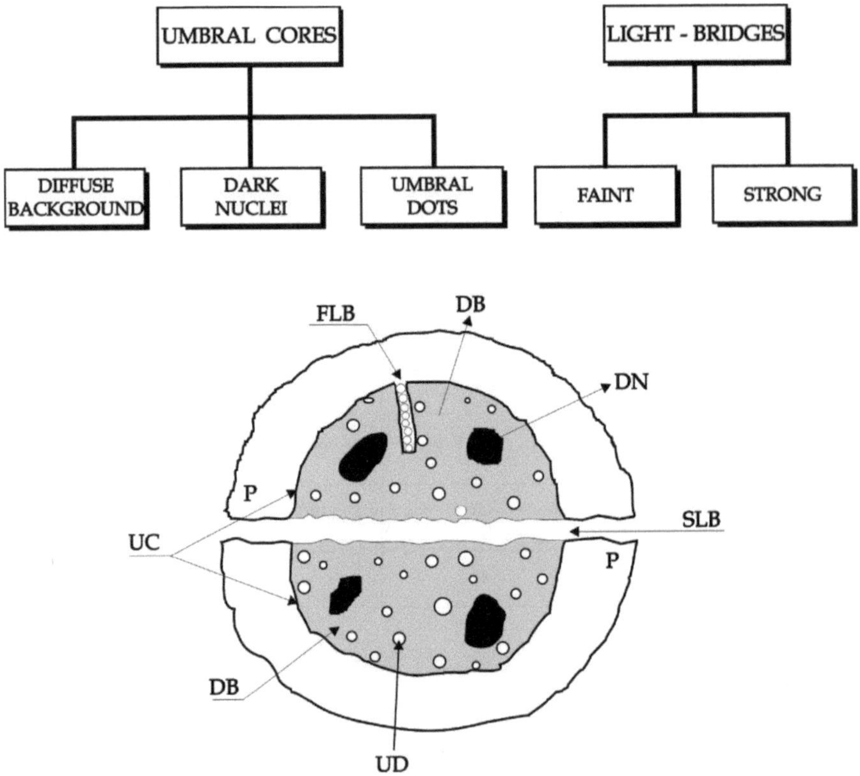

Fig. 1.8. An idealized scheme showing the different fine structures visible in sunspot umbrae. DN (Dark Nuclei), DB (Diffuse Background), UC (Umbral Core), FLB (Faint Light-Bridge) and SLB (Strong Light-Bridge).

Umbral cores have smoothly varying intensities with brighter and darker regions, known as the diffuse background (DB), which has two principal features, the dark nuclei (DN), which correspond to distinctive local intensity minima of the core, and umbral dots (UDs), small bright structures embedded in the diffuse background. Vázquez (1973) presented photographs of the granular structure of light-bridges and their role in sunspot dissolution, obtained with a 15 cm refractor. For modern studies on these structures see Sobotka et al. (1994) and Socas Navarro et al. (2004). Table 1.4 summarizes the main observed properties of umbral fine structures.

The penumbra occupies 85% of the total sunspot area and has on average 75% of the photospheric brightness. Morphologically, it is characterized by dark and bright filaments (see Figure 1.9). Inward proper motions have been observed in the bright elements of the inner penumbra, while in the outer penumbra the proper motions are outwards.

Table 1.4. Characteristics of the main fine structures observed in sunspots (excluding pores). For LBs the size and brightness correspond to the individual grains, and the lifetime to the whole structure. The brightnesses are wavelength-dependent and here are indicated as a fraction of the mean photospheric intensity in the green spectral range

Structure	Size ($''$)	Brightness	Lifetime
SLB	1.2	0.6–1.0	days
FLB	0.5	0.5–0.7	
DN	1.5	0.1–0.4	days
UD	< 0.60	0.2–1.0	minutes

Magnetic Field

The discovery of strong magnetic fields in sunspots by G.E. Hale in 1908 marked a decisive milestone in our understanding of these structures (see Del Toro Iniesta, 1996 for the historical background of this event). A longitudinal magnetic field, B, splits one spectral line at the wavelength λ into two components separated by a distance, $\Delta\lambda_{\mathrm{B}}$. This is the Zeeman effect

$$\Delta\lambda_{\mathrm{B}} = 4.7 \times 10^{-13} \mathrm{g}\, \mathrm{B}\, \lambda^2$$

where g is the Landé factor describing the sensitivity of the spectral line to the magnetic splitting. Its values range between 0 and 3.

Sunspots appear on the solar surface in groups structured as magnetic bipoles. From the magnetic point of view, sunspot groups can be classified as: α, where only a sunspot of one polarity is visible, β, and finally γ, complex

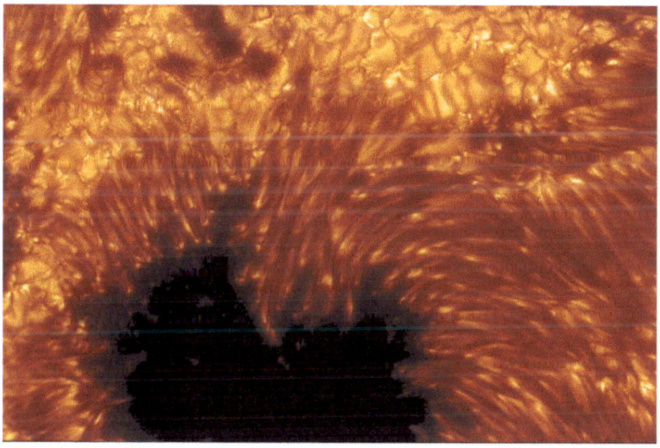

Fig. 1.9. A high-resolution picture of sunspot penumbra obtained at the 1 m Swedish Solar Telescope. Roque de los Muchachos Observatory (La Palma).

active regions in which the positive and negative polarities are so irregularly distributed as to prevent classification as a bipolar group.

Individual spots: The magnetic field strengths range from 1000 to 4000 gauss depending mainly on the sunspot size. It reaches peak values in the darkest part of the umbra where the field lines are generally close to the vertical. A clear correlation exists between magnetic field strength and temperature (Martínez Pillet and Vázquez, 1993).

At the penumbra, the mean field is inclined, becoming almost horizontal at the outer edge. It has long been a topic of debate whether the horizontal magnetic fied is concentrated in the dark or in the bright filaments. Recent measurements indicated a "corrugated" structure of the magnetic field. The magnetic field has an essentially horizontal component that carries the Evershed flow, and a less inclined component. The field strength is weaker where the fields are more horizontal (Thomas, 2000).

Biermann (1941) showed that the strong sunspot magnetic field would impede the convective motions carrying energy from the convective zone. In strong fields, matter can move only along the field lines, thus it is difficult for the material required for convective transport to return.

Parker (1979b) proposed that sunspot cores are composed of individual bundles of magnetic tubes. In order to hold the loose cluster of tubes together, a downdraft beneath the sunspot is needed. Local helioseismology support this hypothesis (Zhao et al., 2001).

Figure 1.10 shows a sketch of the structure of umbral dots within the cluster model. A hot plume of field-free gas penetrates from deep subphotospheric layers up to near the visible surface (shaded area).

Recent reviews on sunspots were given by Martínez Pillet (1997), Sobotka (1999) and Thomas and Weiss (2004).

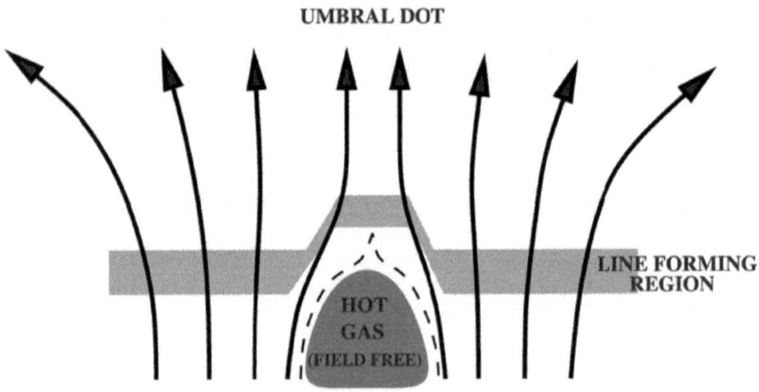

Fig. 1.10. Cartoon representation of an umbral dot in the sunspot cluster model. The solid lines with arrows represent magnetic field lines in the umbra. Adapted from Socas Navarro et al. (2004).

1.2.5 Faculae

Sunspots are usually accompanied by bright structures called faculae (see Figure 1.11). They often precede and considerably outlast the sunspots. The brightening in white light near the disk centre is barely detectable but increases towards the limb. Like sunspots, faculae are associated with strong magnetic fields.

The method of observing faculae near the disk centre is to use narrow-band filters centred on temperature-sensitive lines such as the CN-band at 384 nm (Sheeley, 1969) and the G-band[4] at 430.8 nm (Muller and Roudier, 1984). It was found that these bright structures correspond to small-scale concentrations of the magnetic field (Stenflo, 1966; Livingston and Harvey, 1969). In addition to these magnetic concentrations, there is a diffuse and complex magnetic field that pervades the whole solar photosphere. It is difficult to detect since it leaves almost no brightness signature in images, but it seems to contain a significant part of the solar magnetic flux (see e.g. Sánchez Almeida, 2004).

Table 1.5 shows the relevant parameters of the various magnetic structures observed in the solar photosphere.

A critical point is to understand how brightness is related to magnetic flux, going from bright faculae to dark sunspots. This phenomenon has been simulated numerically by Spruit and Zwaan (1981), who calculated the balance between the inhibition of convective energy transport (strong in large

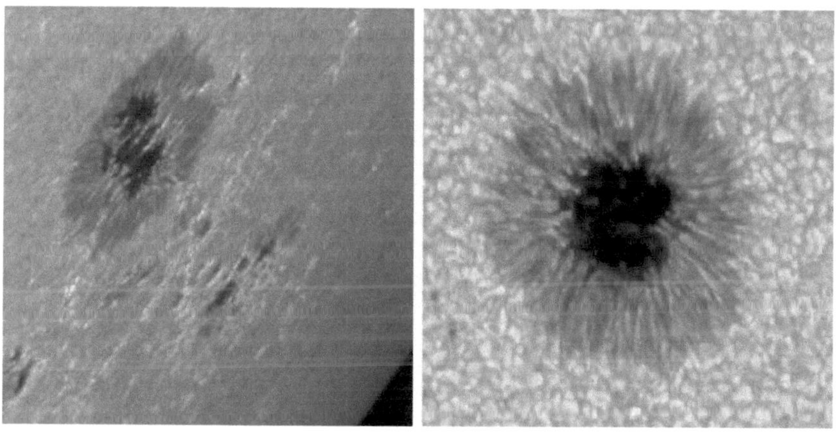

Fig. 1.11. Sunspots near the limb, where also faculae are seen (left) and near the disk centre where the surrounding granulation can be seen. (M. Sobotka, M. Vázquez, J.A. Bonet, A. Hanslmeier, 0.5 m Swedish Vacuum Solar Telescope, La Palma, Observatorio Roque de los Muchachos).

[4] It is called the G-band because it is the "G" feature of the original Fraunhofer spectrum shown in Figure 1.3.

Table 1.5. Hierarchy of magnetic concentrations in the solar photosphere. AR stands for Active Regions and QR for Quiet Regions. Adapted from Schrijver and Zwaan (2000)

	Sunspots	Pores	Faculae	Quiet Sun
Flux (10^{18} Mx)	3×10^4–500	250–25	≤ 20	
Radius (Mm)	28–4	1.8–0.7	~ 0.1	
B (gauss)	2900–2400	2200	1500	0–1500
Cohesion	Compact		In clusters	Very diffuse
Occurrence	Active Regions		QR and AR	Everywhere

magnetic concentrations and in deep layers) and lateral radiative heating from the non-magnetic surroundings, which is substantial in small magnetic structures having less density. They found that the transition between bright and dark structures occurs at sizes around 700 km. More recently, an intermediate family has been found, called dark faculae, which are dark in the centre of the solar disk and bright at the solar limb (see Figure 1.12).

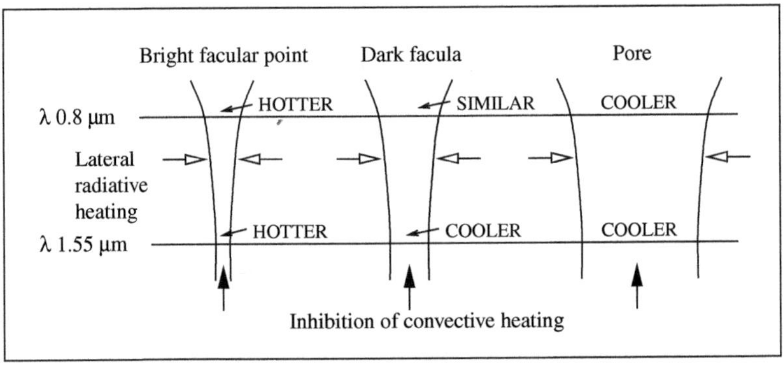

Fig. 1.12. Schematic view of temperature conditions in magnetic features of different sizes. Adapted from Sobotka et al. (2000). At 1.55 and 0.8 μm are located the minimum (deep layers) and the maximum (upper layers) of the absorption continuum coefficient of the solar atmosphere, respectively.

1.3 Observing the Solar Surface

Basic instructions to observe the solar surface are given in Beck et al. (1995), Kitchin (2002) and Macdonald (2003). More specialized monographs are Sánchez et al. (1992), Rimmele et al. (1999), Von der Lühe (2001) and Bhatnagar (2003).

1.3.1 Telescope Basics

A telescope is an instrument that gathers light coming from an object and focuses that light to build up an image. Basically, it consists of a convergent optical system, called the objective. The objective of diameter D forms the image of the object in the focal plane at a distance f (Figure 1.13). The f-ratio = f/D describes the performance of the telescope.

The diameter of the image of the full Sun in the prime focal plane is

$$\mathrm{d} = 2\,\mathrm{f}\mathrm{R}_\mathrm{S}/\mathrm{A}$$

where R_S is the solar radius and A the distance to the Sun. The resulting long focal distances is a major characteristic of many solar telescopes.

The subtended angle in radians is $\phi = d/f$, and since one radian is 206265 arcseconds, then the image scale (i.e. millimetres subtended by one arcsecond) is

$$\mathrm{s} = \frac{f}{206265}$$

Many solar observations were done by the projection method, but soon a second convergent system of lenses, the eyepiece, was added, allowing the enlargement of the primary image of the object for a more detailed study. The angular magnification supplied by the eyepiece is

$$\text{Magnification} = \frac{\text{Focal length of the telescope}}{\text{Focal length of the eyepiece}}$$

If the eyepiece of a telescope is in the right place, the image is "in focus", and will appear sharp. To put the eyepiece in that position, the telescope has a mechanical device called the focuser, which allows you to shift the eyepiece back and forth very precisely, by means of either a couple of focusing knobs using an electric motor, or simply by turning the eyepiece.

The image that comes through the telescope, through the eyepiece and onto the surface of your eye, will appear as a sharply focused disk of light. That disk of light is known as the exit pupil, and its size will vary according to how much magnification the eyepiece/telescope combination is providing. The formula to work out the size of the exit pupil is:

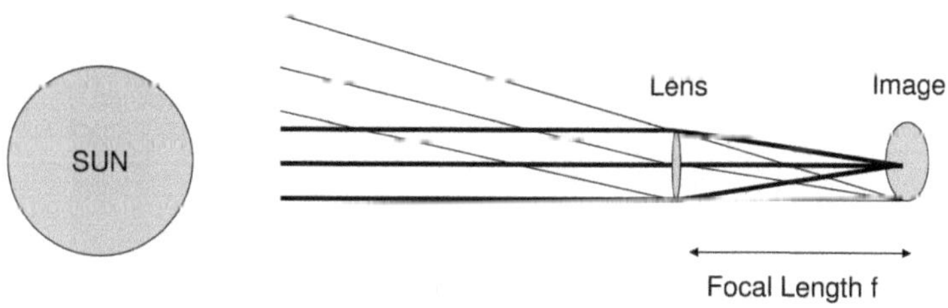

Fig. 1.13. Formation of a solar image by direct projection. Thick lines: rays coming from center of the solar disk. Thin lines: rays coming from the solar limb.

$$\text{Exit Pupil} = \frac{\text{D}}{\text{Magnification}}$$

Two main types of eyepieces have been used during the time covered by this book: (a) Huygens: designed by C. Huygens in 1703, consists of two plano-convex lens with the plane side towards the eye separated by an air gap. The main disadvantages are high image distortion and the narrow field of view. However, they can be very useful for solar projection. (b) Ramsden: designed by J. Ramsden (1735–1800) in 1783, comprises two plano-convex lenses of the same focal length and glass, placed less than one focal length apart. See Rudd (2007) for more details. The telescope field of view, FOV, is given by

$$\text{FOV} = \frac{\text{Eyepiece field of view}}{\text{Magnification}}$$

1.3.2 Image Formation of Extended Objects

Diffraction occurs because of the wave nature of light. The image of a point source is not a point but a disk surrounded by faint concentric rings (Figure 1.14), a pattern called the Airy disk. The size of the Airy disk expresses the maximum angular resolution of the optical system and is given by

$$\Delta\theta = \frac{r_1}{f} = \frac{1.22\,\lambda}{D}$$

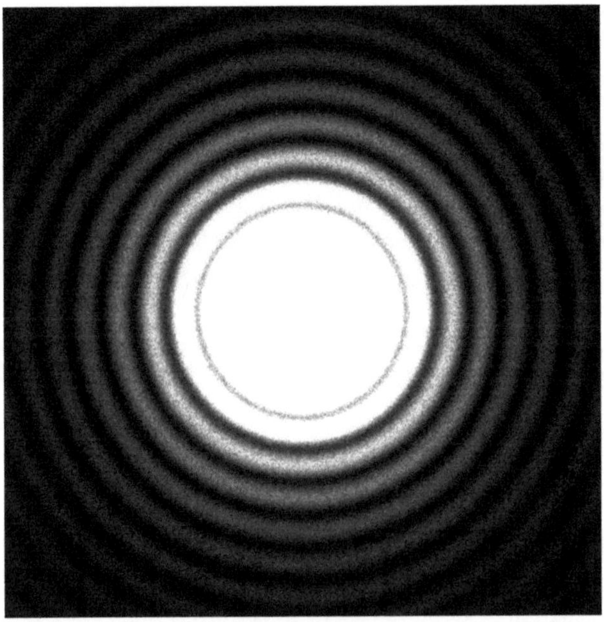

Fig. 1.14. Airy disk.

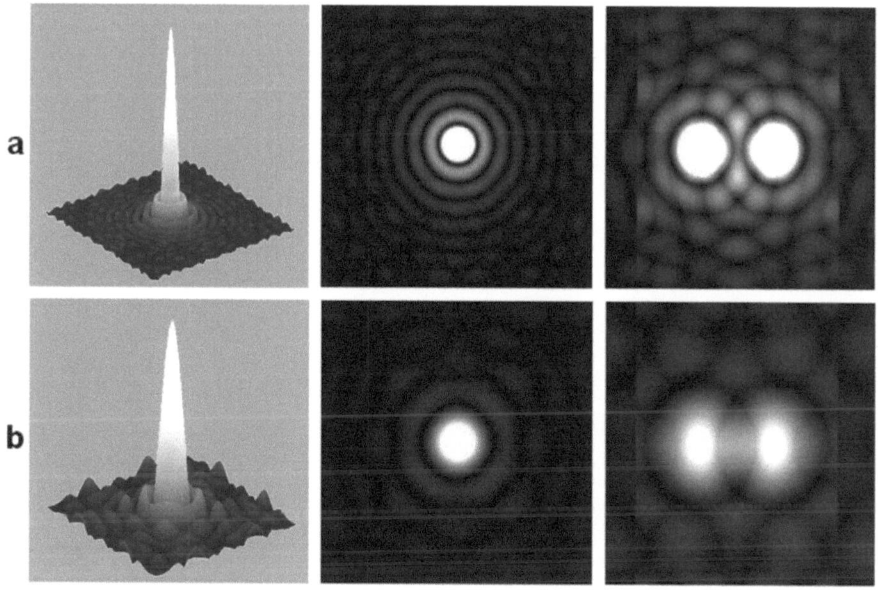

Fig. 1.15. Simulation of the image of a star. A larger telescope aperture (**a**) produces a concentration of light in a smaller space and therefore better spatial resolution. Two stars can be separated whereas with a smaller aperture (**b**) the two stars are seen as a blurred spot. Courtesy: A. Ardanuy (Astronomical Association of Sabadell).

where r_1 is the radius of the first dark Airy ring and $\Delta\theta$ the resolution in radians.

Figure 1.15 shows a simulation of the image of a star and its brightness distribution for two different telescope apertures.

The image of an extended object always suffers a certain degree of degradation when formed through an optical system such as a telescope. Let us imagine a simple case to illustrate how image degradation can be measured. We have as the object a sine wave grating formed by dark and light bars, separated by a distance d, the spatial wavelength (Figure 1.16).

Different parameters are used to describe the spacing of brightness in the objects and images. They are related by the following relations:

$$\mathrm{k} = \frac{\omega}{\mathrm{c}} = \frac{2\pi\nu}{\mathrm{c}} = \frac{2\pi}{\mathrm{d}}$$

where k is the wavenumber, ω is the angular frequency (rad/sec), ν the spatial frequency and c the speed of the propagation.

The modulation M of the light is measured by the function

$$\mathrm{M} = \frac{\mathrm{I_{max}} - \mathrm{I_{min}}}{\mathrm{I_{max}} + \mathrm{I_{min}}}$$

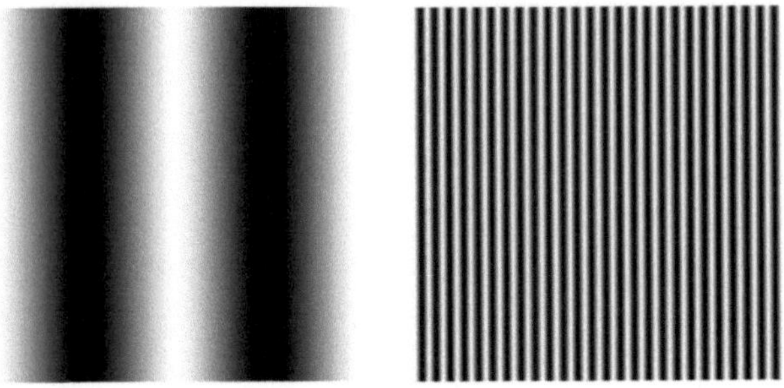

Fig. 1.16. Two sine waves with different spatial scales. The brightness profile across the patterns is a sinusoid.

where I_{max} and I_{min} are the extreme values of the image brightness.

However, we have considered an ideal case and most of the objects can be considered as being composed of an infinite array of spatial frequencies, ν, of all orientations, each with a specific amplitude and phase. The resulting image is the overlapping of different frequencies.

The Modulation Transfer Function (MTF) indicates the ability of an optical system to transfer various levels of detail (spatial frequencies) from the object to the image. It gives the reduction in contrast in the image with respect to the object as a function of spatial frequency

$$\mathrm{MTF}(\nu) = \frac{\mathrm{M(image)}}{\mathrm{M(object)}}$$

The smaller the spacing, the smaller the contrast of the image made by the telescope. Eventually the contrast is so low that we cannot distinguish between bright and dark lines and the structure becomes unobservable. Below is a mathematical illustration of this point.

We can imagine an extended object composed by incoherent point sources with respective intensities $I(\xi, \eta)$. Assuming that the optical system is linear, the intensities of their respective images, I', are added at the focal plane of the telescope and the intensity distribution in the resulting image will be

$$I'(x, y, t) = \int\int I(x', y')S(x, y; x', y')\, dx'dy' = I(x, y) * S(x, y, t)$$

Figure 1.17 illustrates how the image of a point located at (ξ, η) will not be another point but an extended disk whose normalized intensity distribution is denoted by S, the *Point Spread Function* (PSF). In fact, the extended disk is composed by the superposition of individual images of different point sources (see Figure 1.17b).

Fourier theory states that any image or object can be expressed as a sum of sinusoids indicating the variations in brightness across the image:

$$F(x) = a_0 + \sum_{n=1}^{\infty} \cos(\omega_n x - \phi)$$

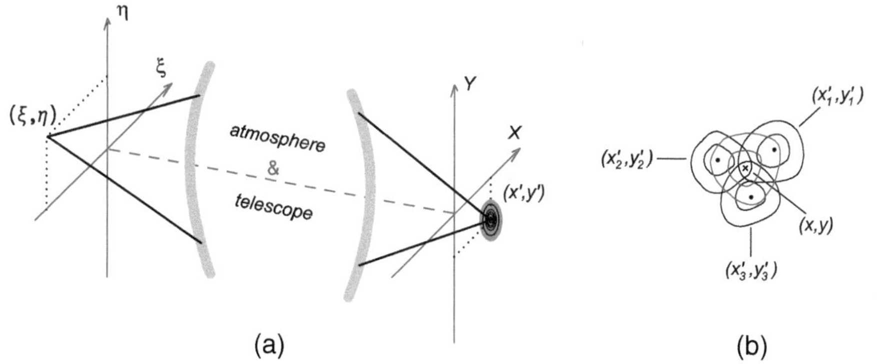

Fig. 1.17. Image formation of an extended object.

A transform is a mathematical operation for transferring information from one domain to another. Jean Baptiste Fourier (1768–1830) applied this technique to transfer from the spatial to the frequency domain. We should point out that high (low) frequencies correspond to small (large) details in the image.

The magnitude of the maximum/minimum peak corresponds to its contrast and the phase represents how the signal is shifted relative to the origin. A Fourier transform encodes all the spatial frequencies present in an image. The Fourier transform of a sine wave as shown in Figure 1.18 is plotted as a single peak at point ν along the spatial frequency axis; the height of the peak is related to the contrast. There is also a "DC term" corresponding to zero frequency, which represents the average brightness across the whole signal. In other words, the Fourier transform is a means of transforming a signal defined in the spatial domain to the spatial frequency domain.

Making use of Fourier analysis, the last expression of the brightness distribution results in a simple product in the spatial Fourier domain

$$\mathrm{i}'(\mathrm{u},\mathrm{t}) = \mathrm{i}(\mathrm{u}).\mathrm{s}(\mathrm{u},\mathrm{t})$$

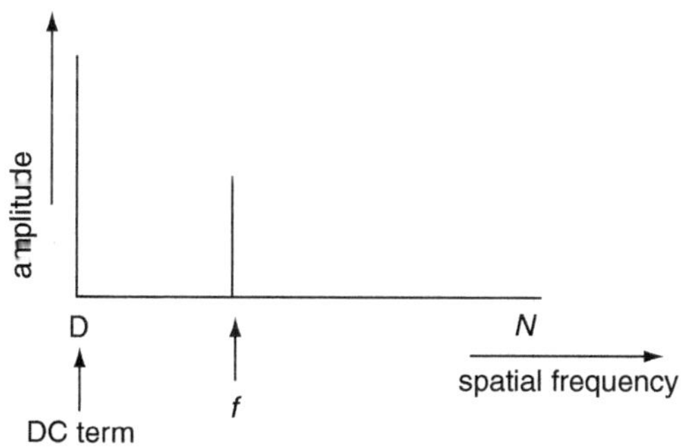

Fig. 1.18. A sine wave in the Fourier domain.

The function s is referred to as the *Optical Transfer Function* (OTF) and its modulus is the *Modulation Transfer Function* (MTF) of the imaging system. The MTF of a perfect telescope is given by

$$\mathrm{MTF_T}(\nu_\mathrm{r}) = \frac{2}{\pi}\left\{ \arccos\left(\frac{\nu_r}{\nu_\mathrm{max}}\right) - \frac{\nu_\mathrm{r}}{\nu_\mathrm{max}}\sqrt{\left[1-\left(\frac{\nu_\mathrm{r}}{\nu_\mathrm{max}}\right)^2\right]}\right\}$$

where ν_max (cycles/arcsec) $= (D/\lambda)(1/2006265)$ is the highest transmitted frequency for the telescope and therefore $\mathrm{MTF}(\nu_\mathrm{max}) = 0$. This resolution expressed in angular distances (arcseconds) is given by

$$\mathrm{a}('') = 1.22\frac{\lambda}{\mathrm{D}}\,206265$$

Figure 1.19 shows radial cuts of the PSF and MTF of two telescopes of different diameters (50 and 100 cm). The corresponding FWHM (full width at half maximum) of the PSF are 0.23 and 0.11 arcsec respectively.

Let us further illustrate the effects on the contrast of the diffraction, discussed above in the basics of telescope optics. For this purpose, we have degraded an image of the solar granulation, assumed to be perfect, with MTFs of telescopes with decreasing apertures. Figure 1.20 shows how the contrast of the granules strongly decreases when observed with smaller telescopes. The granules also become roundish, losing small-scale details (high spatial frequencies).

The collecting area of a telescope increases with the square of the diameter, D^2, while the area of the solar surface covered by a resolution element, the Airy disk, decreases with the same power. Therefore both effects cancel each other out.

In describing the possible loss of light in the optical system, it is relevant to come back to the concept of pupil. The entrance pupil refers to the aperture

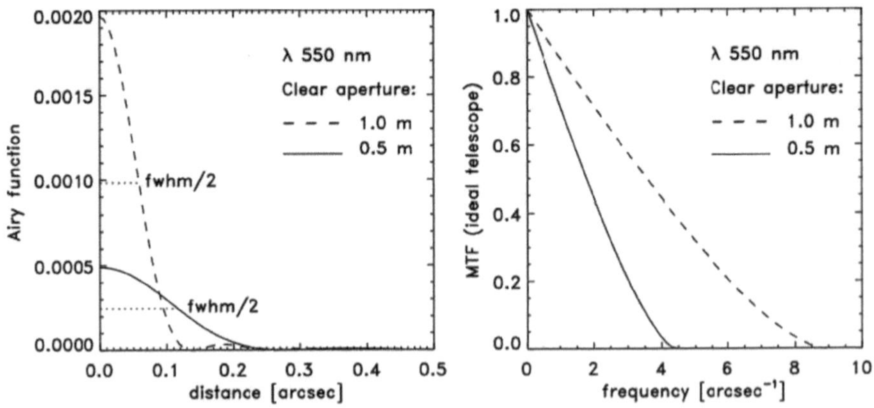

Fig. 1.19. PSF and MTF of telescopes with different diameters (50 and 100 cm) observing at a wavelength of 550 nm. Courtesy of J.A. Bonet (IAC).

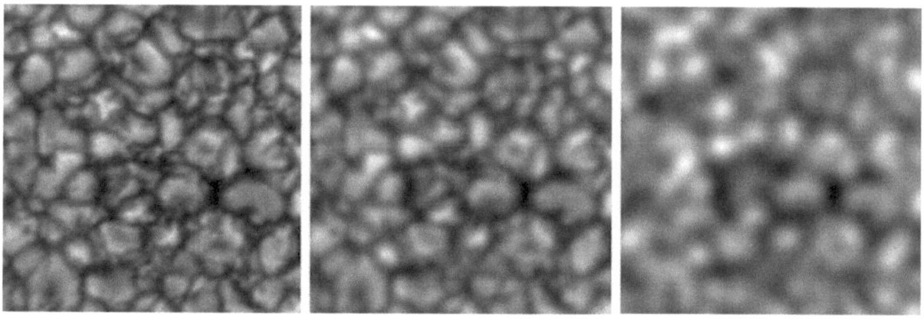

Fig. 1.20. Image of the solar granulation obtained with the 50 cm SVST at the Roque de los Muchachos (left) and subsequently degraded to a telescope of 25 cm (centre) and 10 cm (right). Courtesy of J.A. Bonet (IAC).

through which light enters the telescope, often the lens itself. The exit pupil is a small circle just behind the telescope through which all emerging rays pass. You can see this exit pupil as a little disk of light floating in the air behind the eyepiece. Its size is given by

$$\text{Size Exit Pupil} = \frac{\text{Telescope Aperture}}{\text{Magnification Eyepiece}}$$

In order to avoid loss of light, the eyepiece must be designed in such a way that its exit pupil should have a size similar to the pupil of the human eye, which normally has an opening about 5 mm in diameter in subdued daylight. The pupil may contract to as little as 2.5 mm in bright light, and it may open to 8 mm when the eye has adapted to darkness. Magnifications smaller than values equal to the telescope's diameter will result in a bundle of light larger than the pupil, thus losing image brightness. Magnifications larger than about 10 times the aperture will result in a bundle so small that the image quality will suffer.

See Dainty (1975), Bonet (1999), Von der Lühe (2001) and Schmidt (2000) for a more detailed explanation of this topic. In the next section, we will describe the effects of the rest of the factors influencing the quality of a solar image, starting with the imperfections in the telescope optics.

1.3.3 Telescope Aberrations

The telescope is not a perfect instrument. Apart from diffraction, image aberrations must be taken into account. In 1857 Phillip Ludwig von Seidel (1821–1896) decomposed the first-order aberrations into five constituent aberrations.

Chromatic aberration: In optical systems composed of lenses, the position and size of the image depend upon the refractive indices of the glass. Since the index of refraction varies with the colour or wavelength of the light, it follows that a system of lenses projects images of different colours in somewhat different places and sizes, blurring the images.

Chester Moor Hall (1703–1771) found that by combining lenses of different kinds of glass, the chromatic aberration could be corrected. In 1733 he constructed the first real achromatic telescope.[5]

Spherical aberration: Image imperfection produced by the increased refraction of light rays that occurs when rays strike a lens near its edge, in comparison with those that strike nearer the centre (Figure 1.21). When the image is recorded in a plane, e.g. by the human eye, this causes blurring of the image that changes within the image. Spherical aberration can be minimized by careful choice of the curvature of the surfaces (convex and concave lenses).

The effect is proportional to the fourth power of the diameter and to the third power of the focal length. The criterion to calculate whether a telescope is free of this aberration requires that

$$\mathrm{D} \leq 512\lambda\,\mathrm{N}^3$$

where D is the telescope diameter and N the f-ratio. Spherical main mirrors are used for solar telescopes, which cause spherical aberrations.

Coma: This is encountered if we move off the optical axis of the telescope. The image of a point is distorted in a cometary-like shape. The correction for coma can be included in the multiple lenses that are needed for chromatic and spherical aberration. A lens that is corrected for both spherical aberration and coma is called an aplanatic lens. The angular radius of the usable field (in radians) is

$$\omega_0 = 19.2\,\lambda\,\mathrm{N}^2/\mathrm{D}$$

Astigmatism: This also involves rays coming at an angle to the optical axis of the collector. Now, however, we distinguish whether the rays are off-centre "up-down" or "side-to-side". The "up-down" is called the tangential plane and the "side-to-side" is called the sagittal plane. Astigmatism is present if the rays in the two planes are focused at different points. The angular radius of the usable field (in radians) is now

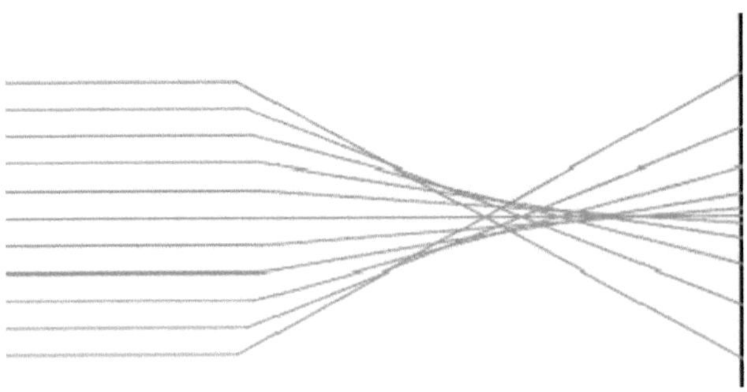

Fig. 1.21. Spherical aberration in an optical system.

[5] Typically, an achromatic system is a doublet consisting of two lenses, one convergent and the other divergent.

$$\omega_0 = (2.7\,\lambda\,\mathrm{N/D})^{1/2}$$

Distortion: The shape of the image is not a true copy of the object, even though it may be in sharp focus. The two kinds of distortion are barrel, in which the image is bowed convex around the centre line, and pin cushion, in which the image is bowed concave around the centre line.

Curvature of field: The image does not fall on a flat plane. Instead, the surface of sharp focus is a curved surface.

1.3.4 Atmospheric Seeing

Fundamentals and Measurement

Solar images observed from the ground are degraded by the Earth's atmosphere. The influence of this medium on image formation is similar in effect to intercalating changeable and poor lenses in the optical path.[6] The atmospheric turbulence produces temporal and spatial fluctuations of the air temperature, ΔT. The refractive index of the air, n, depends on the temperature as

$$\mathrm{n} = 1 + 7.6 \times 10^{-5}\frac{\mathrm{p}}{\mathrm{T}}$$

where the pressure, p, is given in millibars and the temperature in degrees kelvin.

Therefore, light rays coming from the object cross atmospheric bubbles characterized by different refractive indexes. These fluctuations, Δn, are Gaussians to a good approximation:

$$\Delta\mathrm{n} = 7.6 \times 10^{-5}\frac{\mathrm{p}}{\mathrm{T}^2}\Delta\mathrm{T}$$

The resulting image is formed by rays that arrive with different delays (Figure 1.22).

Statiscally, the temperature fluctuation can be described by the so-called structure function (Tatarski, 1961).

$$\mathrm{D_T(r)} = \langle[\mathrm{T}(\mathbf{x},\mathrm{t}) - \mathrm{T}(\mathbf{x}+\mathbf{r},\mathrm{t})]^2$$

where the only dependence of D_T on the magnitude of r reflects the fact that the ΔT are homogeneous and isotropic:

$$\mathrm{D_T(r)} = \mathrm{C_T^2 r^{2/3}}$$

where C_T^2 is the structure coefficient of the thermal field.

---

6 In his *Micrographia* (Observation LVIII), Robert Hooke first suggested the existence of *small, moving regions of atmosphere having different refracting powers which act like lenses.* This book was reprinted in 1961 by Dover.

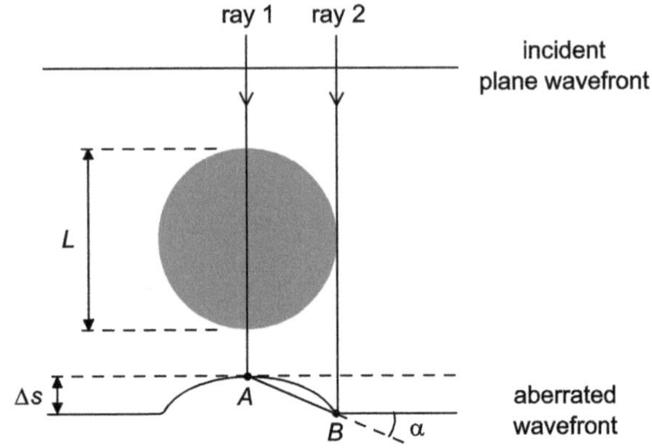

Fig. 1.22. Degradation of a plane wavefront intercepted by a bubble of turbulent air.

Likewise, the variation of the refractive index is described by the corresponding structure function

$$D_n(r) = \left(7.6 \times 10^{-5} \frac{p}{T^2}\right)^2 C_T^2 r^{2/3} = C_n^2 r^{2/3}$$

where p is expressed in millibars and T in kelvins, and C_n is the index structure constant. This dependence of atmospheric turbulence on two-thirds power is known as the Kolgomorov–Obukhov law.

The influence of atmospheric turbulence on image quality is usually represented by the Fried parameter, r_0 (Fried, 1965),

$$r_0 = 0.18\,\lambda^{6/5}\,(\cos z)^{3/5}\left[\int C_N^2(h)dh\right]^{-3/5}$$

The optical performance of a telescope observing through a turbulent atmosphere characterized by C_N^2 is equivalent to that of a diffraction limited telescope with an aperture diameter of r_0. Since the Fried parameter characterizes the resolution in long-exposure images, the resolution in short-exposure images (such as the visual ones) may still be much better than that indicated by these figures.

The effects on the images produced by the atmospheric perturbation is usually described as the result of three separate contributions: (a) a defocusing or loss of sharpness in the observed structures, known as blurring, (b) a global displacement of the image which produces an agitation referred to as image motion, and (c) distortion of the structures caused by differential image motion in distinct parts of the image. These degrading effects are referred to in the literature as seeing. The actual value of the seeing is very variable with atmospheric conditions.

From many site testing campaigns, it is now accepted that the most favourable site for a solar observatory is a high mountain (above the inversion layer) surrounded by a large water surface. The observatories in the Canary Islands (Tenerife and La Palma) are tangible proof of this concept, first proposed by I. Newton in his *Optiks* (1730, Fourth Edition, Book One, part I, pages 110–111).[7]

In short, each component of a complex optical system is described by its own MTF, and the total MTF is the product of each one of them,

$$\mathrm{MTF_{tot}} = \mathrm{MTF_T} \times \mathrm{MTF_S} \times \mathrm{MTF_D}$$

where the indices denote diffraction and telescope aberrations (T), atmospheric seeing (S) and detector (D).

1.4 The Chromosphere

The chromosphere is an irregular layer above the photosphere and it can be observed during a total solar eclipse or in the light of spectral lines formed there. The temperature rises in the chromosphere from the temperature minimum up to 20 000 K and the layer is approximately 2000 km thick. It is highly structured in three dimensions and very dynamic. In the chromosphere, between the thin photospere and the extended corona, the physics passes from being hydromagnetically dominated to magnetically dominated.

1.4.1 Spectral Lines

Line formation and radiative transport in the chromosphere has been reviewed by Kneer and von Uexküll (1999), Uitenbroek (2006) and Rutten (2007). Table 1.6 lists a set of spectral lines formed in the chromosphere. The most prominent ones are:

H_α: the core appears at a wavelength of 656.2797 nm and is formed between 1200 and 1700 km above the solar surface (which is usually defined as the layer where the optical depth at a wavelength of 500 nm becomes unity). The wings are produced between 100 and 300 km (Vernazza et al., 1981).

Ca II doublet (H and K):[8] These are absorption features centred at 396.8 nm (H1) and 393.46 nm (K1), which originate in the photosphere (see Figure 1.23). The so-called plages are hotter than the surrounding areas, and their number and area increase with the level of solar activity. This increase causes the enhancements of the emission cores (H2, K2), where a self-reversal is produced (H3, K3). The formation heights according to

[7] Long telescopes may cause objects to appear brighter and larger than short ones can do, but they cannot be formed so as to take away that confusion of the rays which arise from the tremors of the atmosphere. The only remedy is a most serene and quiet air, such as may perhaps be found on the tops of the highest mountains above the grosser clouds.

[8] H and K as originally labelled by Fraunhofer.

Table 1.6. List of chromospheric indicators

Spectral Line	Wavelength (nm)	Temperature
Continuum	160.00	4.2×10^3
Continuum	200 μ–1 mm	$0.8–1.5\times10^3$
Ca II (H,K)	393.300 ; 396.800	6.0×10^3
Ca II IR	848.200 ; 856.200 ; 866.200	6.0×10^3
H_α	656.300	
He I	1083.000	
He I	587.600	
Mg II (h,k)	279.553 ; 280.270	7.0×10^3
Si II	180.801	6.2×10^3
O I	130.200 ; 130.500	8.0×10^3
C II	133.453 ; 133.566	2.0×10^4
H I (Ly$_\alpha$	121.567	1.5×10^4
Si III	130.100	3.5×10^4
C III	117.5	7.0×10^4
Si IV	139.375 ; 140.277	7.3×10^4
He II	164.000	8.0×10^4

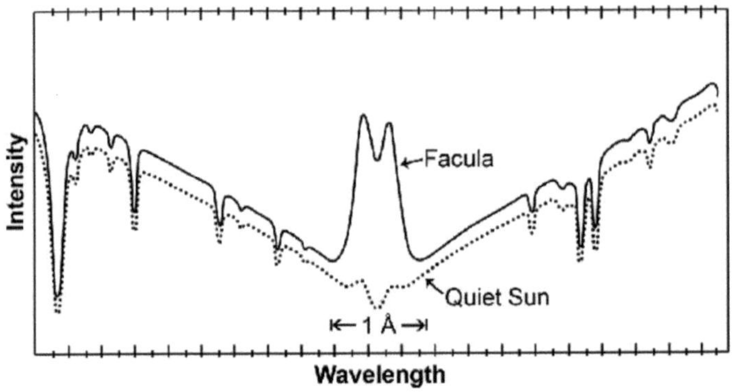

Fig. 1.23. Average profiles of Ca II K for an active region and a nearby quiet region. Adapted from White and Livingston (1981) Fig. 1b.

Vernazza et al. (1981) are 450–650 km (at 0.15 nm from the core, K1), 700–1450 km (K2), and 1800–2000 km (core, K3). This emission is seen in other solar-type stars, thus providing important information about their chromospheres and activity cycles. Figure 1.24 shows full disk images in Ca II K and H_α.

We next describe the most important observable structures in the chromosphere (see Bray and Loughhead, 1974 for a monograph).

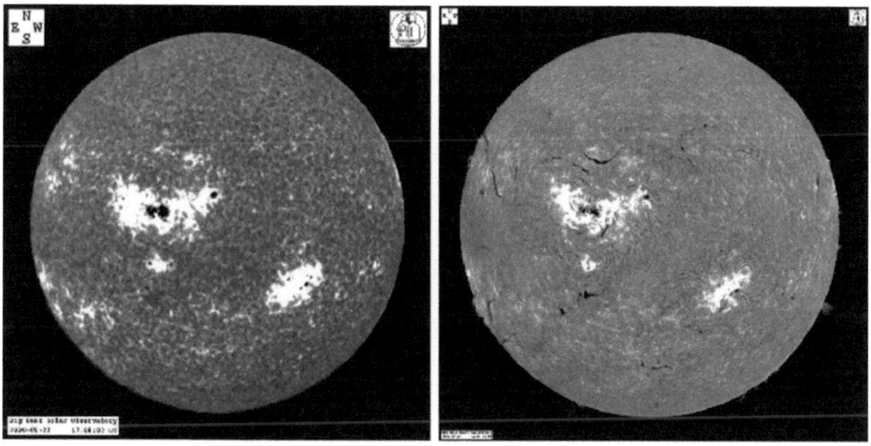

Fig. 1.24. The chromosphere seen in two spectral lines formed in the chromosphere: Ca II K and H_α. Courtesy: Big Bear Solar Observatory.

1.4.2 Plages and the Chromospheric Network

The chromospheric network is a cellular web-like pattern most easily seen in H_α and Ca II K emission. Simon and Leighton (1964) found that there exists a chromospheric emission network that coincides with the borders of supergranules and the photospheric network. The spicules are small, jet-like eruptions seen throughout the chromospheric network. They last for a few minutes and eject material into the corona at 30 km s^{-1} to a height of about 7000 km. They tend to occur at the borders of the supergranules.

Plages are regions hotter and denser than the normal chromosphere. They can be observed in the monochromatic light of spectral lines formed in the chromosphere, H_α and Ca II being used most often. Their typical lifetime is up to 40 days. It is known that they are related to magnetic fields of up to 2000 gauss. They present sizes up to 50 000 km.

In order to keep the equilibrium of pressures, the gas pressure inside magnetic flux tubes, P_g (in), is lower than the pressure in the surrounding photosphere, P_g (out), at all the heights, being compensated by the magnetic pressure, P_m:

$$\mathrm{P_g(out) - P_g(in) + P_m}$$

As the gas pressure decreases with height, the field lines will expand and the magnetic field strength will decrease. Since the magnetic flux ($\Phi = $ B.A) must remain constant, the cross-section of the tube must increase, thereby explaining the larger extension of chromospheric active regions (Figure 1.25).

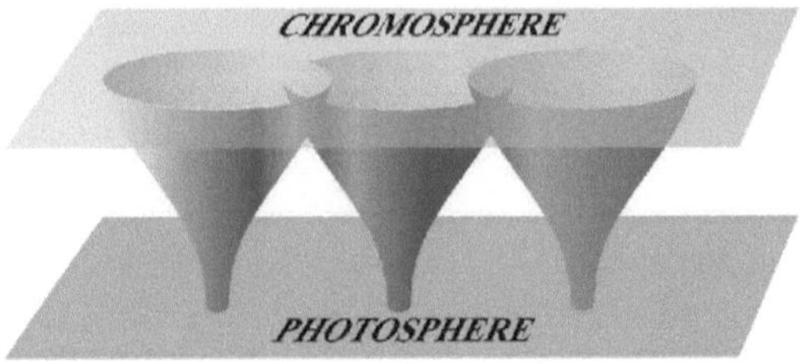

Fig. 1.25. Expansion of magnetic field lines from the photosphere to the chromosphere. The photosphere is covered with flux tubes of intrinsically strong magnetic field, forming a magnetic canopy at chromospheric heights.

1.4.3 Quiet Chromosphere

In principle, the quiet chromosphere coincides with regions free of organized magnetic fields, located in the interior of supergranules. In this view, the heating of these layers occurs mainly through acoustic waves. However, polarimetric observations indicate an average field larger than 20 G, and that an important part has likely remained undetected (Stenflo, 1982; Sánchez Almeida and Lites, 2000; Khomenko et al., 2005). Sánchez Almeida (2007) has presented a model for the distribution of magnetic fields in this region.

Most of the chromospheric lines are broad and therefore magnetic fields in this region are difficult to diagnose with the the Zeeman effect. Other physical mechanisms must be considered to measure the magnetic field (e.g. the Hanle effect).

1.4.4 Prominences

Prominences are cloud-like condensations in the lowest part of the corona with temperatures of about 10 000 K and densities a hundred times that of the surrounding corona. They range in size from a few thousand to two hundred thousand kilometres in length and they are ten-thousand kilometres high and wide. They are observed in emission near the limb (Figure 1.26) and in absorption as dark filaments across the solar disk; therefore, they must be cooler than the surroundings.

Eruptive prominences rise rapidly from the solar surface, and quiescent prominences have a much longer lifetime, typically several solar rotation periods, hanging suspended by magnetic fields in loop-like structures. The magnetic loops that support quiet prominences change slowly, to reach a point where they are not longer stable. Then, filaments and prominences can erupt and rise from the solar surface in minutes or hours. Active prominences are associated with sunspots and flares.

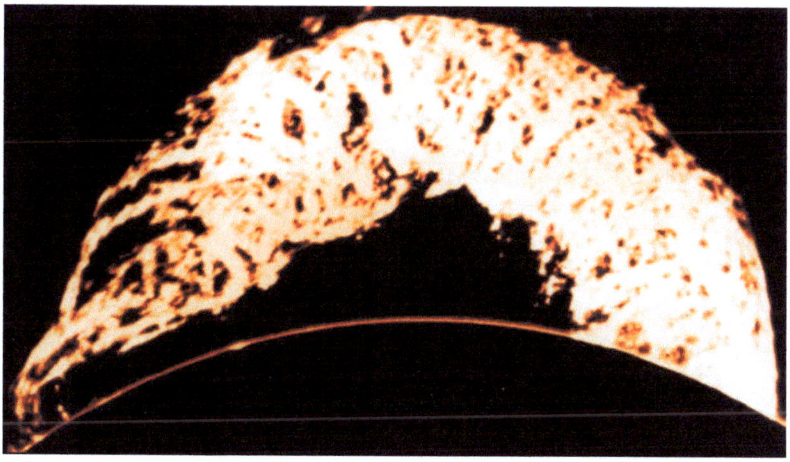

Fig. 1.26. Huge prominence observed on June 4, 1946 by W.O. Roberts (High Altitude Observatory, Colorado).

1.5 The Corona

The corona is the outermost layer of the solar atmosphere. It is visible during total eclipses of the Sun as a pearly white "crown" surrounding the Sun and its shape is strongly dependent on the phase of the solar activity cycle. At solar maximum, the corona appears more symmetric around the occulted solar disk. During activity minimum it is more elongated about the solar equator.

The spectrum of the corona shows that it is formed by three components: (a) *K corona*, the result of light from the solar photosphere being scattered by electrons within the corona, (b) *F corona*, solar light scattered by dust particles and (c) *E corona*, radiation emitted by the corona itself.

The identification of the puzzling coronal spectrum in the visible gave rise to the discovery of a new chemical element apparently unknown on Earth, at first named coronium. It was realized later that the spectrum could be explained by forbidden lines of a high degree of ionization of known elements, such as Fe XIV and Ca XVI, implying temperatures of several 10^6 K (Grotrian, 1939; Edlén, 1943).[9] The density of the corona is extremely low and collisions occur infrequently, thus allowing the observation of forbidden transitions.[10]

[9] Fe XIV means iron 13 times ionized, in other words having lost 13 electrons.

[10] The ion can spend an unusually long time in an excited state.

Visual observations during total eclipses show *coronal streamers*, huge structures covering 150 degrees of solar longitude and spreading over 45 degrees in latitude. They have the shape of a flat blade.

Coronal loops are found around sunspots and in active regions and are associated with the closed magnetic field lines that connect magnetic regions of opposite polarities on the solar surface (Bray et al., 1991). Loops can last for days or weeks but most of them change quite rapidly. Some loops are associated with solar flares and are visible for short periods. These loops contain material denser than their surroundings (see Figure 1.27).

If the corona is observed in X-rays, it does not appear to be uniform: there are dark regions of low temperature called *coronal holes*. These holes are particularly prominent during activity minima and near the solar poles (see Figure 1.28). Coronal holes were discovered by M. Waldmeier, in 1957, who was observing visible coronal lines, as missing coronal material.[11] Their existence was confirmed years later by X-ray instruments on board Skylab. The typical parameters of coronal structures are summarized in Table 1.7.

Fig. 1.27. This image of coronal loops over the eastern limb of the Sun was taken in the TRACE 17.1 nm passband, characteristic of plasma at 10^6 K, on 6 November, 1999, at 02:30 UT. Courtesy: TRACE team and the Lockheed – Martin Solar and Astrophysics Lab.

[11] Waldmeier, M., 1957, *Die Sonnenkorona II*, Verlag Birkhauser, Basel.

Fig. 1.28. Full disk X-ray image obtained by Yohkoh. A polar coronal hole is clearly seen in the upper part. Courtesy: T. Shimizu. The Institute of Space and Astronautical Science, Japan.

Table 1.7. Physical parameters of the principal coronal structures. N_e is the electron density. Source: Lang (2001) *The Cambridge Encyclopedia of the Sun*, Cambridge University Press

Structures	Extension (Mm)	T (10^6 K)	N_e (cm^{-3})
Coronal Holes	700–900	1.0–1.5	4.0×10^{14}
Hot Loops	10	2.0–4.0	$1.0–7.0 \times 10^{15}$
Bright Points	5–20	2.5	1.4×10^{6}

1.6 The Solar Wind

The first hint for a solar wind arose from observations of cometary tails (Biermann, 1951), which are produced when comets are close to the Sun, the tails always pointing away from the Sun. The high temperature of the corona led Chapman and Zirin (1957) to consider the existence of a static solar wind driven by the hot corona. As an alternative, Parker (1958) proposed a thermally driven outflow from the solar corona into interstallar space that accommodates the transition from a subsonic flow near the surface to a supersonic flow further out. In January 1959, the first observation of the solar wind was made by the spacecraft Luna I. Shortly thereafter, an unambiguous measurement was performed by Mariner II (Neugebauer and Snyder, 1962).

The solar wind consists mostly of high-energy electrons and protons. The average speed at the Earth is about 400 km s^{-1} and the total mass loss is some 10^{-14} solar masses per year. However, the solar wind contributes significantly to the loss of angular momentum.

As the Sun rotates, the streams of the wind rotate as well and produce a pattern much like that of a rotating lawn sprinkler. The solar magnetic field is frozen on this wind and its field line follows this motion (Figure 1.29). At 1 A.U. and in the plane of the ecliptic, the interplanetary magnetic field (IMF) shows two or four sectors per solar rotation. In each of them, the polarity of the IMF is directed towards or away from the solar direction.

The solar wind is accelerated in coronal holes (Kohl and Cranmer, 1999), but stronger values are measured in polar regions, as evidenced by the Ulysses spacecraft, which made measurements of the solar wind outside of the ecliptic for the first time.

Fig. 1.29. Structure of the interplanetary magnetic field.

1.7 3-D Topology of the Magnetic Field

The total magnetic flux emerging from the solar surface has two main components, according to the topology of the magnetic field (Figure 1.30):

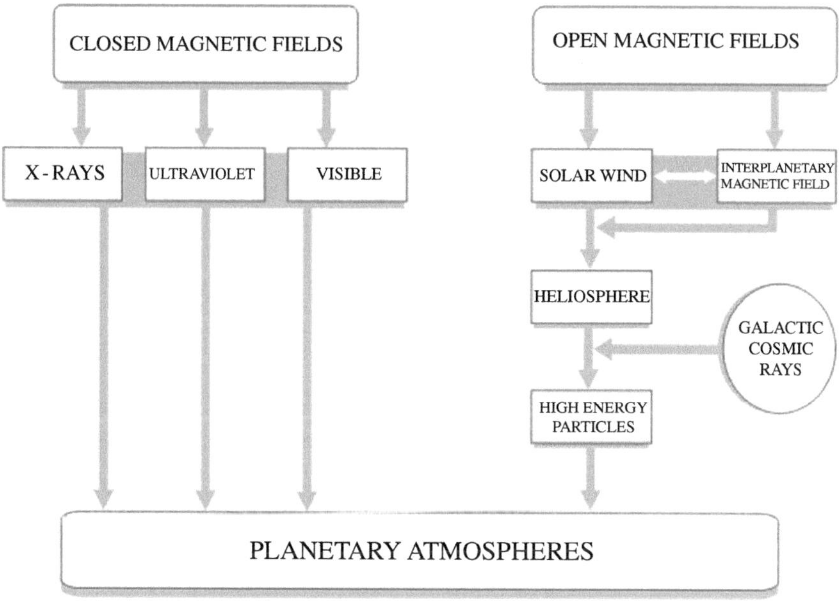

Fig. 1.30. Sources of solar variability.

(a) The magnetic flux of the active regions (hereafter referred to as closed magnetic fields, CMF), characterized by magnetic configurations with closed field lines dominating the variations of the total irradiance and emission in the high-energy range of the solar spectrum (ultraviolet and X-rays). Most of the radiative losses from the outer layers occur in these regions.

(b) Large-scale magnetic regions, which have field lines open toward the interplanetary medium. They are the main source of a continuous outward flow of charged particles (protons, electrons and He nuclei) known as the solar wind. The solar magnetic field of the open regions (OMF) is frozen into this wind, configuring the interplanetary magnetic field (IMF), which produces a huge magnetic region, the heliosphere, that fills the whole Solar System. Galactic cosmic rays (hereafter GCRs) are high-energy particles (mainly protons with energies in the range 1–20 Gev), originating outside our planetary system and striking the Earth from all directions. Both the flux and energy spectrum of GCRs are modulated by the strength of the heliosphere, being stronger when the IMF is weaker.

Figure 1.28 show a full disk X-ray image, in which we can distinguish two structures: (a) *bright regions* corresponding to active regions with a closed topology of the magnetic field and (b) *dark regions* characterized by magnetic lines open towards the interplanetary medium.

The largest coronal holes, showing large solar wind speeds, appear at the solar poles shortly after the maximum of the 11-year sunspot cycle. Recently, the spacecraft Ulysses has mapped the 3-D structure of the solar wind showing clearly that the equatorial regions are the seat of the low-speed solar wind (McComas, 2000).

1.8 Observing the Outer Layers

As seen visually, the outer layers of the solar atmosphere offer very low contrast with respect to the flux from the photosphere. Therefore, total solar eclipses offered the first opportunity to observe them (see Chapter 4 for details).

In 1869 P.J. Janssen (1824–1907) and N. Lockyer (1836–1920) independently developed a solar spectroscope to produce a spectrum and then looked at the solar limb through a slit placed in the red part of the spectrum, permitting the observation of prominences during a solar eclipse. William Huggins wrote in 1869: *Last Saturday, February 13, I succeeded in seeing a solar prominence so as to distinguish its form. A spectroscope was used; a narrow slit was inserted after the train of prisms before the object-glass of the little telescope.... The slit of the spectroscope was then widened sufficiently to admit the form of the prominence to be seen. The spectrum then became so impure that the prominence could not be distinguished. A great part of the light... was then absorbed by a piece of deep ruby glass. The prominence was then distinctly perceived* (Huggins 1869).

The input of G.E. Hale (1868–1938) and H. Deslandres (1853–1948) was decisive for subsequent development with the spectroheliograph, that allowed scans of the solar surface in selected broad spectral lines formed in the chromosphere.[12]

Later, technology made it possible to construct narrow-band filters centred on the main chromospheric lines (H_α and Ca II H and K). They were widely used until the recent development of two-dimensional spectrographs.

G.B. Airy (1801–1892) attempted to observe solar prominences outside a total eclipse by cutting a hole in a screen to match precisely the size of the solar image, so that the brightness of the disk's image would not interfere with the prominences that were expected to appear arrayed around the hole. However, he did not succeed with this method. This was also the case with other attempts to observe the inner corona outside an eclipse, by G.E. Hale and H. Deslandres using a spectroheliograph.

In 1930, B. Lyot (1897–1952) designed an instrument called a coronograph to block out the direct light from the solar photosphere. Basically, it consisted of the lens, which was stopped down with a diaphragm, followed by a screen to block the image of the solar disc, and at the end of the tube, a second field lens and two reflecting prisms cast a picture of the corona on the slit of a

[12] A description of the development of Hale's spectroheliograph is included in the book of A.J. Meadows *Early Solar Physics* pp. 237–249.

spectograph. The first visual observation of the corona outside an eclipse was made at the Pic du Midi Observatory on 25 July 1930 (Lyot 1932).[13]

Balloons, airplanes and finally satellites permitted the observation of the Sun in spectral windows (UV, X-rays) blocked by the Earth's atmosphere. After World War II, R. Tousey (1908–1997) at the Naval Research Laboratory (NRL) designed a small spectrograph, 60 cm height, to be installed in the nose-cone of a V-2 rocket. After a number of failed attempts,[14] on 10 October 1946, the rocket reached an altitude of 173 kilometres. On recovering the capsule from the nose-cone of the rocket, it was possible to observe the first UV spectrum of the Sun in the range between 210–300 nm (Tousey et al., 1947). The spectral resolution was only 0.3 nm, but it was high enough to detect clearly the Mg II doublet at 280 nm (Figure 1.31).

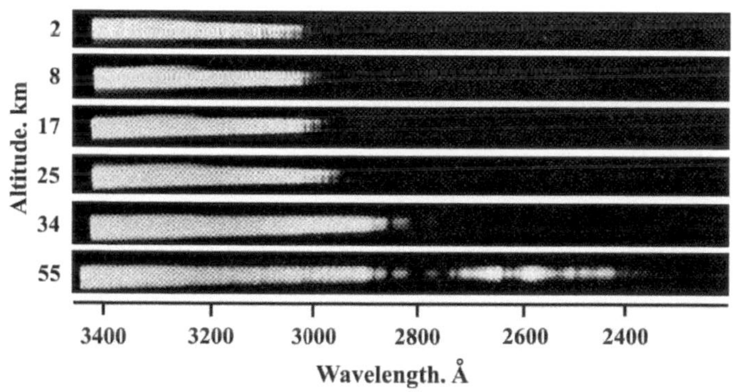

Fig. 1.31. Solar spectra in the ultraviolet region obtained during the ascent of a V-2 rocket. A sequence of exposures shows how more and more ultraviolet radiation of the Sun is recorded as the rocket rises in altitude from 2 to 55 km above sea level (US Navy Photograph. NASM SI 87-8423).

1.9 Time Scales of Solar Variability

The variability of the Sun is linked to the dissipation of the available energies (De Jager, 1972). Different sources are characterized by distinct time-scales. The time-scales we are interested in are those related to the magnetic energy.

[13] Coronagraphs operating within the Earth's atmosphere suffer from scattered light in the sky itself, due primarily to Rayleigh scattering of sunlight in the upper atmosphere. Space-based coronagraphs such as LASCO avoid the sky brightness problem. This technique is also the basis of future telescopes designed with the aim to image extrasolar planets.

[14] A first attempt was made on 28 June, 1946, but the camera was never retrieved from the crater made by the V-2 impact on the desert floor.

1.9.1 The Solar Cycle

S.H. Schwabe (1789–1875) found, in 1844, an 11 year periodicity in the number of sunspots; this is the solar activity cycle and the first clear signal of solar variability. J.R. Wolf (1816–1893), who was professor at the ETH Zürich, defined the sunspot relative number, measuring the sunspot activity independently of the quality of the observations.

The sunspot number (also called the Wolf or Zürich number) is defined as

$$\mathrm{R} = \mathrm{k}\,(10\mathrm{g} + \mathrm{f})$$

where g is the number of spot groups, f is the number of individual spots and k is a correction factor taking into account the observing site, the telescope, atmospheric conditions, and the enthusiasm of the observer (see Figure 1.32). Although the sunspot number contains many uncertainties, it is the oldest known parameter for determining the solar activity directly. The official sunspot number is now given by the Solar Influences Data Analysis Center (SIDC), a group at the Royal Observatory, Belgium, which was founded in 1981 to continue the work of the Zürich Observatory.

The sunspot area, A, is related to the sunspot number by $A \sim 16.7R$, where A is given in millionths of the disk area. Hoyt et al. (1994) proposed using the number of groups to measure solar activity, a parameter less sensitive to calibration uncertainties, especially for earlier epochs.

Sunspot cycles have the following statistical properties (Hathaway and Wilson, 2004): (a) Periods of 131±14 months with a normal distribution, (b) asymmetric with a rapid rise and a slow decay, (c) the rise time from minimum to maximum drecreases with cycle amplitude and (d) although the two hemispheres remain linked in phase, there are significant assymmetries in the activity.

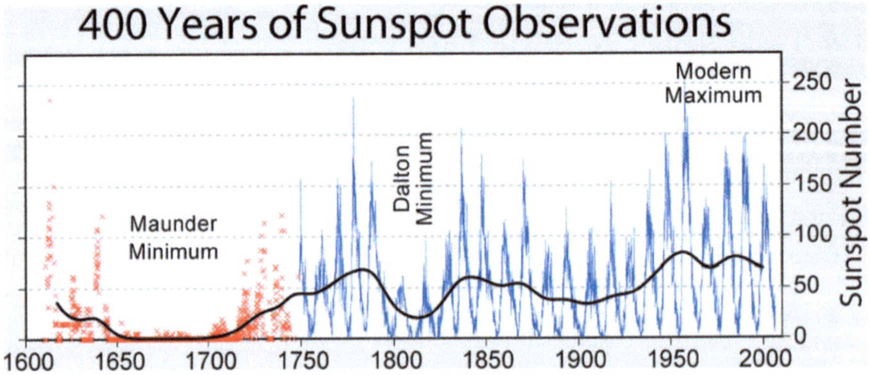

Fig. 1.32. The record of sunspot numbers showing the 11 year cycle. Prior to 1749 only sporadic observations are available. Credit: Robert A. Rohde.

G. Spörer (1822–1895) and R.C. Carrington (1826–1875) found that the Sun does not rotate as a rigid body but differentially and that the sunspots drift gradually to the equator in the course of the sunspot cycle, now known as Spörer's law (Spörer, 1885; Carrington, 1858). Maunder (1904) plotted the first sunspot "butterfly diagram", showing the migration of sunspot groups from higher latitudes at the beginning of a sunspot cycle to lower latitudes at the end. Figure 1.33 shows an updated version of this sunspot distribution.

When sunspots come in pairs, one tends to have a magnetic field polarity which is opposite that of the other. During a given sunspot cycle, the leading sunspots in groups in the northern hemisphere of the Sun all tend to have the same polarity, while the same is true of sunspots in the southern hemisphere, except that the common polarity is reversed from that of sunspots in the northern hemisphere. This behaviour is visualized in Figure 1.34. Two sunspot cycles are needed to return the Sun to the same magnetic state; the resulting 22-yr periodicity is known as the Hale cycle (Hale and Nicholson, 1925).

These charateristics of the solar cycle were explained with the empirical model of Leighton (1969) in the framework of the dynamo theory (Parker, 1979a).

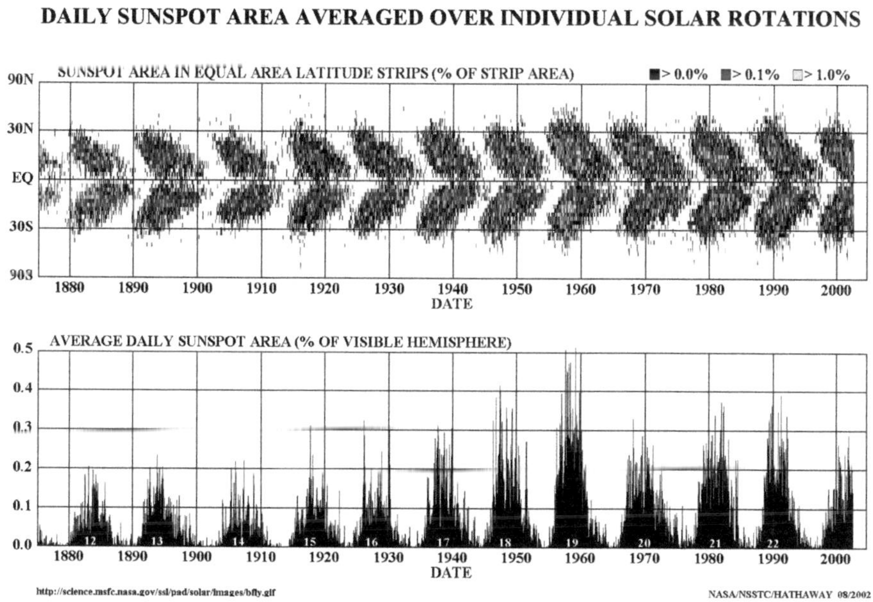

Fig. 1.33. The butterfly diagram (top) and the sunspot area (bottom). Courtesy: D. Hathaway.

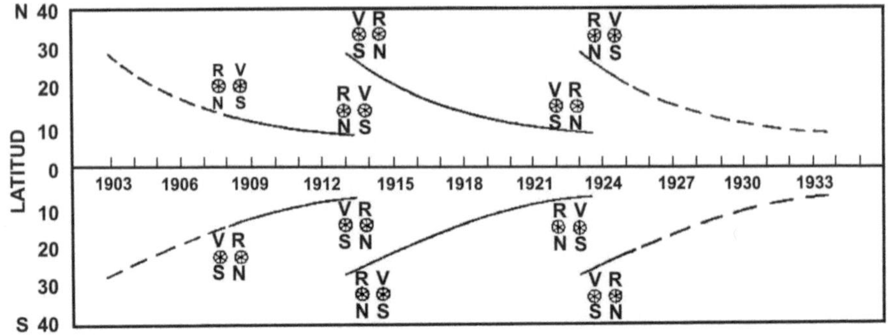

Fig. 1.34. The Hale–Nicholson law for bipolar sunspot groups. V (violet) and R (red) stand for the distinct polarities (positive and negative) of the magnetic field (Hale and Nicholson, 1925).

1.9.2 Long-Term Variations

The variability of both the CMF and OMF is dominated by the 11 year period, but longer periods exist, as is evidenced by the presence of several deep minima of solar activity; for example, the Maunder minimum in the second half of the 17th century. Figure 1.35 shows the variation of CMF and OMF over the last 300 years.

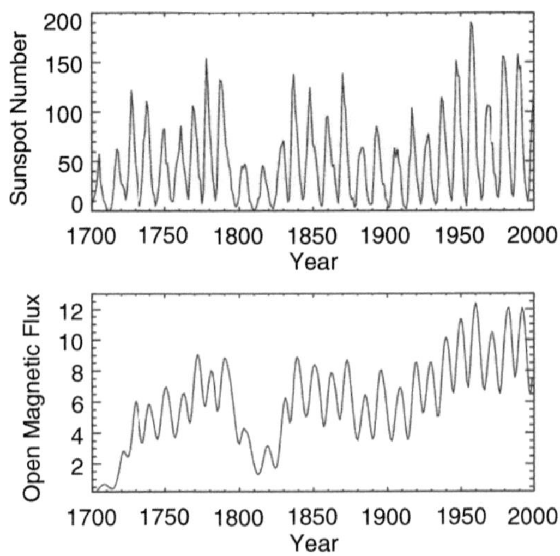

Fig. 1.35. (Top) variation of the closed magnetic field (CMF), indicated by the sunspot number; (Bottom) variation of the open magnetic flux (OMF) calculated by Usoskin et al. 2002 (data courtesy of M. Schüssler, MPIfA, Germany).

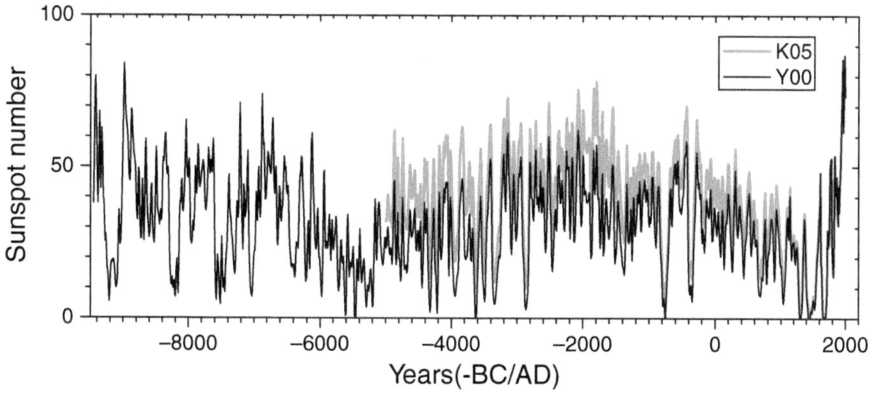

Fig. 1.36. Long-term sunspot-number reconstruction from ^{14}C data (after Usoskin et al., 2007, Figure 1c). Solid (Y00) and grey (K05) curves are based on the paleo-geomagnetic reconstructions of Yang et al. (2000) and Korte and Constable (2005). Courtesy of I. Usoskin. Reprinted with permission of A&A Editorial Office.

Based on dendrochronologically dated radioisotope concentrations, Solanki et al. (2004) and Usoskin et al. (2007) (see also Usoskin, 2008) have reconstructed the sunspot number covering the last 11 400 years. The present epoch appears as a very active period, probably the strongest in the last 8000 years (Figure 1.36).

1.9.3 Flares

A flare[15] is defined as a sudden, rapid, and intense variation in brightness. It occurs when the magnetic energy that has built up in the solar atmosphere is suddenly released. The basic process involved is magnetic reconnection.[16]

Flares are among the most intense phenomena in the solar system. The energy released during a flare is typically of the order of 10^{27} ergs per second. Large flares can emit up to 10^{32} ergs of energy.[17]

Classically, flares have been observed in the chromosphere in H_α, but very intense flares, called white light flares, can also be observed in the photosphere. As a general rule, flares occur above locations in the photosphere where the electric current has a maximum (the highest $\nabla \times B$). They tend to appear

---

[15] For textbooks on this particular subject see: B.V. Somov, 1992, *Physical Processes of solar flares*, Kluwer; E. Tandberg-Hanssen and A.G. Emslie, 1988, *The Physics of Solar Flares*, Cambridge Univ. Press and M.J. Aschwanden, 2002, *Particle Acceleration and Kinematics in Solar Flares: A Synthesis of Recent Observations and Theoretical Concepts*, Kluwer.

[16] Process whereby magnetic field lines from different magnetic domains are spliced to one another, changing the overall topology of a magnetic field.

[17] The amount of released energy is the equivalent of millions of 100-megaton hydrogen bombs.

in those regions in sunspots or groups of sunspots where new and oppositely directed magnetic flux emerges from below. Large gradual flares often occur above the neutral lines in the photosphere, which separate regions of opposite magnetic polarity. Neutral lines are bridged by arcades of loops and, in H_α one sees two bright ribbons formed by the footpoints on each side of the neutral line (Figure 1.37). Flares then occur above the part of the neutral line that has experienced most shear due to different surface motions on both sides. In quiet regions, the most powerful microflares occur at the boundary of supergranular cells.

The access to space data has allowed observations of flares in the upper layers of the solar atmosphere (UV and X-rays).

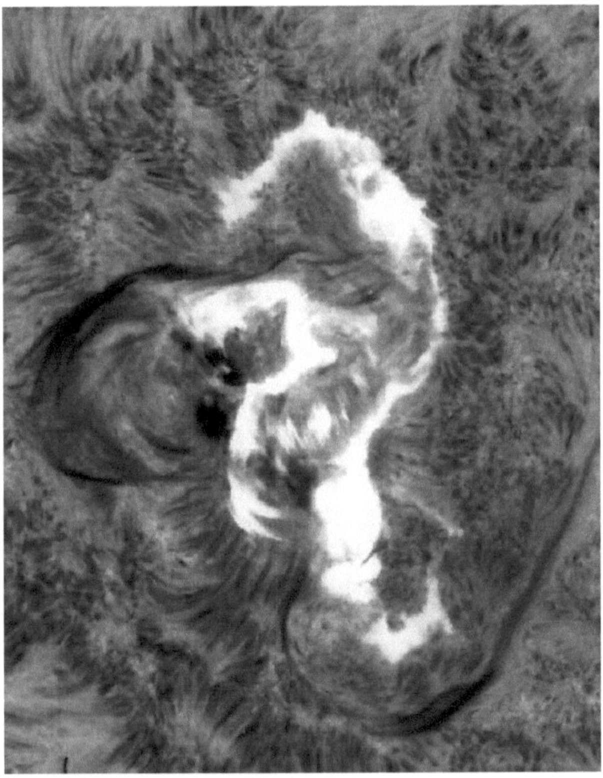

Fig. 1.37. The great "Seahorse Flare" of 7 August, 1972. This image, taken in the blue wing of H_α, shows the two-ribbon structure late in the event, with bright loops connecting the ribbons. Courtesy: Big Bear Solar Observatory.

1.9.4 Coronal Mass Ejections

Coronal mass ejections (CMEs) were first detected in observations made with a coronograph on board the OSO 7 observatory (1971–1973). During a solar

Table 1.8. Variation of solar phenomena during activity cycle

Name	Variation Between Activity Minimum and Maximum
Spots, R	0 − > 150
Flares, gradual	0 −100/yr
Flares, impulsive	0 −1000/yr
Faculae	Contribution during maximum ≫ spot
CMEs	0.5–2.5 per day

eclipse the corona is visible only for a few minutes and no change in the coronal structure can be observed. CMEs are huge bubbles of gas that are ejected from the Sun over the course of several hours. They disrupt the flow of the solar wind.

To summarize this section, Table 1.8 shows the main time-scales of variability of the different phenomena representing the solar activity.

1.10 Solar–Terrestrial Relations

Solar radiation constitutes the main energy source driving all the physical and biological processes taking place in the terrestrial atmosphere. The Sun interacts with our planet through two main channels, namely, radiation and particle flux, the latter remaining unknown until 1962. A change in solar output can be expected from a variation in any of the solar energy sources (see De Jager, 1972, for a summary). We are interested now in solar influences that may affect our civilization; therefore, we shall concentrate on changes associated with solar magnetism.

Using a terrestrial analogy, we divide the solar influences into two time-scales: climate (long-term, decades) and weather (short-term, days-hours).

1.10.1 Sun – Climate

William Herschel (1801) was the first astronomer to suggest a connection between solar radiation and the terrestrial climate. The discovery of the 11-year cycle in sunspot number counts (Schwabe, 1844) marked a crucial moment in solar–terrestrial studies. Numerous studies have tried to find a correlation between the number of sunspots and different terrestrial parameters related to climate (see Hoyt and Schatten, 1997 for a summary). However, several fundamental problems arise when trying to verify the existence of a physical link between solar radiation and terrestrial climate. First, most climate parameters are too deficient in spatial and temporal coverage to be representative of the global climate. Second, it is necessary to identify the real source of solar variability influencing the Earth and the range of its variation. No less important are dating uncertainties, quasi-periodicities related to naturally occurring cycles, and to the Moon, which happen to match solar time scales. Meadows (1975), Pittock (1978) and Burroughs (1992) give critical views concerning the reliability of these relationships.

The average energy density of solar radiation just above the Earth's atmosphere, in a plane perpendicular to the direction of the Sun, is about 1367 W m^{-2}, a parameter called the solar constant. The Earth receives a total amount of radiation determined by its cross-section (πR^2), but as the planet rotates this energy is distributed across the entire surface area ($4\,\pi R^2$). Hence, the average incoming solar radiation (known as "insolation") is 1/4th the solar constant or ~342 W m^{-2}. The climate sensitivity, s_c, is defined as the linear ratio between the cause, solar radiation, and the effect, change of temperatures.

During the past two decades, satellite observations have revealed that the total solar irradiance, $S_\odot$, changes both on a short time-scale (Willson et al., 1981) and on the time-scale of the solar cycle (Willson et al., 1986). These changes in the solar constant with magnetic activity could affect the terrestrial climate, especially if larger excursions occurred in the past. Figure 1.38 show an updated version of these changes, where data from several experiments have been combined.

The simplest calculation indicates that the influence of the solar radiation changes during the solar cycle (0.1% of 239.4 $\sim$ 0.24 W m^2) should be negligible (ΔT = 0.07 °C for s_c = 0.3) . However, there are several facts which indicate that the influence may well be above this simple estimate, and there are feedback mechanisms enhancing the solar signal on the terrestrial climate.

We can easily understand that stationary equilibrium of the atmosphere and instantaneous climate response to an external force is an unrealistic

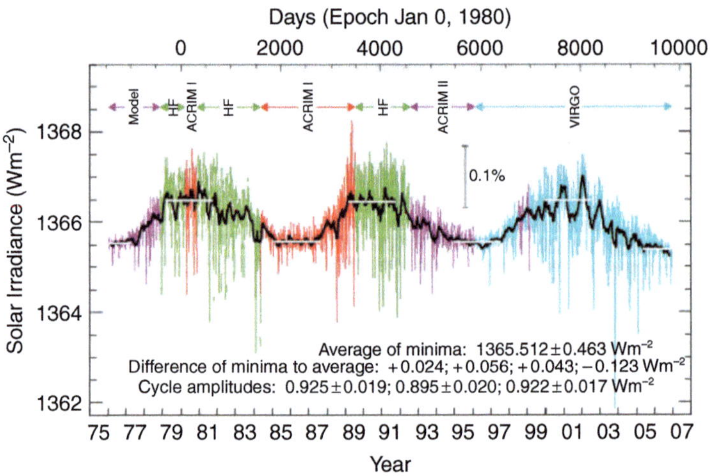

Fig. 1.38. Composite showing the variations of solar irradiance during the last three solar cycles. Courtesy: Claus Fröhlich, World Radiation Center, Davos.

situation. The time response of the climate system, $t_R = C/s_c$, is controlled by the heat capacity, C, of the component considered (e.g. oceans, land, etc.). If C is large, for example in the oceans, the time response will be long. Equally, if the climate sensitivity, s_c, is small, the response time will be long.

Different empirical correlations between solar and terrestrial parameters show that the solar–terrestrial connection is real. This is specially true, when we look at long-term variations, identifying periods of low solar activity (Maunder minima) with cool episodes of the climate. This connection was largely popularized by Eddy (1976) and verified by later work (Mann and Jones, 2003; Jones and Mann, 2004).[18]

In any case, the reconstruction of solar irradiance gives information on irradiance reductions in the order of 0.2 % during these activity minima, too small to produce measurable climate impacts. In order to explain this apparent paradox, the best procedure is probably to look in other channels of the solar output different from the visible range (Table 1.9), where the intrinsic flux is small but the variations with the solar activity are larger.

The solar UV produces considerable heating of the upper layers of the terrestrial atmosphere. Temperatures in the thermosphere change significantly from solar minimum to solar maximum. Figure 1.39 shows how the temperature change with the altitude between two periods with different levels of solar activity (Schmidtke, 2000). The question arises as to how the thermal variations in these upper layers finally influence the climate in the lower layers (see Arnold, 2005).

Other mechanisms of interaction have been proposed such as (a) changes in the propagation of planetary waves; (b) the existence of a global electric circuit and (c) a relation between cosmic rays and cloudiness

Table 1.9. Energy densities in the three main channels for the Sun–Earth connection. From the National Research Council Report "Solar Influences on Global Change", National Academy Press,1994

Source	Energy (W/m^2)	Solar Cycle Change	Terrestrial Deposition (km)
SOLAR RADIATION			
Total irradiance	1366	0.002	Surf. and Trop.
UV (200–300 nm)	15.4	1.3	0–50
UV (0–200 nm)	0.1	0–16	50–500
PARTICLES			
Solar protons	0.002		30–90
Cosmic rays	0.000007		0–90
Solar wind	0.0003		> 500

[18] This connection clearly fails in the last 50 years, when the anthropogenic action on climate becomes dominant.

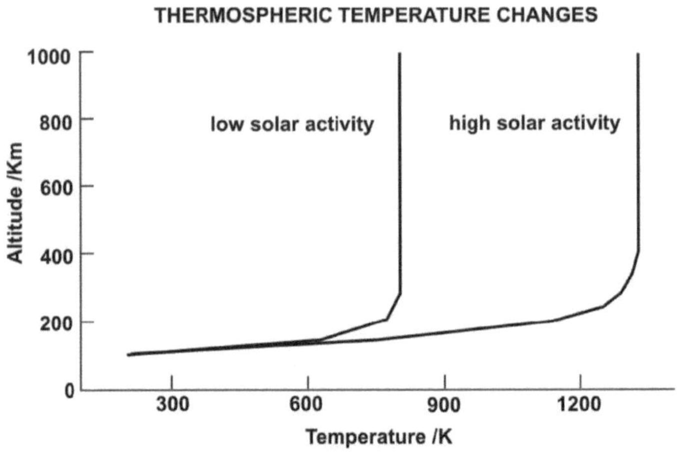

Fig. 1.39. XUV-induced temperature changes in the thermosphere for two levels of solar activity. Adapted from Schmidtke (2000).

Various review papers (Chapman, 1987; Hudson, 1988; Lean, 1997; Vázquez 1999, 2003, 2004; Beer et al., 2000; Fligge et al., 2001; Bard & Frank, 2006; Foukal et al., 2006), monographs (Hermann & Goldberg, 1978; Nesme-Ribes & Thuillier, 2000; Pap and Fox, 2004; Benestad, 2006) and conference proceedings (White, 1977; Sonnett et al., 1991; Donnelly, 1992; Pap et al., 1994; Nesme-Ribes, 1994; Friis-Christensen et al., 2000; Wilson, 2000; Haigh et al., 2005) provide a view of how our knowledge on this topic has changed with time.

1.10.2 Space Weather

The term space weather denotes the change of the environmental conditions in outer space. It is mainly influenced by the speed and density of the solar wind, and the interplanetary magnetic field carried by the solar wind. Sources of space weather variations are CMEs, flares and coronal holes.

The interaction of the solar wind with the Earth's magnetosphere gives rise to a set of phenomena.

Geomagnetic Disturbances

At the beginning of the 19th century, Alexander von Humboldt (1769–1859) initiated a series of measurements of the Earth's magnetic field, studying the orientation of the compass. On 21 December 1806 strong disturbances were noted. It was a solar storm, which caused the value of the terrestrial magnetic field to decrease over the course of several hours.

Soon after this discovery, a positive correlation with the 11 year cycle was established. The *aa* index was the first index representing the value of the magnetic field. Maunder (1905) established a correlation between the solar storms and the solar rotation period supporting the idea of a solar origin, which we now pin down to the coronal holes.

Aurorae

Aurorae are extended sources of light with different forms and colours mainly observed at high latitudes. Their brightness may reach the intensity of the full moon. Aurorae are generated by the interaction of solar particles with atoms and molecules of the Earth's upper atmosphere. The occurrence and brightness of aurorae follows the 11-yr cycle of solar activity (Figure 1.40). The historical background is described in Chapter 6. For monographs on aurorae see Earther (1980) and Bone (1997).

Ionosphere

The first ideas about an electric conducting layer in the terrestrial atmosphere were advanced in 1839 by C.F. Gauss (1777–1855), who proposed that

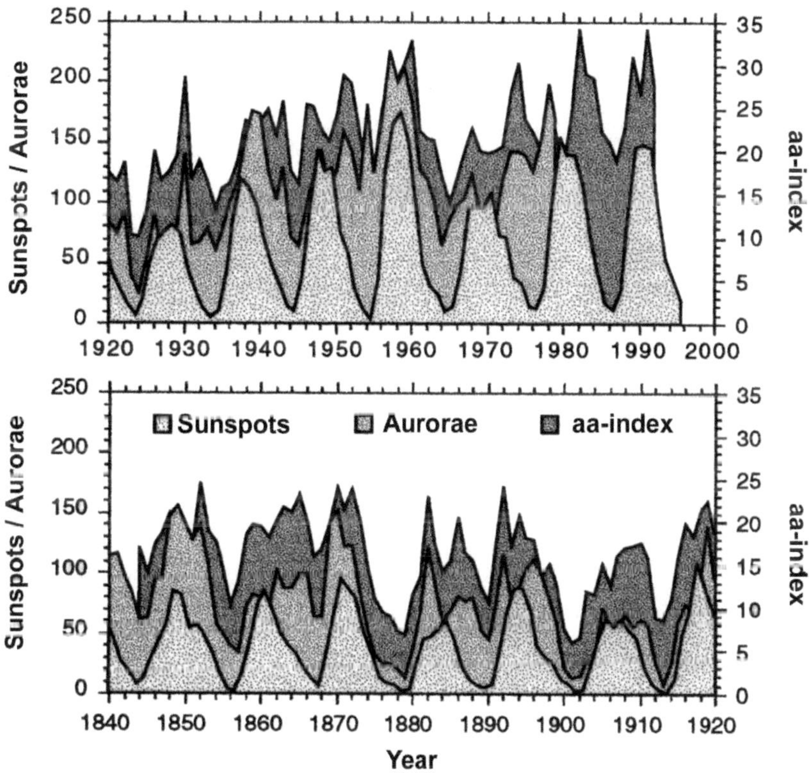

Fig. 1.40. Temporal variation of the *aa* index and the number of aurorae and sunspots.

small daily variations in the geomagnetic field could be explained by electric currents flowing in such a layer. Belfour Stewart (1828–1887) described such currents, around 1886, as arising from electromotive forces generated by periodic motions of the electric layer across the terrestrial magnetic field.

In 1901, G. Marconi (1874–1937) was able to establish transatlantic radio-communication.[19] In 1902, Oliver Heaviside (1850–1925) and Arthur E. Kennelly (1861–1939), independently, explained this in terms of a reflection of the radio waves by free charges in the high atmosphere. Fleming (1906) proposed that solar UV radiation generates such free charges.

The first observational evidence of a new layer in the atmosphere, the ionosphere,[20] came from E. Appleton (1892–1965) and M.A.F. Barnett (1901–1979) in 1926. This finding was later verified by Gregory Breit (1899–1981) and Merle Anthony Tuve (1901–1982) by using pulsed radio waves.

On 29 June 1927, an eclipse provided an opportunity to study the effect of the Sun on the ionosphere. Appleton verified that as soon as the rays of light were cut off, the height of the reflecting layer changed. In 1928, E.O. Hulburt proposed that the ultraviolet radiation shortwards of 123 nm might be the source of the ionosphere. After World War II, soundings with V-1 and V-2 rockets showed that this radiation shapes the bottom of the ionosphere.

The components of the atmosphere may be ionized by capturing photons whose energy ($hc/\lambda_{\rm crit}$) exceeds the corresponding ionization potential (see Table 1.10). Thus, only radiation with $\lambda < \lambda_{\rm crit}$ produces ionization, which corresponds to the X-ray and EUV regions of the solar spectrum.

The combination of the amount of incident radiation and the density of the atmospheric particles configures the extension of this layer (Figure 1.41), which is subdivided into different regions, also called Chapman layers.

Table 1.10. Ionization potential of components of the terrestrial atmosphere. From Hargreaves (1979) *The Upper Atmosphere and Solar–Terrestrial Relations*, Van Nostrand Reinhold

	I (ev)	λ_{max} (nm)		I (ev)	λ_{max}
NO	9.25	134	CO_2	13.79	89.9
O_2	12.08	102.7	N	14.54	85.3
H_2O	12.60	98.5	H_2	15.41	80.4
O_3	12.80	97.0	N_2	15.58	79.6
H	13.59	91.2	Ne	21.56	57.5
O	13.61	91.1	He	24.58	50.4

[19] On 12 December 1901, he transmitted a Morse code signal from Cornwall (England) to Newfoundland (Canada), at a distance of 2900 km, baffling scientists at the time, as they were unable to explain how the waves propagating as straight lines could bend over to follow the curvature of the Earth.

[20] This term was first used in 1926 by Robert Watson-Watt (1892–1973), the father of radar.

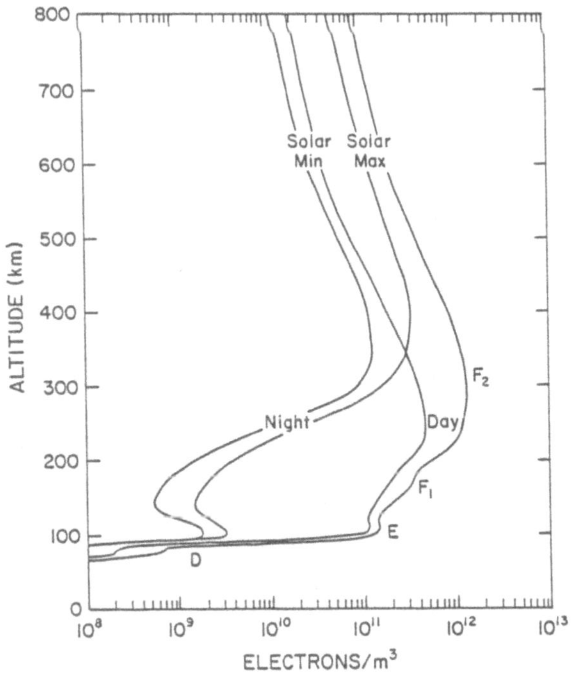

Fig. 1.41. Variation of the electron density, N_e, with altitude during day and night for two different levels of solar activity.

- *D layer*: The lowest, at altitudes between 50 and 80 km. It disappears at night, when solar radiation is blocked by the Earth. The main sources of ionization are Lyman-α (121.5 nm) on NO; EUV (102–111 nm) on oxygen O, and hard X-rays and cosmic rays on all atmospheric constituents.
- *E layer*: Found between 100 and 125 km. Also practically disappears at night. Ions in this region are mainly O_2^+ and NO^+, produced through the following reactions:

$$
\begin{aligned}
O^+ + N_2 &\longrightarrow NO^+ + N \\
N_2^+ + O &\longrightarrow NO^+ + N \\
O^+ + O_2 &\longrightarrow O_2^+ + O
\end{aligned}
$$

- *F layer*: The ionization production is smaller than in the previous lower level layers, but the much longer electron lifetime permits large values of electron density to be reached. Its strength varies according to the time of day, the season, and the level of solar activity. During the daytime, it is split into two sublayers, with F1 placed around 180 km, where NO^+ and O_2^+ ions dominates, and the F2 at 400 km or more, with O^+ as the main contributor.

The ionosphere shows transient perturbations that have been attributed to solar activity in the form of variable emission of particles. We can reasonably expect variations of solar magnetism to change the EUV radiation and therefore to cause variable levels of electronic density, and ionospheric properties. Figure 1.41 illustrates the change of N_e between day and night and for two phases of solar 11-yr cycle.

High-Frequency communication depends on radio waves reflected from the ionosphere, which enable communications from one part of the globe to another, as well as from spacecraft to the ground and between spacecraft.

H.A. Lorentz (1853–1928) investigated the reflection and refraction of a radiowave in a layer of plasma with refractive index n. When n^2 is less than unity, the incoming waves are bent away from the normal to the layer, and radio waves are consequently reflected. This behaviour is expressed as the Appleton–Hartree formula, obtained after observations made during the solar eclipse of June 29, 1927:

$$\mathrm{n}^2 = 1 - \left(\frac{\omega_\mathrm{p}{}^2}{\omega}\right)$$

with the plasma frequency, ω_p, given by

$$\omega_\mathrm{p} = \sqrt{\frac{\mathrm{N_e e^2}}{\mathrm{m_e}\epsilon_0}}$$

where $\mathrm{N_e}$ and $\mathrm{m_e}$ are the density and mass of the electron, respectively, e the electron charge, ϵ_0 the dielectric constant of free space, and ω the angular frequency of the transmitted signal.

As the wave penetrates into the ionosphere, the refractive index becomes smaller. If $\omega_p > \omega$, the wave cannot propagate because the refractive index is imaginary. The energy carried by the radio wave is therefore reflected back to the ground.

The limiting frequency, ω_0, above which a radio wave is no longer reflected by an ionized layer depends on the square root of the electronic density, N_e. The radio transmission is characterized by the Maximum Usable Frequency (MUF) and the Lowest Usable Frequency (LUF). MUF values change progressively in the D–(16 MHz), E–(28 MHz) and F–(16 MHz) layers.

Neither the MUF nor the LUF is a practical operating frequency. While radio waves at the LUF can be refracted back to Earth at the desired location, the signal-to-noise ratio is still much lower than at the higher frequencies, and the probability of multipath propagation is much greater. Operating at or near the MUF can result in frequent signal fading and dropouts when ionospheric variations alter the length of the transmission path. The Optimum Usable Frequency (OUF) is roughly about 85 % of the MUF, but the actual percentage varies and may be either considerably more or less than this value.

The MUF depends on the maximum of $\mathrm{N_e}$ in the F region and the angle of incidence of the emitted radio wave. The LUF is controlled by the amount of absorption of the radio wave in the lower D and E layers.

The maximum distance over the Earth's surface that can be reached by a single ionosphere reflection[21] is

$$2\mathrm{R}\arccos\left(\frac{\mathrm{R}}{\mathrm{R}+\mathrm{h_i}}\right)$$

where R is terrestrial radius and h_i the altitude of the different ionospheric layers involved in the transmission.

Variations in the total electronic density of the ionosphere are of great importance for the precision of satellite-based positioning systems such as the Global Positioning System (GPS), Very Long Baseline Interferometry (VLBI) and Satellite Laser Ranging (SLR), as illustrated in Figure 1.42. In the near future these will be followed by the European system Galileo.

For general monographs on space weather see Hanslmeier (2002) and Schwenn (2006).

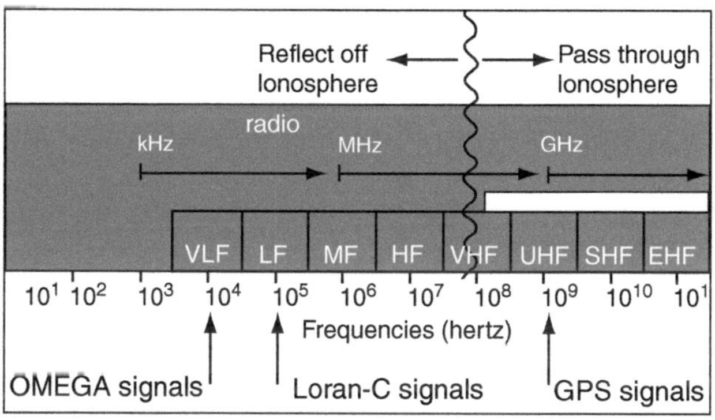

Fig. 1.42. Working range of various communication systems.

1.11 Monographs and Textbooks on Solar Physics

Abetti, G., 1952, *Il Sole*, Editore Ulrico Hoepli, Milano (English Translation 1957, McMillan Co.)

Abbot, C.G., 1911, *The Sun*, Appleton, New York

Abbot, C.G., 1929, *The Sun and the Welfare of Man*, Smithsonian Scientific Series

Antia, H.M., Bhatnagar, A., Ulmschneider, P. (Eds.), 2003, *Lectures on Solar Physics*, Springer

Bhatnagar, A., Livingston, W., 2005, *Fundamentals of Solar Astronomy*, World Scientific

Brester, A., 1924, *Le Soleil*, W.P. Van Stockum & Fils, La Haye

[21] This distance is not sufficient for transatlantic communications, for which at least two reflections are required.

Bruhat, G., 1951, *Le Soleil*, Presses Universitaires de France, Paris
Bruzek, A., Durrant, C.J. (eds.), 1977, *Illustrated Glossary for Solar and Solar Terrestrial Physics*, D. Reidel.
Foukal, P., 1990, *Solar Astrophysics*, Wiley
Frazier, K., 1978, *Our Turbulent Sun*, Prentice-Hall Inc.
Gamow, G., 1940, *The Birth and Death of the Sun*, The Viking Press
Gamow, G., 1964, *A Star called the Sun*, The Viking Press
Gibson, E.G., 1973, *The Quiet Sun*, NASA SP–303
Giovanelli, R.G., 1984, *Secrets of the Sun*, Cambridge University Press
Golub, L., Pasachoff, J.M., 2002, *Nearest Star: The Surprising Science of Our Sun*, Harvard University Press
Goodwin, J., Doster, A., Page, J.K., Bourne, R. (eds.), 1981, *Fire of Life: The Smithsonian Book of the Sun*, Smithsonian Exposition Books
Gribbin, J.R., 1991, *Blinded by the Light: The secret life of the Sun*, Harmony Books, New York
Kippenhahn, R., 1994, *Discovering the Secrets of the Sun*, John Willey Sons
Kuiper, G.P. (ed.), 1953, *The Sun*, University of Chicago Press
Landi Degl' Innocenti, E., 2007, *Fisica Solare*, Springer
Lang, K., 2001, *The Sun from Space*, Springer
Lang, K., 2001, *The Cambridge Encyclopedia of the Sun*, Cambridge University Press
Lockyer, N., 1897, *The Sun's Place in Nature*, MacMillan, New York. Reprinted by Kessinger Publishing in 2004
Lockyer, N., Norman, J., 1874, *Contributions to Solar Physics*, Mc Millan, London
Mattig, W., 1940, *Die Sonne*, C.H. Beck Wissen
Meadows, A.J., 1970, *Early Solar Physics*, Pergamon Press
Menzel, D.H., 1959, *Our Sun*, Harvard University Press
Newton, H.W., 1958, *The Face of the Sun*, Penguin Books Ltd
Noyes, R.W., 1982, *The Sun, Our star*, Harvard University Press
Phillips, K.J.H., 1992, *Guide to the Sun*, Cambridge University Press
Priest, E.R., 1984, *Solar Magnetohydrodynamics*, D. Reidel Pub. Co.
Proctor, R.A., 1876, *The Sun: Ruler, Fire and Life of the Planetary System*, Longmans, Green and Co., London
Schrijver, C.J., Zwaan, C., 2000, *Solar and Stellar Magnetic Activity*, Cambridge University Press
Secchi, A., 1875, *Le Soleil*, Gurthier-Villiars, Paris
Severny, A., 2004, *Solar Physics*, University Press of the Pacific
Stix, M., 2000, *The Sun* (2nd ed.), Springer
Sturrock, P.A., 1986, *Physics of the Sun*, three volumes, D. Reidel
Tandberg-Hansen, E., 1967, *Solar Activity*, Blaisdell
Tayler, R.J., 1997, *The Sun as a star*, Cambridge University Press
Young, C., 1895, *The Sun*, D. Appleton, New York
Vázquez Abeledo, M., 2003, *El Sol, algo más que una estrella*, Editorial Sirius, Madrid

Whitehouse, D., 2004, *The Sun: A Biography*, Wiley
Zirker, J.B., 2002, *Journey from the Center of the Sun*, Princeton University Press

References

Arnold, N. *Solar Variability, Coupling Between Atmospheric Layers and Climate Change*, p. 391–417. Advances in astronomy, Vol. I., Imperial College Press, J.M. Thompson (ed.), 2005.

Athay, R. G. (ed.). *The Solar Chromosphere and Corona: Quiet Sun*, volume 53. D. Reidel, 1976.

Bard, E. and Frank, M. Climate change and solar variability: What's new under the sun? *Earth and Planetary Science Letters*, 248:480–493, 2006.

Beck, R., Hilbrecht, H., Reinsch, K., and Völker, P. *Solar Astronomy Handbook*. Willmann Bell Inc., 1995.

Beer, J. Long-term indirect indices of solar activity, *Space Sciences Reviews*, 94: 53–66, 2000.

Benestad, R. E. *Solar Activity and Earth's Climate* (2nd ed.). Springer/Praxis, 316 pp. 2006.

Bhatnagar, A. Instrumentation and observational techniques in solar astronomy. In H. M. Antia, A. Bhatnagar, and P. Ulmschneider (eds.), *Lectures on Solar Physics*, volume 619, pp. 27–79. Springer Verlag, 2003.

Biermann, L. Der gegenwärtige Stand der Theorie konvektiver Sonnenmodelle. *Vierteljahresschrift Astron. Gesellschaft*, 76:194–200, 1941.

Biermann, L. Kometenschweife und solare Korpuskularstrahlung. *Zeitschrift für Astrophysik*, 29:274–286, 1951.

Bone, N. *The Aurora*. New York, Wiley, 1997.

Bonet, J. A. High Spatial resolution imaging in solar physics. In A. Hanslmeier and M. Messerotti (eds.), *ASSL Vol. 239: Motions in the Solar Atmosphere*, pp. 1–12, 1999.

Bray, R. J., Cram, L. E., Durrant, C., and Loughhead, R. E. *Plasma Loops in the Solar Corona*. Cambridge, Cambridge University Press, 1991.

Bray, R. J. and Loughhead, R. E. *The Solar Chromosphere*. The International Astrophysics Series, London, Chapman and Hall, 1974.

Bray, R. J., Loughhead, R. E., and Durrant, C. J. *The Solar Granulation* (2nd ed.). Cambridge and New York, Cambridge University Press, 270 pp. 1984.

Burroughs, W. J. *Weather Cycles: Real or Imaginary?* Cambridge University Press, 1992.

Carrington, R. C. On the distribution of the solar spots in latitudes since the beginning of the year 1854, with a map. *Monthly Notices Royal Astronomical Society*, 19:1–3, 1858.

Chapman, G. A. Variations of solar irradiance due to magnetic activity. *Annual Review of Astronomy and Astrophysics*, 25:633–667, 1987.

Chapman, S. and Zirin, H. Notes on the Solar Corona and the terrestrial ionosphere. *Smithsonian Contributions to Astrophysics*, 2:1–12, 1957.

Cox, A. (ed.). *Astrophysical Quantities*, Springer, 2000.

Dainty, J. C. (ed.). *Laser speckle and related phenomena*, Springer, 1975.

de Jager, C. Solar energy sources. In C. de Jager and E. R. Dyer, (eds.), *ASSL Vol. 29: The Sun. Part 1 of Solar-Terrestrial Physics/1970*, pp. 1–8, 1972.

del Toro Iniesta, J. C. On the discovery of the Zeeman effect on the sun and in the laboratory. *Vistas in Astronomy*, 40:241–256, 1996.

Donnelly, R. F. (ed.). *Proceedings of the Workshop on the Solar Electromagnetic Radiation Study for Solar Cycle 22*, National Technical Information Service, Springfield, 1992.

Earther, E. H. *Majestic Lights.* Washington, American Geophysical Union, 1980.

Eddy, J. A. The Maunder Minimum. *Science*, 192:1189–1202, 1976.

Edlén, B. Die Deutung der Emissionslinien im Spektrum der Sonnenkorona. *Zeitschrift fur Astrophysics*, 22:30–64, 1943.

Elste, G. and Gilliam, L. Limb darkening measurements free of instrumental stray light. *Solar Physics*, 240:9–15, 2007.

Fleming, J. A. On the electric radiation from Bent Antennæ. *Proceedings of the Physical Society of London*, 20:409–426, 1906.

Fligge, M., Solanki, S. K., Pap, J., Fröhlich, C., and Wehrli, Ch. *Journal of Atmospheric and Solar-Terrestrial Physics*, 63:1479–1487, 2001.

Foukal, P., Fröhlich, C., Spruit, H., and Wigley, T. M. L. Variations in solar luminosity and their effect on the Earth's climate. *Nature*, 443:161–166, 2006.

Fried, D. L. Statistics of a geometric representation of wavefront distortion. *Journal Optical Society of America*, 55:1427–1435, 1965.

Friis-Christensen, E., Fröhlich, C., Haigh, J. D., Schüssler, M., and von Steiger, R. (eds.). *Solar variability and Climate*, Kluwer, 2000.

Grotrian, W. Zur Frage der Deutung der linien im Spektrum der Sonnenkorona. *Naturwissenschaften*, 22:20, 1939.

Hargreaves, J. K. *The Upper Atmosphere and Solar-Terrestrial Relations*, Van Nostrand Reinhold, 1979.

Haigh, J. D., Lockwood, M., Giampapa, M. S., Rüedi, I., Güdel, M., and Schmutz, W. (eds.) *The Sun, Solar Analogs and the Climate*, 2005.

Hale, G. E. and Nicholson, S. B. The Law of Sun-Spot Polarity. *Astrophysical Journal*, 62:270–300, 1925.

Hanslmeier, A. *The Sun and Space Weather.* ASSL Vol. 277, Dordrecht, Kluwer Academic Publishers, 2002.

Hathaway, D. H. and Wilson, R. M. What the Sunspot Record Tells Us About Space Climate. *Solar Physics*, 224:5–19, 2004.

Herschel, W. Observations Tending to Investigate the nature of the Sun, *Philosophical Transactions Royal Society London* Series I, 91, 265, 1801.

Hermann, J. R., Goldberg, R. A. *Sun, Weather and Climate*, Dover, New York, 1978.

Hoyt, D. V., Schatten, K. H. *The role of the Sun in Climate Change*, Oxford University Press, 1997.

Hoyt, D. V., Schatten, K. H., and Nesmes-Ribes, E. The one hundredth year of Rudolf Wolf's death: Do we have the correct reconstruction of solar activity? *Geophysical Research Letters*, 21:2067–2070, 1994.

Hudson, H. S. Observed variability of the solar luminosity. *Annual Review of Astronomy and Astrophysics*, 26:473–507, 1988.

Huggins, W. Note on a method of viewing the solar prominences without an eclipse. *Proceedings of the Royal Society London*, 17:302–303, 1869.

Hulburt, E. O. Ionization in the upper atmosphere of the Earth. *Physical Review*, 31:1018–1037, 1928.

Jones, P. D. and Mann, M. E. Climate over past millennia. *Reviews of Geophysics*, 42:1–42, 2004.

Khomenko, E. V., Martínez González, M. J., Collados, M., Vögler, A., Solanki, S. K., Ruiz Cobo, B., and Beck, C. Magnetic flux in the internetwork quiet Sun. *Astronomy and Astrophysics*, 436:L27–L30, 2005.

Kirchhoff, G. and Bunsen, R. Chemische analyse durch spectralbeobachtungen. *Annalen der Physik und Chemie*, 110:161–189, 1860.

Kitchin, C. *Solar Observing Techniques.* Patrick's Moore Practical Astronomy Series, Springer, 2002.

Kneer, F. and von Uexküll, M. Diagnostics and dynamics of the solar chromosphere. In A. Hanslmeier and M. Messerotti, (eds.), *ASSL Vol. 239: Motions in the Solar Atmosphere*, pp. 99–118, 1999.

Kohl, J. L. and Cranmer, S. R. (eds.). *Coronal Holes and Solar Wind Acceleration, Proceedings SOHO-7 Workshop.* Kluwer Academic Publishers, 1999.

Korte, M. and Constable, C. G. The geomagnetic dipole moment over the last 7000 years – new results from a global model. *Earth and Planetary Science Letters*, 236:348–358, 2005.

Lean, J. The Sun's variable radiation and its relevance for Earth. *Annual Review of Astronomy and Astrophysics*, 35:33–67, 1997.

Livingston, W. and Harvey, J. Observational Evidence for Quantization in Photospheric Magnetic Flux. *Solar Physics*, 10:294–296, 1969.

Lyot, B. Étude de la couronne solaire en dehors des éclipses. *Zeitschrift für Astrophysik*, 5:73–95, 1932.

Macdonald, L. *How to Observe the Sun safely.* Patrick Moore's Practical Astronomy Series, London, Springer, 2003.

Mann, M. E. and Jones, P. D. Global surface temperatures over the past two millennia. *Geophysical Research Letters*, 30:5–1, 2003.

Martínez Pillet, V. Polarimetric Measurements of Sunspots. In J. C. Del Toro, B. Schmieder and M. Vázquez (eds.), *Advances in the Physics of Sunspots*, ASP Conf. Series 118:212–236, 1997.

Martínez Pillet, V. and Vázquez, M. The continuum intensity-magnetic field relation in sunspot umbrae. *Astronomy and Astrophysics*, 270:494–508, 1993.

Maunder, E. W. Note on the distribution of sun spots in heliographic latitude, 1874-1902. *Monthly Notices Royal Astronomical Society*, 64:747–761, 1904.

Maunder, E. W. Magnetic disturbances and associated sun-spots. *Monthly Notices Royal Astronomical Society*, 65:538–556, 1905.

McComas, D. J., Barraclough, B. L., Funsten, H. O., Gosling, J. T., Santiago-Muñoz, E., Skoug, R. M., Goldstein, B. E., Neugebauer, M., Riley, P., and Balogh, A. Solar wind observations over Ulysses' first full polar orbit. *Journal of Geophysical Research*, 105:10419–10434, 2000.

Meadows, A. J. A hundred years of controversy over sunspots and weather. *Nature*, 256:95–97, 1975.

Muller, R. The Solar granulation. In A. Hanslmeier and M. Messerotti (eds.), *ASSL Vol. 239: Motions in the Solar Atmosphere*, pp. 35–70, 1999.

Muller, R. and Roudier, T. Variability of the quiet photospheric network. *Solar Physics*, 94:33–47, 1984.

Nesme-Ribes, E. (ed.). *The Solar Engine and its Influence on Terrestrial Atmosphere and Climate.* Springer, 1994.

Nesme-Ribes, E., Thuillier, G. *Histoire solaire et climatique*, Belin, Paris, 2000.

Neugebauer, M. and Snyder, C. W. Solar plasma experiment. *Science*, 138:1095–1097, 1962.

Nordlund, Å. Solar photosphere and convection. In B. N. Dwivedi (ed.), *Dynamic Sun*, pp. 148–164. Cambridge University Press, 2003.

Pap, J., Fröhlich, C., Hudson, H. S. and Solanki, S. K. (eds.). *The Sun as a variable star: Solar and stellar irradiance variations*, IAU Coll. 143, Cambridge University Press, 1994.

Pap, J. M. and Fox, P. (eds.). *Solar Variability and its Effects on Climate.* American Geophysical Union, Geophysical Monograph 141, 2004.

Parker, E. *Cosmic Magnetic Fields.* Clarendon Press, 1979a.

Parker, E. N. Dynamics of the interplanetary gas and magnetic fields. *Astrophysical Journal*, 128:664–676, 1958.

Parker, E. N. Sunspots and the physics of magnetic flux tubes. I – The general nature of the sunspot. II – Aerodynamic drag. *Astrophysical Journal*, 230:905–923, 1979b.

Pittock, A. B. A critical look at long-term sun-weather relationships. *Reviews of Geophysics and Space Physics*, 16:400–420, 1978.

Rimmele, T. R., Balasubramaniam, K. S., and Radick, R. R. (eds.). *High Resolution Solar Physics: Theory, Observations, and Techniques*, 1999b.

Rudd, M. E. Chromatic aberration of eyepieces in early telescopes. *Annals of Science*, 64:1–18, 2007.

Rutten, R. J. Observing the solar chromosphere. In P. Heinzel, I. Dorotovič, and R. J. Rutten (eds.), *The Physics of Chromospheric Plasmas*, volume 368 of *Astronomical Society of the Pacific Conference Series*, pp. 27–48, 2007.

Sánchez, F., Collados, M., and Vázquez, M. (eds.). *Solar Observations: Techniques and Interpretation.* Cambridge University Press, 1992.

Sánchez Almeida, J. The Magnetism of the Very Quiet Sun. In T. Sakurai and T. Sekii (eds.), *The Solar-B Mission and the Forefront of Solar Physics*, volume 325 of *Astronomical Society of the Pacific Conference Series*, pp. 115–124, 2004.

Sánchez Almeida, J. A simple model for the distribution of quiet-Sun magnetic field strengths. *Astrophysical Journal*, 657:1150–1156, 2007.

Sánchez Almeida, J. and Lites, B. W. Physical Properties of the Solar Magnetic Photosphere under the MISMA Hypothesis. II. Network and Internetwork Fields at the Disk Center. *Astrophysical Journal*, 532:1215–1229, 2000.

Schmidt, W. Solar telescopes and instruments: Ground. *Encyclopedia of Astronomy and Astrophysics*, P. Murdin (ed.), 2000.

Schmidtke, G. Solar Variability in Ionizing Radiation (UV, X-rays). In *ESA SP-463: The Solar Cycle and Terrestrial Climate, Solar and Space weather*, pp. 61–68, 2000.

Schwabe, M. Sonnenbeobachtungen im Jahre 1843. *Astronomische Nachrichten*, 21:233–236, 1844.

Schwenn, R. Space weather: The solar perspective. *Living Reviews in Solar Physics*, 3(2), 2006.

Sheeley, N. R. Jr. The evolution of the photospheric network. *Solar Physics*, 9:347–357, 1969.

Simon, G. W. and Leighton, R. B. Velocity fields in the solar atmosphere. III. Large-scale motions, the chromospheric network, and magnetic fields. *Astrophysical Journal*, 140:1120–1147, 1964.

Sobotka, M. Fine Structures in Sunspots. In A. Hanslmeier and M. Messerotti (eds.), *ASSL Vol. 239: Motions in the Solar Atmosphere*, pp. 71–97. Kluwer Academic Publishers, 1999.

Sobotka, M., Bonet, J. A., and Vazquez, M. A high-resolution study of the structure of sunspot light bridges and abnormal granulation. *Astrophysical Journal*, 426:404–413, 1994.

Sobotka, M., Vázquez, M., Cuberes, M. S., Bonet, J. A., and Hanslmeier, A. Infrared photometry of solar photospheric structures. I. Active regions at the center of the disk. *Astrophysical Journal*, 544:1155–1168, 2000.

Socas-Navarro, H., Pillet, V. M., Sobotka, M., and Vázquez, M. The thermal and magnetic structure of umbral dots from the inversion of high-resolution full stokes observations. *Astrophysical Journal*, 614:448–456, 2004.

Solanki, S. K., Usoskin, I. G., Kromer, B., Schüssler, M., and Beer, J. Unusual activity of the Sun during recent decades compared to the previous 11,000 years. *Nature*, 431:1084–1087, 2004.

Sonnett, C. P., Giampapa, M. S., and Mathews, M. S. *The Sun in Time*, University of Arizona Press, 1991.

Spörer, G. Über die Periodicität der Sonnenflecken seit dem Jahre 1618 . *Nova Acta der ksl. Leop.-Carol Deutschen Akademie der Naturforscher*, 53:283–324, 1885.

Spruit, H. C. and Zwaan, C. The size dependence of contrasts and numbers of small magnetic flux tubes in an active region. *Solar Physics*, 70:207–228, 1981.

Stenflo, J. O. On the influence of the slit size on measurements with a solar magnetograph. *The Observatory*, 86:73–75, 1966.

Stenflo, J. O. The Hanle effect and the diagnostics of turbulent magnetic fields in the solar atmosphere. *Solar Physics*, 80:209–226, 1982.

Stewart, B. On the cause of the solar-diurnal variations of terrestrial magnetism. *Proceedings of the Physical Society of London*, 8:38–49, 1886.

Tatarski, V. F. *Wave propagation in a turbulent media.* New York, Dover, 1961.

Thomas, J. Sunspot penumbra: Structure and activity. *Encyclopedia of Astronomy and Astrophysics*, P. Murdin (ed.), 2000.

Thomas, J. H. and Weiss, N. O. Fine structure in sunspots. *Annual Review of Astronomy and Astrophysics*, 42:517–548, 2004.

Tousey, R., Strain, C. V., Johnson, F. S., and Oberly, J. J The solar ultraviolet spectrum from a V-2 rocket. *Astronomical Journal*, 52:158–159, 1947.

Uitenbroek, H. Chromospheric diagnostics. In J. Leibacher, R. F. Stein, and H. Uitenbroek (eds.), *Solar MHD Theory and Observations: A High Spatial Resolution Perspective*, volume 354 of *Astronomical Society of the Pacific Conference Series*, pp. 313–323, 2006.

Usoskin, I. G. A history of solar activity over millennia. *Living Reviews in Solar Physics*, 5(3), 2008.

Usoskin, I. G., Solanki, S. K., and Kovaltsov, G. A. Grand minima and maxima of solar activity: new observational constraints. *Astronomy and Astrophysics*, 471:301–309, 2007.

Vázquez, M. A morphological study of the light-bridges in sunspots. *Solar Physics*, 31:377–387, 1973.

Vázquez, M. Long-term changes in the total solar irradiance. In A. Hanslmeier and M. Messerotti (eds.), *ASSL Vol. 239: Motions in the Solar Atmosphere*, pp. 169–195, 1999.

Vázquez, M. The influence of solar variability on terrestrial climate. In S.G. Pandalai (ed.), *Recent Research Developments in Astrophysics, Vol. 1, Research Signpost*, pp. 1–26, 2003.

Vázquez, M. Solar variability and climate change. In A. Marquina (ed.), *Environmental Challenges in the Mediterranean 2000–2050, Nato Science Series, Kluwer*, pp. 29–46, 2004.

Vernazza, J. E., Avrett, E. H., and Loeser, R. Structure of the solar chromosphere. III - Models of the EUV brightness components of the quiet-sun. *Astrophysical Journal Supplement*, 45:635–725, 1981.

von der Lühe, O. Solar instrumentation. An introduction. In A. Hanslmeier, M. Messerotti, and A. Veronig (eds.), *The Dynamic Sun*, volume 259 of *Astrophysics and Space Science Library*, pp. 43–67, 2001.

Waldmeier, M. *Ergebnisse und Probleme der Sonnenforschung.* Leipzig, Geest Portig, 2. erweiterte Aufl., 1955.

White, O. R. (ed.), *The Solar Output and its Variation.*, 1977.

White, O. R. and Livingston, W. C. Solar luminosity variation. III – Calcium K variation from solar minimum to maximum in cycle 21. *Astrophysical Journal*, 249:798–816, 1981.

Willson, R. C., Gulkis, S., Janssen, M., Hudson, H. S., and Chapman, G. A. Observations of solar irradiance variability. *Science*, 211:700–702, 1981.

Willson, R. C., Hudson, H. S., Frohlich, C., and Brusa, R. W. Long-term downward trend in total solar irradiance. *Science*, 234:1114–1117, 1986.

Wilson, A. (ed.). *The Solar Cycle and Terrestrial Climate*, 2000.

Wollaston, W. A method of examining refractive and dispersive powers, by prismatic reflection. *Philosophical Transactions Royal Society London*, 92:365–379, 1802.

Yang, S., Odah, H., and Shaw, J., Variations in the geomagnetic dipole moment over the last 12000 years. *Geophysical Journal International*, 140:158–162, 2000.

Zhao, J., Kosovichev, A. G., and Duvall, T. L. Jr. Investigation of mass flows beneath a sunspot by time-distance helioseismology. *Astrophysical Journal*, 557:384–388, 2001.

2

Naked-Eye Sunspots

Humans have been looking at the Sun for millennia. Nevertheless, on most occasions one only manages to be dazzled. But there are some situations in which the intensity of the light of the Sun diminishes and we can observe the solar disk with the naked eye. Some examples are the Sun observed through fog or during a dust-storm, or even the Sun observed through the smoke during a forest fire. On these occasions, we can see the solar disk without being dazzled. If there was a sunspot of great size just at that moment, then we might distinguish it inside the solar disk (see Figure 2.1).

There are several myths concerning the vision of spots or blemishes on the Sun in antiquity. For example, a supposed Aztec pre-Columbian myth has a god-Sun (creator of the world) with dirty marks on his face. This seems to be slightly more convincing than some supposed drawings of sunspots in Egyptian civilization. One can also mention a beautiful story from the area of the Zambesi in Africa that seems to be more credible. The Moon envies the Sun and throws mud at its face. Luckily, this does not happen very often since the Sun is vigilant. But every ten years approximately, the Sun loses concentration and gets dirtied by the mud (Brody, 2002).

Undoubtedly written observations are more trustworthy than myths and orally transmitted legends.[1] A 3000-year-old Babylonian tablet might contain a reference to observations of sunspots (Sayce, 1877). Surviving Late Babylonian astronomical diaries are very detailed (Sachs and Hunger, 1988), but a clear allusion to sunspots has not been found. It should be emphasized that: (1) no more than 5% of the original texts have survived and (2) systematic observations such as eclipses or planetary movements were recorded in abundance, but other sporadic events such as aurorae or meteors are very scarce. One can speculate that naked-eye sunspot observations in

[1] Early writing systems are dated in the late 4th millennium BC. However, they were not a sudden invention and were based on previous traditions of symbol systems (Houston, 2004).

J.M. Vaquero, M. Vázquez, *The Sun Recorded Through History*,
Astrophysics and Space Science Library 361, DOI 10.1007/978-0-387-92790-9_2,

Fig. 2.1. Large sunspot groups in the solar disk on 27 October 2001 (courtesy of Catania Astrophysical Observatory). The amateur astronomer J. Ruiz saw, that day, four naked-eye sunspots using a filter.

China go back at least to the twelfth century BC (Hsü, 1972) and, probably, the classical Greeks also observed sunspots.

Bicknell (1968) has suggested that a naked-eye sunspot was observed by Anaxagoras of Clazomenai (500–428 BC) in 467 BC, but it is just a supposition. Sarton (1947) noted that there are no clear references to sunspots in classical literature. However, there are some texts that might suggest systematic solar observations for weather prediction. There are several references to possible naked-eye sunspot observations in the surviving fragments of *De Signis Tempestatum* ("On weather signs") by Theophrastus (371–287 BC). The work consists of several chapters describing signs of different weather meteors. Three fragments have references to possible sunspots (Hardy, 1991): (1) "If the Sun has a black mark when it rises, or if it rises out of clouds it is a sign of rain" (in the section "Signs of Rain"), (2) "If the Sun rises with a burning heat but does not shine brilliantly, it is a sign of wind. If the Sun has a hollow appearance, it is a sign of wind or rain... also black spots on the Sun or Moon indicate rain, red spots wind" (in the section "Signs of Wind") and (3) "If the Sun rises brilliantly but without scorching heat and without showing any special sign on his orb, it indicates fair weather" (in the section "Signs of Fair Weather"). We consulted the *Opera* of Theophrastus (Theophrasti Eresii, 1866), but an English translation of *De Signis Tempestatum* can be consulted (Theophrastus, 1916).

Nevertheless, Theophrastus might not be an "observer of sunspots" in the strict sense since he only expounded a simple rule of weather prediction using sunspots. Later, we can find other similar texts in Greek literature. Examples may be verses 822–824 of *Phaenomena* (Appearances) by Aratus (315–240 BC) or some verses of *Georgics* by Virgil (70–19 BC). Modern knowledge of major maxima and minima of solar activity (Usoskin et al., 2007) can help us. During the 4th century BC solar activity was low. Thus, it is improbable that Theophrastus's contemporaries saw naked-eye sunspots. However, solar activity during the 5th century BC was very high and it is quite probable that ancient Greeks observed sunspots (including Anaxagoras). Moreover, it is well known that the pre-Socratics were keen observers.

This chapter is devoted to naked-eye sunspot observations, and especially to their possible use for the reconstruction of solar activity during the last two millennia. Firstly, we will study the human eye as a detector of light. Secondly, we will study the criteria of visibility of sunspots. Then, we will be ready to present the historical observations that have survived and the modern programmes of observation carried out by amateur astronomers. Finally, we will show the interest of this kind of observation for the reconstruction of the history of the Sun during the last two millennia.

2.1 The Human Eye as a Detector of Light

In the context of this book, the human eye is the available detector. Therefore it is of interest to review its main properties. *Oculus hoc est fundamentum Opticum* by C. Scheiner (1619) is probably the first work about visual optics and anatomy. It includes detailed observations of the pupil during accommodation and of the refractive power of the lens and the aqueous and vitreous humour (Southall, 1922). He was impressed by the analogy between the eye and the camera obscura (Daxecker, 2004).

Sunlight enters the eye by passing through the cornea, where the image is focused. The brightness of the image is controlled by the varying diameter of the pupil, the aperture of the optical system. The average size of the exposed pupil varies throughout life, but averages between 6 and 7 millimetres. The iris regulates the amount of light passing through the pupil, acting like a camera shutter. As the amount of light entering the eye diminishes, the iris muscle pulls away from the centre, causing the pupil to dilate and allowing more light to pass. The image is given a fine focus by the *lens*. Finally the image falls on the retina, where the real power of the eye is located (Figure 2.2). At any given instant, the retina can resolve a contrast ratio of 100:1.

The retina is a thin layer composed of five types of cells lining the back of the eye. The cells are arranged in four layers. Using the outermost two, the retina can turn parts of itself on and off as it analyses the image. The photoreceptors are located behind. The ones that allow us to see colour are the cones, with their concentration moving outwards from an area of the retina called fovea centralis through the central part of the retina, called the *macula*. A few degrees from the fovea appear the rods, outnumbering

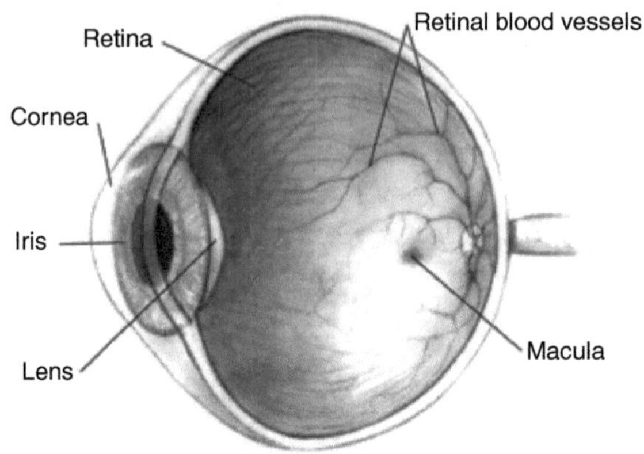

Fig. 2.2. Cross-section of the human eye.

the cones by 20 to 1, and acting as light collectors. These elements respond to light by generating electrical impulses that travel out of the eye through the optic nerve to the brain.

The sensitivity of the eye ranges over about 14 orders of magnitude from a minimum threshold to a light level that could possibly cause damage. The photopic (cone) threshold is almost four magnitudes above the minimum. The next two magnitudes are called the mesopic range and it is here that both rods and cones contribute to vision. The scotopic peak sensitivity (of rod cells) is at about 500 nm, while photopic sensitivity peaks at around 550 nm (Figure 2.3). Wavelengths shorter than 315 nm are absorbed by the cornea (causing injury) and do not reach the retina.

As in the case of telescopes, eyes are not perfect optical systems. So the relative intensity of the object is distributed across the retina as shown in

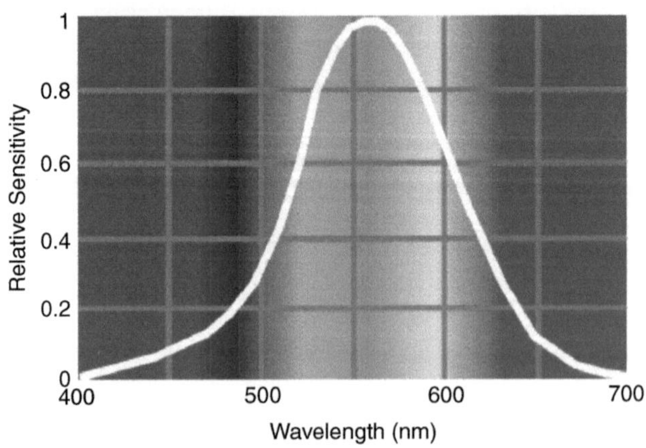

Fig. 2.3. Photopic sensitivity of the human eye.

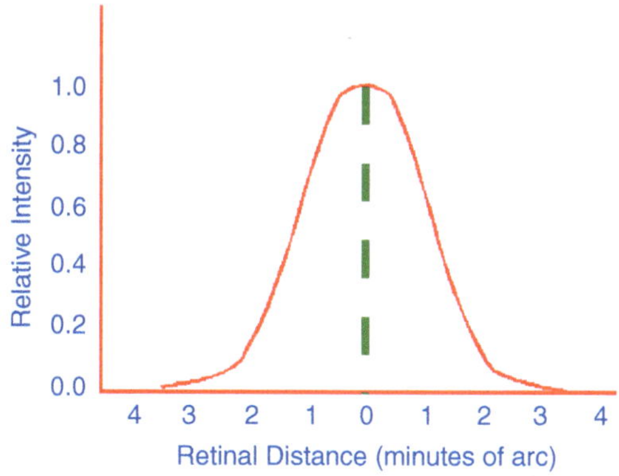

Fig. 2.4. The Point Spread Function of the retina of the eye.

Figure 2.4. At any given instant, the retina can resolve a contrast ratio of 100 to 1.

2.1.1 Solar Damage to the Eye

Solar retinopathy is a kind of damage to the eye's retina, particularly to the macula, induced by prolonged exposure to solar radiation. It usually occurs due to staring at the Sun or viewing a solar eclipse. The main damage to the eye is photochemical rather than thermal (Ham et al., 1976). Young people are much more likely to suffer damage than their elders, because the eye gradually becomes yellower with age, filtering out the harmful UV photons (Istock, 1985).

In a letter to British philosopher John Locke (1632–1704), Isaac Newton (1643–1727) describes the effects of looking at the reflection of the Sun in a mirror, while standing in a darkened room.[2] Westfall (1980) comments that "Newton left the sun alone after that". Newton's description of his symptoms

[2] "The observation you mention [. . .] I once made upon my self with the hazzard of my eyes. The manner was this. I looked a very little while upon ye sun in a looking-glass with my right eye and then turned my eyes into a dark corner of my chamber and winked to observe the impression made by the circles of colours which encompassed it & how they decayed by degrees & at last vanished. This I repeated a second and a third time. At the third time when the phantasm of light & colours about it were almost vanished, intending my phansy upon them to see their last appearance I found to my amazemt that they began to return & by little & little to become as lively & vivid as when I had newly looked upon the sun. But when I ceased to intende my phansy upon them they vanished again. After this I found that as often as I went into the dark & intended my mind upon them as when a man looks earnestly to see any thing which is difficult to be seen, I could make the phantasm return without looking any more upon the sun. And the oftener I made it return, the more easily I could make it return again. And at length by repeating this without looking any more upon the sun I made such

closely agrees with a typical mild solar retinopathy characterized by a yellow foveolar dot and a central scotoma.[3] The damage was not evident immediately, but after several hours; the symptoms were most visible for a few days, and gradually subsided over a long time (several months), eventually disappearing entirely, or nearly so. Photophobia (in which the victim avoids the light) is also a common symptom.

2.2 Visibility Criteria

When is a sunspot visible by using an instrument? We are especially interested in the response when the instrument is only the human eye. Counting sunspots to establish the sunspot number is of great importance for astronomers, climatologists, and space engineers. Even at present, the sunspot counts are made using small telescopes. For this, several authors have calculated the conditions that must be fulfilled in order for a sunspot to be visible. In particular, their calculations serve to indicate when a sunspot can be visible to the naked eye.

The simplest visibility criteria are based on the spatial resolving power of optical instruments. Rayleigh's criterion establishes that the power of an optical system to distinguish two structures separated by an angular distance

---

an impression on my eye that if I looked upon the clouds or a book or any bright object I saw upon it a round bright spot of light like the sun. And, which is still stranger, though I looked upon the sun with my right eye only & not with my left, yet my phansy began to make the impression upon my left eye as well as upon my right. For if I shut my right eye and looked upon a book or the clouds with my left eye I could see the spectrum of the sun almost as plain as with my right eye, if I did but intend my phansy a little while upon it. For at first if I shut my right eye and looked with my left, the spectrum of the Sun did not appear till I intended my phansy upon it; but by repeating this, appeared every time more easily. And now in a few hours time I had brought my eys to such a pass that I could look upon no bright object with either eye but I saw the sun before me, so that I durst neither write nor read but to recover the use of my eyes shut myself up in my chamber made dark for three days together & used all means to divert my imagination from the Sun. For if I thought upon him I presently saw his picture though I was in the dark. But by keeping in the dark and imploying my mind about other things I began in three or four days to have some use of my eyes again & by forbearing a few days longer to look upon bright objects recovered them pretty well, th not so well but that for some months after the spectrum of the sun began to return as often as I began to meditate upon the phaenomenon, even tho I lay in bed at midnight with my curtains drawn. But now I have been very well for many years, tho I am apt to think that if I durst venture my eyes I could still make the phantasm return by the power of my fansy." The quotation is taken from "The Correspondence of Isaac Newton" (Turnbull, 1961, pp. 153–154). The matter is discussed more briefly by R. Westfall (1980, pp. 93–94).

[3] An area of lost or depressed vision within the visual field, surrounded by an area of less depressed or of normal vision.

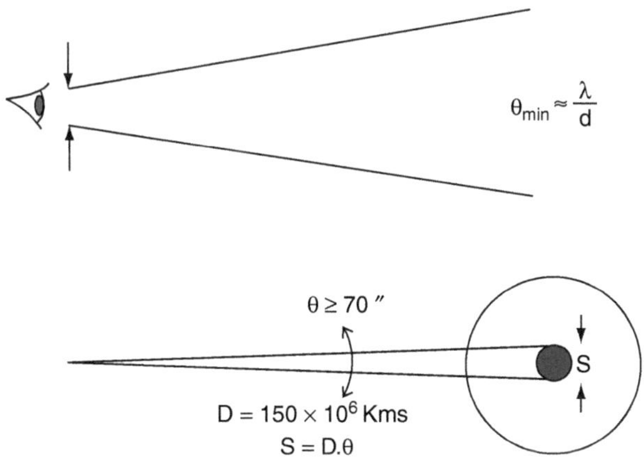

Fig. 2.5. The simplest visibility limit is based on the spatial resolving power of the human eye.

ϑ_{min} can be expressed as

$$\vartheta_{min} \approx (1.22\lambda)/d$$

where λ is the light's wavelength and d is the diameter of the optical system (see Figure 2.5). The diameter of the pupil of our eyes changes between 1 and 5 mm, depending on the illumination conditions. We can assume a value of 1.5 mm for a diurnal observation and a value of 500 nm for the wavelength of the sunlight. Using these values, we obtain $\vartheta_{min} \sim 70''$. Spots of this size are not too rare (3%) especially close to a maximum of the 11-year cycle of solar activity. It is obvious to think that a sunspot smaller than the value given for resolution of the human eye must be invisible. However, this logic is naive and contradicted by observation. For example, MacRobert (1989) observed sunspot groups with penumbral diameters from 22″ to 26″. It is important to note that since modern observers can use filters, the diameter of the pupil could be greater than 1.5 mm and, therefore, $\vartheta_{min} < 70''$. The sunspot visibility problem should be approached as a problem of calculating contrast thresholds for the human eye (Schaefer, 1991). Schaefer (1993) developed a theoretical model of sunspot visibility that can be applied to naked-eye observation, direct vision through a telescope or pinhole camera, and telescope projection. In the following paragraphs, we shall show this theory applied to naked-eye sunspot visibility.

The threshold contrast ratio C_{th} is the lowest contrast (with respect to the photosphere) which makes a sunspot visible, and can only be established through physiological experimentation.[4] This threshold depends on the

[4] In a general form, the contrast ratio is a measure of a display system defined as the ratio of the luminance of the brightest zone to that of the darkest zone that the system can produce.

background brightness, the size scale of the source, and the brightness distribution of the object. Blackwell (1946) presented half a million observations over eight orders of magnitude in brightness and nearly three orders of magnitude in size scale from 19 observers. This work is the definitive study on C_{th} from the physiological experimentation point of view.

Blackwell (1946) defined the critical visual angle θ_{cva} as the effective size over which the eye integrates, or the diameter of the effective pixel of vision. Using Blackwell's data, for diurnal vision one can write

$$\theta_{cva} = (40''/S)[10^{(B^{0.3}/60)}]$$

where B is the effective brightness in lamberts[5] of the photosphere as perceived by the eye and S is the Snellen ratio. This last parameter is a measure of the visual acuity of an observer compared with a standard observer.

We can write C_{th} as a function of B using Blackwell's data for B brighter than 0.001 lamberts using the equation

$$C_{th,B} = 0.0028 + [0.3 - 0.133\log(B)](\theta_{cva}/\zeta)^2 \tag{2.1}$$

where ζ is the angular diameter of the circular sources used by Blackwell.

There are two problems when we try to model sunspot visibility from Blackwell's work. First, sunspots do not have a simple shape (such as circular). Moreover, their shape is smeared out by diffraction and atmospheric seeing. Second, the experience of the observer is an important factor in the detection of sources with lower contrast. See, for example, Schaefer (1990) and Doggett and Schaefer (1994). The experience effect can be included in the model using

$$C_{th} = eC_{th,B} \tag{2.2}$$

where e is a factor based on the experience of the observer. We have e = 1 and e = 4 for experienced and novice observers, respectively.

The total solid angle covered by an idealized sunspot on the solar disk will be

$$\Omega_{sunspot} = \pi R^2_{sunspot}\cos\theta \tag{2.3}$$

where $R_{sunspot}$ is the angular size of the semimajor axis of the idealized sunspot and θ is its heliocentric angle. Moreover, the relation between the sunspot area and the solid angle of the sunspot can be written as

$$A_{sunspot} = (0.173 \text{ millionths per square arcsec})\Omega_{sunspot}/\cos\theta$$

The surface brightnesses for the photosphere, umbra, and penumbra can be calculated from the surface brightness of the centre of the Sun as viewed above the atmosphere (B_{centre}). This surface brightness is equal to 7.8×10^5

[5] The lambert (symbol L) is a unit of luminance named after Johann Heinrich Lambert (1728–1777). It is equal to $10^4/\pi$ candela per square metre.

lamberts. These surface brightnesses depend on the heliocentric angle from the centre of the disk according to the equations

$$\mathrm{B_{photo}} = B_{centre}(0.41 + 0.59 cos\theta)$$

$$\mathrm{B_u} = B_{photo}(0.17 - 0.09 cos\theta)$$

and

$$\mathrm{B_p} = \mathrm{B_{photo}}(0.76 - 0.02\cos\theta)$$

using visual wavelengths of 5.5×10^{-5} cm (Allen, 1976). Moreover, the contrast ratios of the umbra and penumbra when compared to the background light of the photosphere will be $\mathrm{C_u} = 0.83+0.09\cos\theta$ and $\mathrm{C_p} = 0.24+0.02\cos\theta$. However, these values will vary from sunspot to sunspot. Thus, the result of the visibility model will be expressed with the contrast ratios as free parameters.

The sunspot images are smeared out by turbulence in the atmosphere. In addition, diffraction effects must be taken into account. Thus, we should calculate the two-dimensional convolution of the sunspot structure with smearing functions. However, this calculation is difficult and it is impossible to present the result in a simple and general form. Schaefer (1993) obtained the characteristic area of the convolution as an alternative. It can be proved that the second moment of a convolution of circularly symmetric normalized distributions is the sum of the second moments of the functions being convolved. For this reason, we must calculate the second moments of all the relevant functions.

The second moment of the normalized contrast ratio as viewed by the observer will have a different functional form for every observing method, and will be

$$\mu = \mu_{\mathrm{seeing}} + \mu_{\mathrm{diff}} + \mu_{\mathrm{aper}} + \mu_{\mathrm{sunspot}} \tag{2.4}$$

For the case of naked-eye observations, μ_{seeing}, μ_{diff}, and μ_{aper} can be neglected. For μ_{sunspot}, we can write:

$$\mu_{\mathrm{sunspot}} = (\mathrm{MR_{sunspot}})^2 \left(\frac{0.5[(\mathrm{C_u} - \mathrm{C_p})(\mathrm{A_u}/\mathrm{A_{sunspot}})^2 + \mathrm{C_p}]}{(\mathrm{C_u} - \mathrm{C_p}(\mathrm{A_u}/\mathrm{A_{sunspot}}) + \mathrm{C_p})}\right) \tag{2.5}$$

We can compare the visibilities of the observed light distribution with a uniform circular source. The second moment of a uniform circular source is $\mu_{\mathrm{circ}} = \zeta^2/8$ where ζ is the angular diameter. If we equate the second moments of the uniform circular source and the observed normalized brightness distribution, then the size scale of the equivalent normalized circular source will be

$$\zeta = (8\mu)^{0.5} \tag{2.6}$$

Then, we can write an equation for the contrast ratio for the sunspot as viewed by the observer:

$$\mathrm{C_{obs}} = (\mathrm{MR_{sunspot}}/\zeta)^2 \cos\theta[(\mathrm{C_u} - \mathrm{C_p})\mathrm{A_u}/\mathrm{A_{sunspot}} + \mathrm{C_p}] \tag{2.7}$$

Now, we should compare this contrast ratio against the threshold referenced in Equation (2.2). For naked-eye observations, only the sunspot size is relevant and from Equations (2.4), (2.5), and (2.6), we can write

$$\zeta = 1.74\mathrm{R}_{\mathrm{sunspot}} \tag{2.8}$$

According to our reasoning, the sunspot will be visible if $C_{th} < C_{obs}$. The limiting solid angle for a visible sunspot from Equations (2.1), (2.2), (2.3), (2.7), and (2.8) can be expressed as follows:

$$\Omega_{\mathrm{lim}} = \frac{\pi[0.3 - 0.133\log(\mathrm{B})]\theta_{\mathrm{cva}}^2}{[(\mathrm{C_u} - \mathrm{C_p})\mathrm{A_u}/\mathrm{A_{sunspot}} + \mathrm{C_p}]/\mathrm{e} - 0.0085/\cos\theta}$$

This general result shows that the visibility has a strong dependence on the observer and a weak dependence on the apparent brightness of the Sun and the position of the sunspot. We can estimate Ω_{lim} using some typical values. When a naked-eye sunspot is observed, the apparent brightness of the photosphere will depend on the observation conditions that can vary widely. However, typical values for B and θ_{cva} are 0.6 lambert and $42''\mathrm{S}^{-1}$, respectively. We can take $\mathrm{A_u}/\mathrm{A_{sunspot}}$ equal to 0.17 (see Chapter 1). In fact, larger sunspots are intrinsically darker than smaller ones (e.g. Collados et al., 1994). Using these values and for a sunspot located near the centre of the solar disk ($\theta \approx 0$), the limiting sunspot size is given by

$$\Omega_{\mathrm{lim}} = (35''\mathrm{S}^{-1})^2\mathrm{e}$$

If we assume typical values for $\mathrm{S} = 1.0$ and $\mathrm{e} = 1$ (for experienced observers), we have a reference value for Ω_{lim} equal to 1225 square arcsec. Schaefer (1993) compared his theoretical results with observations. However, he pointed out that the comparison of a single observation is dangerous. The model's limits are statistical in nature. Blackwell's results give the thresholds for the 50% detection level, and the theoretical model should be compared only against large data sets. Moreover, Schaefer made an observational test of his mathematical model. He observed naked-eye sunspots for 124 days (years 1990 and 1991). The observed solid angle subtended by a sunspots with 50% probability of detection was 2000 arcsec^2 and the theoretical values was $\Omega_{\mathrm{lim}} = 2500$ arcsec^2. Figure 2.6 shows the sunspot visibility fraction as a function of sunspot area, with the observed Ω_{lim} being where the probability of detecting a sunspot is 50%. The value of the model Ω_{lim} strongly depends on the adopted values of S and e. Comparison between the modeled and the observational values shows good agreement. The differences between the two quantities are comparable with the uncertainties in both the sunspot areas and the acuity of the observer.

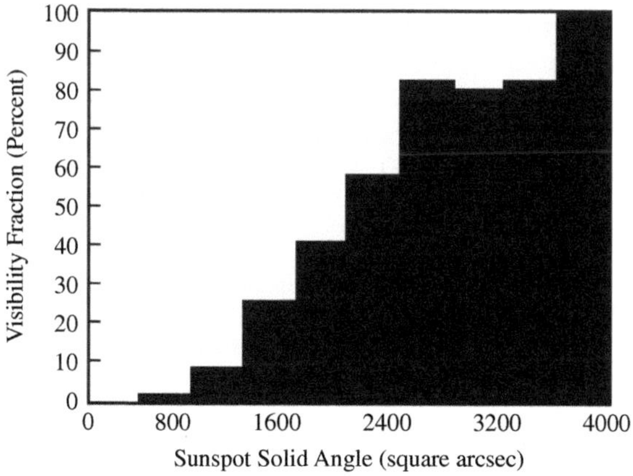

Fig. 2.6. Percentage of naked-eye sunspots visible as a function of sunspot area. Observations by Schaefer (1993) were made for 1990 and 1991. Figure adapted from Schaefer (1993).

2.3 Naked-eye Sunspot Observations

In this section, we will try to provide a general review of naked-eye sunspot records. The historical Oriental observations are the most complete set. However, a few examples of Arabic, European, Mayan, and Indian historical observations are available. Moreover, a considerable number of naked-eye observations were made during the telescopic epoch.

2.3.1 Historical Oriental Observations

The Sun was worshipped in ancient China, and there were daily sacrificial rites for welcoming and seeing off the Sun (Xu, 1990). Bone inscriptions of the Shang Dynasty (ca. 1500–1050 BC) are clear evidence for this. It is known that the ancient Chinese observed and recorded at least four solar phenomena, probably because Sun worship was the cause of spontaneous observations. They are (1) *Ri Hui* (the Sun was dark and gloomy), (2) *Ri Yun* (solar halo), (3) *Ri Zhi* (no definite translation), and (4) *Ri Shi* (solar eclipses). According to Xu (1990), the earliest sunspot observations should be identified by the expression *Ri Zhi* because sunspot is the most reasonable translation, although the words can be interpreted in different ways.

The great majority of the records of naked-eye sunspot observations (95% approximately) come from the East (especially China and Korea). The Oriental record covers a period of time four times longer than the period of telescopic observations. Unfortunately, only approximately 200 records survive. The records suggest the presence of the undecennial solar cycle, but only in the periods where a great density of observations exist (Yunnan Observatory, 1977; Ding et al., 1983; Wittmann and Xu, 1987).

There is information available about the astrological bases of this type of observations and about the differences of philosophy and religion that seem to explain the almost total absence of similar records from the Western world (Needham, 1959). In addition, one also knows how important are the effects of the persistent seasonal haze of the Asian continent on the Oriental record of naked-eye sunspots (Willis et al., 1980).

Nevertheless, our ability to interpret the record is determined by three factors: (1) the observational techniques used, (2) the frequency of the observations, and (3) the capacity of the record to reflect the state of solar activity during that epoch.

Of course, sunspots can be seen with the unaided eye if the sunlight is attenuated by haze, smoke, clouds, or similar, or when the Sun is near the horizon (sunrise and sunset) and the atmosphere becomes another natural filter. However, Needham (1959) has suggested that the Oriental astrologers may have used attenuating filters manufactured with rock-crystal or polished jade. Another possibility is solar observation by reflection on a pool of coloured liquid (Chu Wen-Hsin, 1934; Wang and Siscoe, 1980; Bo Shu-ren, 1983).

Chinese official histories and the systematic record of naked-eye sunspots in China started to be compiled in the Han dynasty (206 BC to AD 220). Moreover, there are numerous local topographies or histories called *Fang Zhi* (local gazettes) that formed an important supplement to the official histories. However, the *Fang Zhi* were ignored by the catalogue compilers until Xu and Jiang (1982).

In the mid-19th century, when Humboldt presented in his *Kosmos* Schwabe's discovery of the solar cycle, and solar studies were revitalized, several authors started to make compilations of naked-eye sunspot records. The most complete pioneering study was done by Kanda (1933), who compiled a catalogue of Far Eastern observations. There are more detailed works by the Yunnan Observatory (1977), Clark and Stephenson (1978) and Chen and Dai (1982). However, the two catalogues most used by the solar physics community are the works of Wittmann and Xu (1987) and Yau and Stephenson (1988). These two catalogues are very similar. There are a few differences in the translation of foreign languages. But the main difference is that Yau and Stephenson (1988) is restricted to records from East Asia extending down to as late as 1918. The most modern and detailed catalogue is the "*Catalog of Large Sunspots (165 BC–1992)*" by A. Wittmann which, although unpublished, can be consulted.[6]

The accounts of sunspots and a great number of astronomical observations such as eclipses, lunar and planetary movements, comets , and novae are mainly cited in astronomical treatises that form important sections of the official dynastic histories of China and Korea. The origin of this great interest in astronomical observation was astrological. An interesting aspect of this interest is that the terminology used in the reports remained almost unchanged

[6] http://www.astro.physik.uni-goettingen.de/~wittmann/.

over the last two millennia. With the aid of the catalogues, an overview can be given of the temporal coverage and geographical distribution of the records, and the most usual descriptions of sunspots (Stephenson, 1990).

There are very few reports before AD 300. The earliest reliable and well-dated report of an Oriental naked-eye sunspot dates from 165 BC. Until the year AD 1150, all the reports come from China except for one Japanese observation in AD 851. After that date, there are also reports from Korea but reports from China are more numerous. In AD 1276, 1593 and 1603, there are occasionally reports from Vietnam. Nevertheless, the number of sunspots recorded by ancient observers is very small in terms of modern sunspot observations (Mossman, 1989; Eddy et al., 1989).

With regard to the terms used by Oriental astronomers in their descriptions, there are two basic forms. The first is "Within the Sun there was a black spot" (*hei-tzu*) and the second is "Within the Sun there was a black vapour" (*hei-ch'i*). Occasionally, there are other descriptions using miscellaneous objects appearing within the Sun such as birds or stars. An important fact is that practically all the descriptions are written in Classical Chinese and there is no grammatical plural in this language. Thus, the best is to assume that all sunspot accounts refer to a single spot. Moreover, observations of double or multiple sunspots are recorded on several occasions. The earliest observation of two different spots in the Sun occurs in AD 355.

Stephenson (1990) summarized the time distribution of naked-eye sunspot records in Oriental sources in five characteristics (see Figure 2.7): (1) there are only sporadic reports before AD 1100, (2) the reports are relatively frequent in China and Korea during the 12th Century, (3) there are very few reports in the period AD 1200–1350, (4) there is a marked peak around AD 1370, and (5) there are relatively few reports of sunspots in China and Korea until AD 1600.

Figure 2.7 shows a record with poor uniformity. There are historical explanations for some of the gaps in this figure. For instance, the gap between approximately AD 600 and 800 is due to the sack of the T'ang Dynasty capital

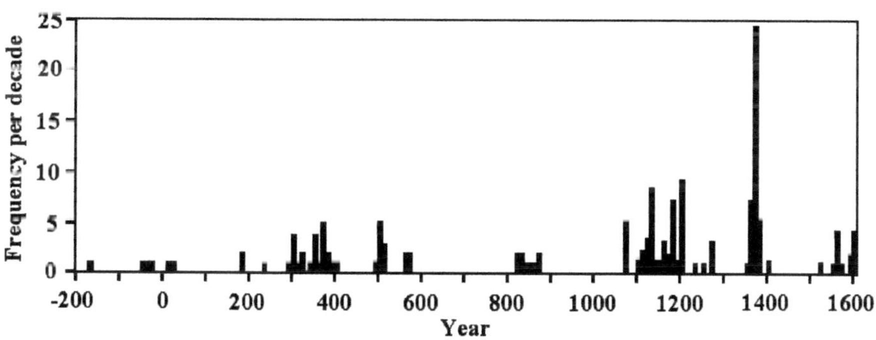

Fig. 2.7. Decadal distribution of naked-eye sunspots recorded in East Asia from 165 BC to AD 1610 (adapted from Yau, 1988).

of Ch'ang-an in AD 755. This situation meant that a large number of ancient records were destroyed. Important documentation was probably lost due to similar facts in other periods (as during the fall of a dynasty).

It is important to point out that sunspots were very often detected near the day of a new moon because at these periods observations were more frequent and intensive due to the determination of the new moon date. This kind of astronomical observation was important for calendar purposes (Wittmann and Xu, 1987).

Another important factor that can explain the poor uniformity of the record is astrology. Stephenson (1990) indicates that more than a half of all Chinese naked-eye sunspots during the pre-telescopic period were reported during the reigns of only six emperors. However, there were approximately 150 emperors during the period covered in Figure 2.7. Park (1977) cites another interesting fact from Korea: in February AD 1204, a sunspot was seen for three days but the Korean Royal astronomer tried to suppress the report. The cause was that he knew from Chinese history that it was a presage of the Emperor's death. Finally, the report was registered and the Korean king died a month later. Curiously, there are no reports of naked-eye sunspots in Korean historical sources during the next 54 years, although several observations were recorded in China! The conclusion is that the influence of astrology in the uniformity of the sunspot record is far from negligible.

Figure 2.8 is a plot of the decadal frequency of astronomical events of all kinds recorded in China from 200 BC to AD 1610. These astronomical observations may be comets, eclipses, lunar and planetary conjunctions, sightings of Venus in daylight, meteors, aurorae, etc.). The figure illustrates the great

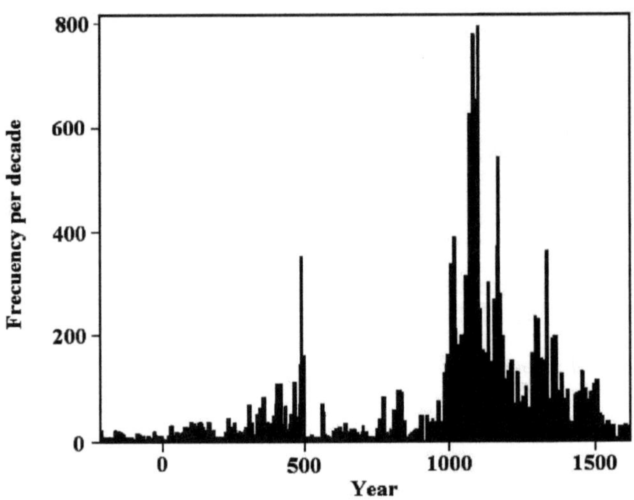

Fig. 2.8. Decadal distribution from 200 BC to AD 1610 of astronomical observation recorded in China (adapted from Yau, 1988).

variability in the number of astronomical phenomena recorded. According to Stephenson (1990), this complex variability must be largely of artificial origin.

Yau (1988) and Stephenson (1990) compared the variability of the naked-eye sunspot record with the variability of solar halos reported in East Asian sources. Although we know that the solar halo is an atmospheric phenomenon (Lynch and Livingston, 2001), it is recorded in the same sections of dynastic histories as sunspots ("solar changes" section). Figure 2.9 shows the decadal variability of solar halos recorded in Chinese sources during the period AD 1–1610. A great variability can be observed. Moreover, some of the main features of Figure 2.9 are similar to those of Figure 2.7. Peaks occurred approximately in AD 400, 1200 and 1400. This result suggests that the long-term variations observed in the sunspot record could be an artifact.

In 2004, a group of Korean solar physicists reviewed historical Korean sources looking for naked-eye sunspots and aurorae (Lee et al., 2004). They used the followings historical sources:

(1) *Koryo-Sa*: The Annals of the Koryo Dynasty (918–1391).
(2) *Choson Wangjo Sillok*: The Annals of the Choson Dynasty (1392–1910).
(3) *Jeungbo Munheon Bigo*: an encyclopaedia of 250 volumes, which was published in 1770 and revised later in 1908.
(4) *Daedong Yaseung*: unofficial historical book of 71 volumes, during the Choson Dynasty.

Table 2.1 presents the dates and descriptions of the naked-eye sunspots that they found. We have added some notes specifying the reference number in the Yau and Stephenson (1988) catalogue or the "new record" character. Lee et al. (2004) indicate that auroral sightings are recorded much more frequently than naked-eye sunspots (788 aurorae with respect to 60 sunspots). The sunspots observed in 1105, 1152, 1608 (17 May) and 1720 have not been included in previous works. Also, the sunspot on 16–19 March 1361 was

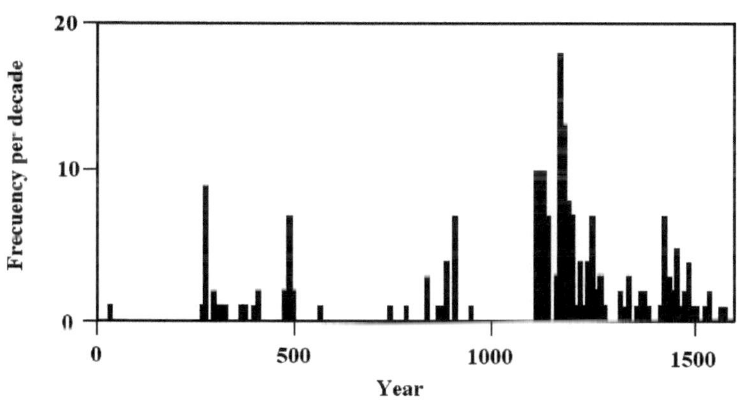

Fig. 2.9. Decadal distribution of solar halos recorded in China (adapted from Yau, 1988).

Table 2.1. Sunspot records compiled by Lee et al. (2004) using Korean historical sources

Year	Month	Day	Brief Description	Notes
1105	2	7	Black light	New record
1151	3	21	Black spot as large as hen's egg	Yau and Stephenson (1988) [79]
1151	3	31	Black spot as large as hen's egg	Yau and Stephenson (1988) [79]
1152	4	1	Black spot as large as hen's egg	New record
1160	9	29	Black spot	Yau and Stephenson (1988) [82]
1171	10	20	Black spot as large as peach	Yau and Stephenson (1988) [83]
1171	11	16	Black spot as large as peach	Yau and Stephenson (1988) [84]
1183	12	4–5	Black spot	Yau and Stephenson (1988) [85]
1185	2	11	Black spot as large as pear	Yau and Stephenson (1988) [86]
1185	3	27	Black spot as large as pear	Yau and Stephenson (1988) [88]
1185	4	18	Black spot	Yau and Stephenson (1988) [89]
1185	11	14	Black spot	Yau and Stephenson (1988) [90]
1200	9	19	Black spot as large as plum	Yau and Stephenson (1988) [93]
1201	4	6	Black spot as large as plum	Yau and Stephenson (1988) [96]
1202	8	23	Black spot as large as pear	Yau and Stephenson (1988) [97]
1204	2	3–5	Black spot as large as plum	Yau and Stephenson (1988) [99]
1258	9	15	Black spot as large as hen's egg	Yau and Stephenson (1988) [103]
1258	9	16	Black spot like a doll	Yau and Stephenson (1988) [103]
1278	8	31	Black spot as large as hen's egg	Yau and Stephenson (1988) [106]
1356	4	4–5	Black spot	Yau and Stephenson (1988) [107]
1361	3	16–19	Black spot	Yau and Stephenson (1988) [108]
1362	10	5	Black spot	Yau and Stephenson (1988) [109]
1371	1	2	Black spot	Yau and Stephenson (1988) [122]
1371	11	21	Black spot	Yau and Stephenson (1988) [125]
1372	5	8	Black spot	Yau and Stephenson (1988) [129]
1373	4	26-27	Black spot	Yau and Stephenson (1988) [132]
1373	10	23	Black spot	Yau and Stephenson (1988) [133]
1375	3	20–21	Black spot	Yau and Stephenson (1988) [136]
1381	3	23	Black spot	Yau and Stephenson (1988) [139]
1382	3	9–11	Black spot as large as hen's egg	Yau and Stephenson (1988) [140]
1387	4	15	Black spot	Yau and Stephenson (1988) [143]
1402	11	15	Black spot	Yau and Stephenson (1988) [144]
1520	3	9	Black gas	Yau and Stephenson (1988) [145]
1556	4	17	Black spot as large as hen's egg	Yau and Stephenson (1988) [146]
1603	4	16	Black spot like a coin	Yau and Stephenson (1988) [154/155]
1604	10	24	Black spot as large as bird's egg	Yau and Stephenson (1988) [156]
1604	10	25	Black spot as large as hen's egg	Yau and Stephenson (1988) [156]
1608	5	10	Black spot as large as pear	Yau and Stephenson (1988) [157]
1608	5	17	Black spot as large as pear	New record
1648	1	16	Black spot	Yau and Stephenson (1988) [183]
1660	5	22	Black gas	Xu (1983) gives a record from China
1720	5	8	Black gas	New record
1720	6	1	Black gas	Yau and Stephenson (1988) [194]
1726	10	21–22	Black gas	Yau and Stephenson (1988) [195]
1743	10	19–21	Black gas	Yau and Stephenson (1988) [197]

compiled by Yau and Stephenson (1988) (catalogue number 108) but only for day 16 and using a Chinese source. Only one naked-eye sunspot was recorded during the Maunder minimum in Korean sources according to the table. It was observed on 22 May 1660. This sunspot was also observed in China (see Xu, 1983). In Europe, Robert Boyle (1627–1691) saw this sunspot with a telescope on 25 May 1660 (Boyle, 1671).

2.3.2 Historical Occidental Observations

Though the majority of the observations contained in the catalogues come from the East, one can find some western observations (including observations from North-Africa, Europe, the Middle East, and Russia).

The most famous European observation of a naked-eye sunspot was related to the death of Charlemagne, King of the Franks and Emperor, by his secretary and adviser Einhard, who wrote *The Life of Charlemagne* (Einhard, 1880). He was not very careful dating the sunspot event. However, there is no doubt that a spot was seen on the Sun in 807 over eight days. In the *Annales Regni Francorum*, a Mercury transit observation appears during the days 17–24 March 807, but it is evidently a sunspot observation that confirms Einhard's account.

Another famous European observation was recorded by John of Worcester on December 8, 1128. He was not an Emperor, but he did make the earliest drawing of a sunspot (Figure 2.10). The text surrounding the drawing reads: *In the third year of Lothar, emperor of the Romans, in the twenty-eight year of the King Henry of the English, in the second year of the 470th Olympiad, seventh indiction, twenty-fifth moon on Saturday, 8 December there appeared from the morning right up to the evening two black spheres against the sun. The first was in the upper part and large, the second in the lower and small, and each was directly opposite the other as this diagram shows*

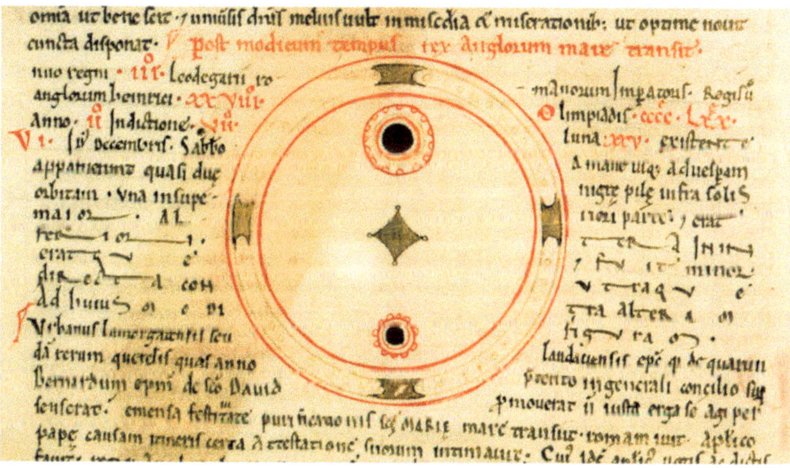

Fig. 2.10. The earliest sunspot drawing by John of Worcester (by permission of the President and Fellows of Corpus Christi College, Oxford, England, MS 157, Folio 380, lower half).

(Darlington and McGurk, 1995; McGurk, 1998). Curiously, the sunspots observed at Worcester were not observed elsewhere. However, this epoch coincides with a period of enhanced solar activity ("The Medieval Maximum") and Willis and Stephenson (2001) presented evidence of recurrent geomagnetic storms and associated aurora from AD 1127 to 1129.

Really remarkable in the sketch is the clear distinction between umbra and penumbra. Stephenson and Willis (1999) estimated that the two sunspots would have been within the latitude belt 25–35 degrees, in northern and southern latitude, with angular diameters of 2 and 3 arcmin, respectively, well within the resolution of the human eye.

Just eleven years later, a naked-eye sunspot was recorded in Bohemia. In the chronicle *Cosmae chronicon boemorum cum continuatoribus. Canonici Wissegradensis continuatio Cosmae*, the following description, dated in 1139, appears: *Fuerunt etiam nonulli, qui diceband se quasi fissuram in sole vidisse*[7] (Emler, 1874, p. 230). Chinese sources indicate a number of naked-eye sunspots during the interval 1136–1139. Křivský (1985) mentioned that a probable date for the sunspot observed in Bohemia is 24 July 1139, assuming 27 days for the Sun's rotation and relating this record with the Chinese observational dates.

Reports of Venus and Mercury transits stand out among the Occidental observations of naked-eye sunspots. Probably, the Aristotelian idea of a faultless Sun contributed to associating any sign or mark on the solar disk as a planetary transit, the only possible explanation for this phenomenon in the context of the astronomical knowledge of this epoch. Goldstein (1969) published the most complete study on mediaeval reports of planetary transits observed by Ibn Ishaq al-Kindi (800–873) also known as Alkindus, Ibn Seena (980–1037) or Avicenna, Ibn Rushd (1126–1198) or Averroes, and Ibn Bajja (1095–1138) or Avempace. Moreover, some transits are mentioned in non-astronomical texts. All these transit reports have to be identified as naked-eye sunspot observations.

In this context, it is interesting to note that Stephenson (1990) compared the list of computed dates for Venus transits (Meeus, 1958) with the dates of naked-eye sunspot reports compiled by Wittmann and Xu (1987) and Yau and Stephenson (1988), indicating that no coincidences or near coincidences of the records appear, and that Jeremiah Horrocks maintains his status of first observer of a transit of Venus in 1639 (see the second half of Chapter 5 for more details).

We have a description of the al-Kindi observation because Ibn al-Qifti wrote the event in his book *Ta'rikh al-hukama* (Ibn al-Qifti, 1903). He said that there appeared a black spot close to the middle of the Sun on 19 Rajab 225 (25 May 840), and al-Kindi mentioned that the spot lingered on the Sun for 91 days and that this spot was due to Venus occulting the Sun (Ibn al-Qifti, 1903, p. 156). This passage may also be found in Casiri (1760, pp. 422–423). It is

[7] Some people said they had seen a fissure in the Sun.

evident that a Venus transit cannot last for 91 days. Moreover, astronomical calculation shows that Venus was near its greatest elongation on 25 May 840. Thus, al-Kindi saw a large sunspot.

Goldstein (1969) consulted an Arabic text identified as *Compendium of the Almagest* by Avicenna from the Bibliotheque National (Paris). Avicenna says that he observed Venus as a spot on the surface of the Sun but he does not mention the date of observation. Avicenna died in the year AD 1037. According to this date, the only Venus transit during Avicenna's life was on 24 May 1032. However, this transit was not visible in his geographical region.

The transit reported by Averroes is the most famous because it is cited by Copernicus (1543) in *De Revolutionibus*. Averroes reports that two black spots were seen on the Sun at the time of Ibn Mu'adh by Ibn Mu'adh's nephew. Moreover, Averroes says that he computed the positions of Venus and Mercury and both planets were in conjunction with the Sun. Averroes probably ignored the planetary latitudes in the calculation because transits of Mercury and Venus do not occur simultaneously. According to Goldstein (1969), a probable computation date is 15 May 1068.

The last transit reported in Goldstein (1969) is attributed to the Spanish philosopher Ibn Bajja (d. 1139), also known as Avempace. Unfortunately, no astronomical works by Avempace survive. However, the astronomer Qutb al-Din al-Shirazi (d. 1311) cites a Venus transit observed by Avempace: *At sunrise one day I was standing on the roof of my house, and I saw two spots on the surface of the Sun. I calculated the positions of Venus and Mercury at that time from the zij, and I found them both near the position of the Sun. Therefore I concluded that the two spots were Venus and Mercury* (Goldstein, 1969, p. 54, translated from British Museum Ms. Add. 7482, fol. 21b).

These notices on sunspot were written by Arabic astronomers and philosophers. However, it is possible to find notices on sunspots in Arabic historical books. Vaquero and Gallego (2002) presented an observation of a sunspot recorded in an Arabic source by Ibn Hayyan called *al-Muqtabis V*. The document narrates diverse facts during the epoch of the caliph an-Nasir from the year AD 912 to the year 942. There are modern translations in Arabic (Ibn Hayyan, 1979) and Spanish (Ibn Hayyan, 1981). The text describing the naked-eye sunspot observation is the following: *At the end of this year (327 h) a strange and unknown prodigy occurred in the solar disk, being covered by a patent spot, visible by eye, that removed part of its light and switched off its rays, situation that continued 7 complete days, 4 in Dhu-al-Hijjah*[8] *at the end of this year (October 14–17, AD 939) and 3 at the beginning of Muharram*[9] *following, at the beginning of the year 328 h. (October 18–20, AD 939). At the end of the week that spot on the Sun disappeared, coming back its rays and light to the normal state after the 7 days, because of a copious rain fallen*

[8] Twelfth and final month in the Islamic Calendar.
[9] First month of the Islamic Calendar.

down during the night of the Thursday before to the morning when it became resplendent (Vaquero & Gallego, 2002, pp. 207–208).

Other non-Oriental sunspot observations are available from medieval Russia (Dodd & Schaefer, 2002). The Sun appeared red and sunspots were visible due to the smoke caused by forest fires in the years 1365 and 1371. These accounts are described in the *Nikonovsly Chronicle* (Vol. II of the St. Petersburg series of Russian Chronicles published in 1897). Vyssotsky (1949) published an English version of the most interesting fragments from the astronomical point of view. The description of the two records is the following:

> "During this year (AD 1365) there was a sign in the sky. The Sun was like blood and there were dark spots on it, and haziness lasted for half of the year. The heat was very intense; the forest, the marshes and the earth itself burned, the rivers dried up, some water-covered lowlands dried up completely, and there was terror, dread and sorrow among men".
>
> "During this year (AD 1371) there was a sign on the Sun. There were dark spots on the Sun, as if nails were driven into it, and the murkiness was so great that it was impossible to see anything for more than seven feet. (...) Woods and forest were burning and the dry marshes began to burn, and the earth itself burned, and great fright and terror spread among men".

The number of naked-eye observations in Occidental sources increases during the early 17th century. Vaquero (2004) pointed out that a naked-eye sunspot was observed by Galileo. Probably, this is the first sunspot that was seen by naked-eye and by telescope at the same time. It is surprising that this case has been forgotten by the compilers of the naked-eye sunspot catalogues in spite of Galileo's fame. Galileo presented his sunspot telescopic observations in his work *Istoria e dimostrazioni intorno alle macchie solari* (1613). This work is organized in the form of letters from Galileo to Mark Welser (see Chapter 3). In the postscript, in the second letter to Mark Welser (Galileo, 1613, p. 56), one can read that on days 19, 20, and 21 August 1612 a naked-eye sunspot was seen by Galileo and other people.

Moreover, Galileo published in his *Istoria* the drawings of the solar disk observed by telescope (see the next chapter for more details). Figure 2.11 shows the solar disk (29 August 1612). One can see a large sunspot near the central meridian of the Sun. An approximate value of the area of this large sunspot from Galileo's sketch can be measured. Its area was some 2000 millionths of a solar hemisphere, large enough to be visible by naked-eye.

Another observation overlooked by the compilers of catalogues is the naked-eye sunspot observed by the poet and painter Raffael Gualterotti (1544–1639) on the day 25 September 1604. This observation is reported by him in his *Discorso sopra l'apparizione de la nuova stella* (1605) published in Florence: "The year 1604, the day 25 September, [...] I saw when the Sun was setting that a spot appeared on his body" (Gualterotti, 1605, p. 28).

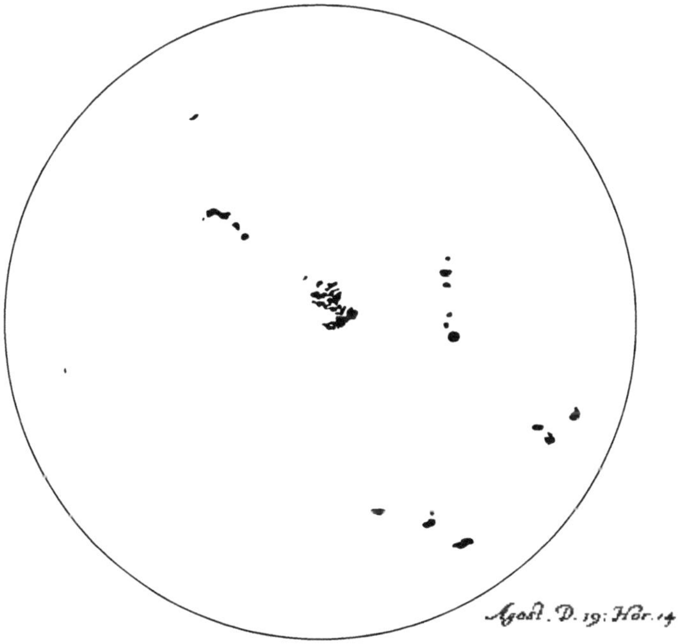

Fig. 2.11. Drawing of the solar disk showing naked-eye sunspots included in the *Istoria e dimostrazioni intorno alle macchie solari* (Galileo, 1613, p. 94). The original figure caption is "Disegni della Macchia grande Solare, veduta con la semplice vista dal Sig. Galilei, e similmente mostrata a molti; nelli giorni 19.20.21. d'Agosto 1612" [Drawing of the large sunspot seen by naked-eye by Galileo, and shown in the same way to everybody during the days 19, 20, and 21 August 1612]

Gualterotti explains that it cannot be Venus (during a transit) because that would have been much smaller and gives a tentative explanation for the observed sunspot. According to Gualterotti, the conjunction of Mars and Saturn attracted vapours and exhalations. Then these vapours were rarefied and taken to the Sun where they became sunspots. Moreover, Gualterotti described an aurora borealis during the same time. The Yau and Stephenson (1988) catalogue lists a naked-eye sunspot observed in Korea on the day 24 October 1604. It is probable that the sunspot observed in Korea is the same as that observed by Gualterotti but in the next solar rotation.

Other Occidental observations were made during ship voyages. Brody (2002) presents two nice examples. In 1590, the master of the ship *Richard of Arundel* registered: *On the 7 [December] at the going downe of the Sunne, we saw a great blacke spot on the Sunne, and on the 8 day, both at rising and setting, we saw the like, which spot to our seeming was about the size of a shilling* (Welsh, 1904, vol. 6, p. 450). Another spot was seen on the Sun, during setting, on the day 16 December. However, the bad weather did not allow other observations (Hosie, 1879; Schove, 1982). The second case from aboard ship is the crew of Henry Hudson's *Half Moon* that reported the Sun having a "slake"[10] in May 1609 when they were near to the Faroe Islands (Brody, 2002, p. 25).

As a final note, one can cite Sarton (1947) in which some possible Western observations of naked-eye sunspots are reviewed. He includes the sunspot observation by Guido Carrara and his son, Giovanni, in Bergamo about 1450. The latter described it in his *De constitutione mundi*: "At a certain time two drops of blood were seen on the Sun and the populace was terrified [...] However, my father, Guido of Carrara, whom I venture to describe as the outstanding figure of his age in every branch of letters, observed and compared the [position of] planets and found that Venus and Mercury were the cause" (Sarton, 1947, p. 70). However, the low solar activity during the 15th century is intriguing.

2.3.3 Mayan and Indian Observations

We can also find examples of observations (or possible observations) of naked-eye sunspots during the pre-telescopic epoch in other regions of the world such as Central America and India.

The city of Mayapan was the most important urban and military centre of the Yucatan Peninsula during the Mayan post-classic period (AD 1000–1519). In the middle of the 15th century Mayapan was destroyed as a result of a civil war. The Fresco Hall is situated at the southern end of the Central Plaza of Mayapan. It is a rectangular structure and the frescoes are still visible. Galindo Trejo and Allen (2005) suggested that descending personages depicted inside the sun-circles in the various panels of the mural might represent a Venus transit or sunspots. Mayan priests could have seen a naked-eye sunspot but it would have been interpreted as Venus inside the solar disk.

Another beautiful example comes from Varanasi, the paradigmatic holy city of Northern India where the Sun is honoured. Specific attributes of the Sun are represented by the *adityas*, the fourteen forms into which the Sun divided himself when he took residence in the city according to the Hindu tradition. The aditya shrines mark the sites of Sun temples destroyed during the Mughal occupation of the city after AD 1192. The *Kashi Khanda* is a fourteenth century Hindu text that contains the Varanasi spiritual traditions. This text describes the aditya temples and details about mytho-historical events related with *adityas*. Malville and Singh (1995) showed that some of those events seem to have involved astronomical phenomena. The presence of coiled and dark snakes on the Sun in a fascinating and complex story in the *Kashi Khanda* is almost certainly a reference to naked-eye sunspots probably dated during the Mediaeval Maximum of sunspot activity. Moreover, references to a "leprous" Sun may also have been related to this period of high sunspot activity.

[10] Expression for mud or slime.

2.3.4 Naked-Eye Observations During the Telescopic Era

With the widespread use of the telescope, naked-eye sunspot observations turned into a mere curiosity for the majority of astronomers. In the European astronomy of the 18th and 19th century no systematic work exists on sunspots visible to the naked eye. Nevertheless, there were astronomers who observed them. We will discuss here two paradigmatic cases. W. Herschel (1738–1822) and E.W. Maunder (1851–1928) were two great astronomers who did major work concerning the Sun. Both observed sunspots without the aid of the telescope.

Herschel made important contributions to solar physics although he is famous for his discovery of the planet Uranus (see, for example, Hoskin, 1963). Herschel published three papers on the Sun, including observations, in the *Philosophical Transactions of the Royal Society of London* (Herschel 1795, 1801a,b). However, the bulk of Herschel's sunspot observations is in his unpublished notebooks. Hoyt and Schatten (1992a,b) transcribed and reproduced the fragment on solar observations of Herschel's notebooks. The original manuscripts are in Churchill College in Cambridge, England. In Herschel's notebooks, one can read that he observed naked-eye sunspots on two occasions:

> "April 19, 1779 (2^{h} $30'$) common time. There is a very large spot on the Sun visible to the naked-eye. By the telescope it appears to be divided into two. The length of the largest of them is by the micrometer 8.06". 7 ft. telescope."
> "September 2, 1792 (9^{h} $55'$) I saw two spots on the Sun with the naked-eye. In the 7 foot telescope, I found two cluster of spots besides many scattered ones; every one of them was certainly below the surface of the Sun."

Another paradigmatic case is Edward Walter Maunder, a pioneer in solar terrestrial physics and "discoverer" of the minimum of long-term solar activity during 1645–1715, which was posthumously named for him (Maunder, 1894). In his book *The Heavens and their Story* (1909) he relates his experience when he was fourteen years old: "*In February, 1866, as I was returning home from school one evening, I saw the Sun, low down in the West, shining red through the mist. The Sun was dim and red enough for me to look at him without blinking, and I saw plainly on him a round black spot*" (Maunder and Maunder, 1909, p. 103). Maunder saw other naked-eye sunspots sixteen years later. A very violent magnetic storm occurred on 17 November 1882. He wrote one day later: "*On November 18, 1882, Queen Victoria was holding a review in Hyde Park. The morning was somewhat foggy, and the Sun shone dull and red through the thick air, so that it was easy to look at him. On this occasion there was a great spot on the Sun; so big that it caught the attention of the soldiers who were marching across Blackheath [. . .]*" (Maunder and Maunder, 1909, p. 106).[11]

If one reads with attention the scientific literature of the 18th and 19th centuries, it is possible to find other observations. We shall present some examples.

The case of Richard Lewis is very interesting. He published a report on the aurora seen at Maryland in October 22, 1730 (Lewis, 1731). The report finished with a comment on naked-eye sunspot observation: "*Dr. Samuel Chew of Maidstone tells me, that he has for some Days past, at Morning and evening, observed several Spots, on the Sun, very plainly with the naked Eye, some of which seemed very large*". To the best our knowledge, this is the first account of naked-eye sunspots published from the American colonies.

Another interesting observation in America is due to John Winthrop (1714–1779), a contemporary and friend of Benjamin Franklin (1706–1790), in 1739. He was the most important American pioneer in mathematics and astronomy. Kilgour (1938) found some sunspot observations in Winthrop's unpublished manuscripts. The observations were made after the appearance of a great sunspot observed without the aid of the telescope:

> "1739 April 19th at Boston. Walking on the Common a little before sunset, the air being so hazy that I was able to look on the sun, I plainly saw with my naked eye a very large and remarkable spot. Its shape was oblong and the length of it was perpendicular to the horizon. I observed it several minutes till the sun was actually set. It was like wise seen by several persons in the company of Messrs. Skinner and Read. [. . .] I am since informed that several persons in the country saw them like wise with their naked eye, particularly some at Medford and the ferrymen at the Charlestown ferry" (cited by Kilgour, 1938, pp. 358–359).

Moreover, the following day, at night, a considerable aurora was seen by Winthrop, and he believes that there was some relation between large sunspots and aurorae (Kilgour, 1938; Eather, 1980), a suggestion usually first attributed to Mairan.

There are some observations published in periodical journals of the 19th century. There is no study that records systematically the information on naked-eye sunspots appearing in astronomical journals. However, there are very many records by (especially amateur) astronomers in this sense. We shall present three examples from *Monthly Notices of the Royal Astronomical Society*.

An extract of a letter written by A. Weld (1823–1890), Director of the Observatory at Stonyhurst College, was published. He says: "*On September 20, I observed a large spot on the sun with our equatoreal, and found that it consisted of several dark nuclei enveloped in one large penumbra*". Then, he

[11] E.W. Maunder's name appears on the book, but he states in the preface that it was "almost wholly" the work of his wife Annie Maunder (1868–1947).

presented the results of his measurements: 2′41″.1 for the greatest diameter of the sunspot, 1′7″.2 for the greatest diameter of the nucleus, 2′14″.1 for the equatorial diameter of the sunspot, 0′49″.2 for the equatorial diameter of the nucleus and 2′14″.1 for the meridian diameter of the sunspot. Weld (1848) finished by saying: "*The spot was distinctly visible to the naked eye before sunset*".

A very brief notice (eight lines only) was published recording the observation by Weld (1851) on the evening of 12 March 1851. The Sun was setting in the midst of a thick haze and Weld observed a sunspot with the naked eye. Another two persons saw it with facility. Weld observed the Sun the next day with a telescope and found a single large sunspot. According to Weld's comments, "Its greatest measured diameter parallel to the equator was $4^s.05$, that of the nucleus $1^s.60$, and its diameter measured along the meridian circle was 52″.53". This observation is in accordance with the naked-eye sunspot recorded in the Oriental source "Ting-nan Hsien-chih 6" corresponding to number 203 in the Yau and Stephenson (1988) catalogue.

One more example could be the record by A.R. Hill published in 1870. He explains that the light of the Sun was "*obscured by a peculiar scud drifting over it, and giving the whole disk a reddish appearance, with the borders less luminous than the centre*" on Sunday, 22th May. He observed three large spots (Figure 2.12, left). The next day, the meteorological conditions were more advantageous for naked-eye observations. Hill saw four spots, but the fourth was only visible at the most advantageous intervals (Figure 2.12, right).

Another interesting case is the record of naked-eye sunspots by William Dawson. He was an active sunspot observer during the period 1867–1890. There are 1623 daily records by Dawson in the Hoyt and Schatten (1998) database. He published three papers with sunspot observations (Dawson, 1888, 1889, 1890). The first (Dawson, 1888) lists his sunspot observations made

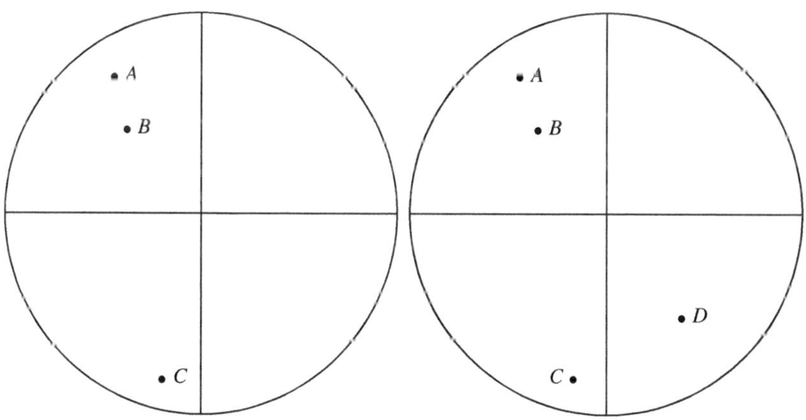

Fig. 2.12. Drawing by A. R. Hill (1870) showing the positions of naked-eye sunspots observed on 22 (left) and 23 (right) May 1870.

during 1884–1886 with a telescope with an aperture of 4.6 inches and about 70 inches focus made by A. Clark & Sons. The 100-power eyepiece was generally used. With respect to the naked-eye sunspot observation, he says: "*In 1871–78, I left off counting the spots on account of failing eyesight, simply noting the number of groups, large and small. Observations were much neglected in 1874–77. I estimate that I have seen about 40 different sun-spots with the naked eye—their diameters (umbrae) ranging from about 8,000 to 35,000 milles*". In this paper, Dawson mentions the naked-eye sunspot observed in the "notes" column. During the period covered in the paper (1884–1886), Dawson observed naked-eye sunspots on the following days: 6 Oct 1884; 3 Feb 1885; 9, 11, 19, and 22 June 1885; 7 and 24 July 1885; 11, 14, and 15 Sep 1885; and 15 Nov 1885. It is interesting to note that the sunspot observed on 7 July 1885 by Dawson could be the same sunspot reported in a Chinese source on 5 July (reference number 226 in the Yau and Stephenson (1988) catalogue).

Other naked-eye sunspot records appear when one is looking for other kinds of observation. Moore (2003) was intrigued to discover that T.W. Webb (1807–1885) observed a transit of Mercury on 9 November 1848. When he consulted Webb's notebooks in the Royal Astronomical Society library, he found a nice naked-eye sunspot observation by Revd Webb on 26 October 1852: "*...a large naked-eye sunspot on the disc of the Sun, dimmed by the morning fog of London about 10 am.*" (Moore, 2003, p. 306). A biography of T. Webb can be consulted (Robinson and Robinson, 2006).

2.3.5 Modern Observations

The low number of historical naked-eye observations could discourage potential modern observers. However, some amateur astronomers have been trying to see sunspots without the aid of a telescope for approximately the last 30 years.

One of the most famous observational campaigns was conducted by Mossman (1989), an experienced English amateur observer. He observed systematically the Sun using a dark filter or through clouds for thirteen months, near the maximum of the solar cycle number 22. Probably, the main conclusion of this campaign was that Mossman saw more sunspots in 13 months than are found in 18 centuries of pre-telescopic observations. Thus, it is clear that the sunspots that were recorded in historical sources are a very small fraction of those that could have been seen. Mossman (1989) showed that he could see features on the solar disk as small as 0.3 arcmin, distinguish roughly the shapes of sunspots, and identify as many as five separate sunspots in a day.

The longest series of naked-eye sunspot observations was made by Heath (1994). He observed naked-eye sunspots from 1959 to 1993 (35 years) during solar cycles 20–22 using a dark Sun filter with a Sun diagonal (Herschel Wedge), or No. 14 welder's glass. Figure 2.13 shows the results obtained by Heath. The total number of naked-eye sunspots observed during a year is well

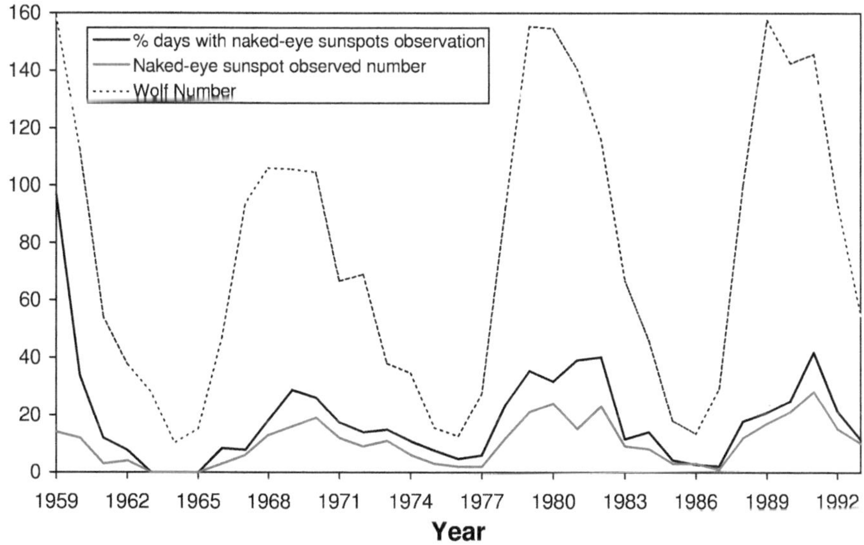

Fig. 2.13. Annual values of total number of naked-eye sunspots and percentage of days with at least one naked-eye sunspot observable during the period 1959–1993 by Alan W. Heath. The Wolf number is also included in the figure for comparisons.

correlated with the Wolf number (correlation coefficient r = 0.885) in spite of the possible influence of the days without observation. Similar behaviour corresponds to the percentage of days with at least one naked-eye sunspot observable on the solar disk (r = 0.778). During the period considered, a total of 357 naked eye spots were recorded, which averages 10.2 per year. Heath's observations suggest that the peak of naked-eye sunspots occurs after the peak of the 11-year solar cycle.

P. Wade was other important modern naked-eye sunspot observer. From February 1980 to December 1992, he made 2876 systematic sightings (61% of the days in all the period). He noted the date and time and each report included details about the number and rough location of the sunspots and other points of interest. Figure 2.14 shows the annual mean daily frequency[12] (MDF) for his naked-eye sunspot observations. The 11-year cycle is clearly visible. The smallest sunspot reported by Wade (1994) was approximately 25 arc seconds in size (only 80 millionths of the solar hemisphere). A comparison between the naked-eye sunspot MDF and telescopic sunspot MDF shows the two quantities to be linearly related.

An interesting task carried out by Wade (1994) was recording the latitude of the naked-eye sunspots he observed. Wade used Stonyhurst disks to estimate the sunspot latitudes from 152 mm diameter images projected from a 73 mm refractor. According to his observations and measurements, the

[12] Mean daily frequency is the average of the number of sunspot observed once per day during a month.

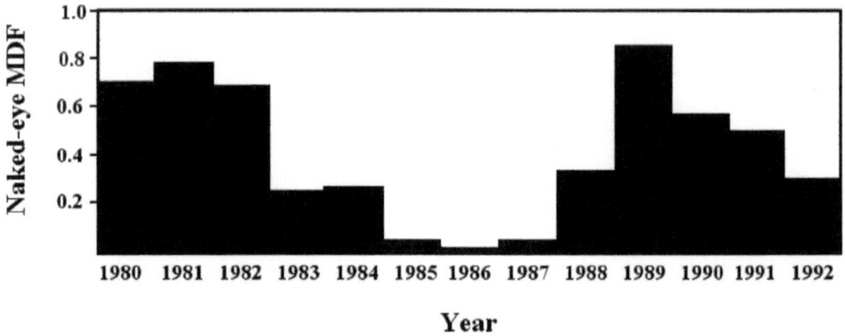

Fig. 2.14. Annual MDF for naked-eye sunspots observed by P. Wade during the period 1980–1992 (adapted from Wade, 1994).

distribution of the latitudes of naked-eye sunspots is similar to the latitude distribution for all sunspot groups observed telescopically. Thus, he concluded that naked-eye sunspots and the general sunspot population have the same main characteristics with respect to frequency and latitude distribution.

Keller and Friedli (1992) calculated an empirical visibility limit for naked-eye sunspot observations. Based on Keller's observations made during the period 1976–1986 (solar cycle number 21), they found basically that the number of naked-eye sunspots varies with solar cycle, that there is a linear relationship between naked-eye sunspot number and relative sunspot number, and that the visibility limit for naked-eye sunspots is 31 arcsec for the penumbral diameter of sunspots (for the acuity of Keller's eyes). They proposed a more general result using the data of an international network of naked-eye solar observers that was observing daily from 1984 to 1989. Keller and Friedli (1992) showed that an average observer is able to detect by naked eye a sunspot whose mean umbral and penumbral diameters are at least 15 and 41 arcsec when the spot is near the centre of the solar disk.

During recent years, a network of amateur astronomers has been observing the Sun without the aid of telescopes. Monthly naked-eye sunspot numbers recorded by the Spanish amateur astronomer Javier Ruiz ($R_{naked-eye}$) and sunspot numbers (R_Z) are plotted in Figure 2.15 from November 1998 (before the maximum of solar cycle 23) until December 2006. It is interesting that the Gnevyshev gap[13] between the principal and the secondary maxima of solar cycle 23 is more noticeable in the naked-eye series. The two data sets are closely related, as is illustrated in Figure 2.16 where the sunspot number is plotted against naked-eye sunspot number. The best-fit straight line, determined by the method of least squares, is also shown. This line is described by the equation

$$R_Z = (56 \pm 7)R_{naked-eye} + (46 \pm 5),$$

and the correlation coefficient is 0.625 (99.9% statistical significance).

[13] The relative minimum between two consecutive peaks in a solar cycle is known as the Gnevyshev Gap (Gnevyshev, 1967, 1977; Kane, 2005).

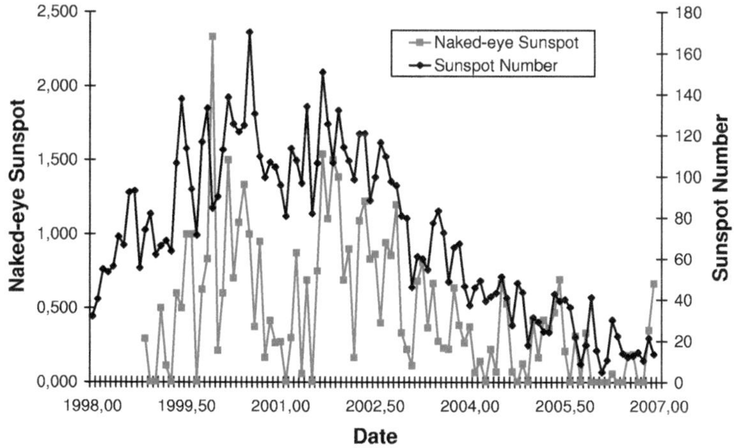

Fig. 2.15. Monthly International Sunspot Number and monthly average of naked-eye sunspots recorded by the Spanish amateur astronomer Javier Ruiz during 1998–2007. Data courtesy of J. Ruiz (available at http://www.astrocantabria.org/parhelio/).

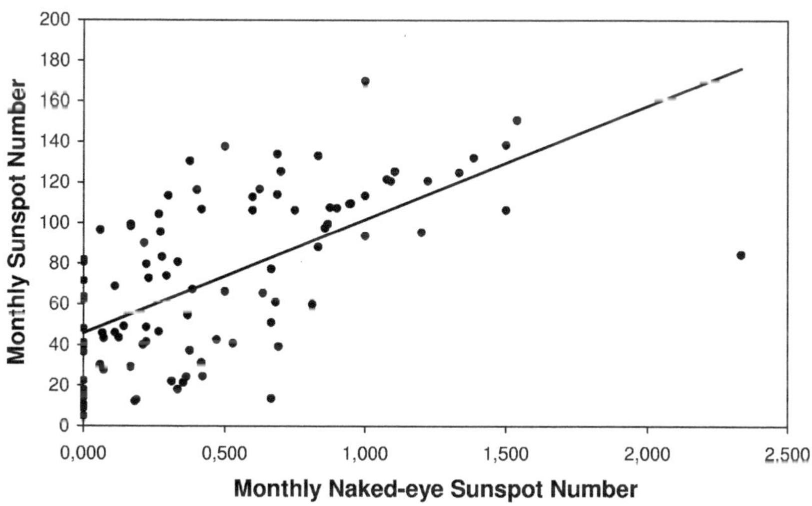

Fig. 2.16. The International Sunspot Number *versus* monthly average of naked-eye sunspots recorded by the Spanish amateur astronomer Javier Ruiz during 1998–2007. Data courtesy of J. Ruiz (available at http://www.astrocantabria.org/parhelio/).

2.4 Naked-Eye Sunspots and Temporal Evolution of Solar Activity

After this review of the historical and modern naked-eye sunspot observations, which showed the great differences between them, the key question is now: Can the historical naked-eye sunspot observations be used to study the Sun during the last two millennia? Usoskin and Kovaltsov (2004) indicate that the naked-eye sunspot record is not straightforward to interpret because: (1) the record is influenced by meteorological phenomena, (2) the record depends on the predominant traditions at the time, and (3) a direct comparison of Oriental naked-eye and European telescopic data shows important discrepancies. Usoskin and Kovaltsov (2004) conclude that the most reliable and useful proxy of long-term solar activity is formed by the data set on cosmogenic radionuclides produced by cosmic rays in the atmosphere of Earth (Stuiver and Quay, 1980; Beer et al., 1990; Bard et al., 1997; Beer, 2000; Solanki et al., 2004).

In the section 2.3.1 devoted to Oriental records, we saw some of the historical problems in interpreting the observations as a series that represents solar activity. The most important problems are the loss of information due to historical periods with political turbulence (such as the gaps existing between AD 600 and 800) and the role of certain political and sociological factors (such as astrology).

Another problem is the different type of documentation used. Hameed and Gong (1991) indicated that the series of the number of naked-eye sunspot observations correlates badly with ^{14}C from 1620. This is probably due to historical motives since during this epoch local historical sources were used. These sources show less reliability in the record of astronomical events. However, some naked-eye sunspots were recorded during the Maunder minimum (see, for example, Clark and Stephenson, 1980).

The influence of atmospheric conditions on the possibility of naked-eye sunspot observations is unquestionable. Willis et al. (1980) established that the dates of naked-eye sunspot sightings occur more frequently in late winter and spring. Figure 2.17 shows the number of unaided-eye sunspot sightings recorded in each month according to the Clark and Stephenson (1978) catalogue during the period from 28 BC to AD 1604. The dark part of the histogram shows the distribution of those dates specified without reservation and the grey part includes 12 uncertain dates. A marked spring maximum and a broad summer minimum are the main characteristics of the figure. The probability of obtaining the seasonal distribution shown in the dark part of the figure by chance is less than 1 in 25 000 using the chi-squared test. Of course, this probability is even smaller if one considers the whole histogram. It is possible to explain this seasonal distribution of naked-eye sunspot observations considering the general characteristics of East Asia's climate. The high number of observations in spring could be attributed to dust storms and sand storms originating over central China frequently during late winter and early

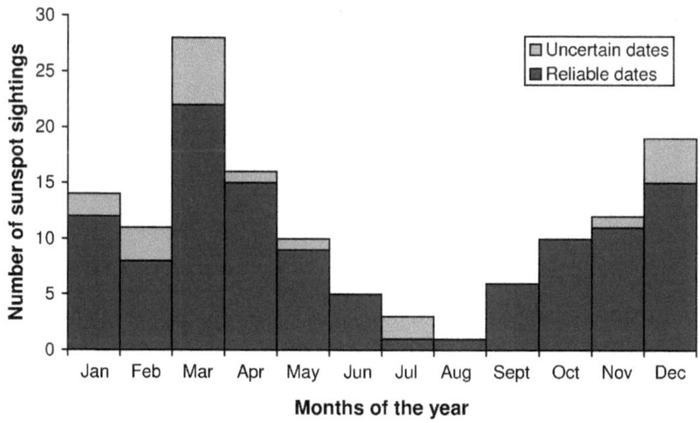

Fig. 2.17. Number of sunspots sightings recorded in each month by Clark and Stephenson (1978), from Willis et al. (1980).

spring. The summer minimum in the figure could be explained by the cloudiness and precipitation during the rainy season in East Asia. Willis et al. (1980) concluded that atmospheric conditions make a simple interpretation of Oriental naked-eye sunspot observations very difficult as a direct index of historical solar activity. In the same context, Hameed and Gong (1991) found a strong correlation between Oriental naked-eye reports and dust/rain events in China during the period 1621–1920.

Many naked-eye sunspot records mention reduction in sunlight intensity. This fact may be related to observation of the Sun very near the horizon or through fog, haze, or dust clouds. Moreover a similar effect could be the result of attenuation of sunlight by volcanic dust and aerosols in the stratosphere. Thus, it is reasonable to think that after a major volcanic eruption the probability of observing a naked-eye sunspot is greater, since the volcanic aerosol reduces the intensity of the sunlight, favouring the observation. Scuderi (1990) made a statistical analysis of Oriental naked-eye sunspot records including descriptions of "vapours" and related phenomena. This analysis shows that these records could be related to the atmospheric dust veil resulting from large volcanic eruptions.

Figure 2.18 shows a comparison between the 50-year moving average of annual naked-eye sunspots and Group Sunspot Numbers. It is evident that, quantitatively, the historical naked-eye observations must be interpreted with caution (Willis et al., 1996; Usoskin and Kovaltsov, 2004). Thus, a time series constructed using naked-eye sunspot observations cannot be presumed to represent the quantitative variation of long-term solar activity. Naked-eye observations cannot be interpreted quantitatively because only observed sunspots are reported and thus we cannot distinguish between no-spot and no-observation cases. Moreover, only some 67% of Oriental naked-eye observations have been confirmed by telescopic observations during the period 1874–1918 (Willis et al., 1996).

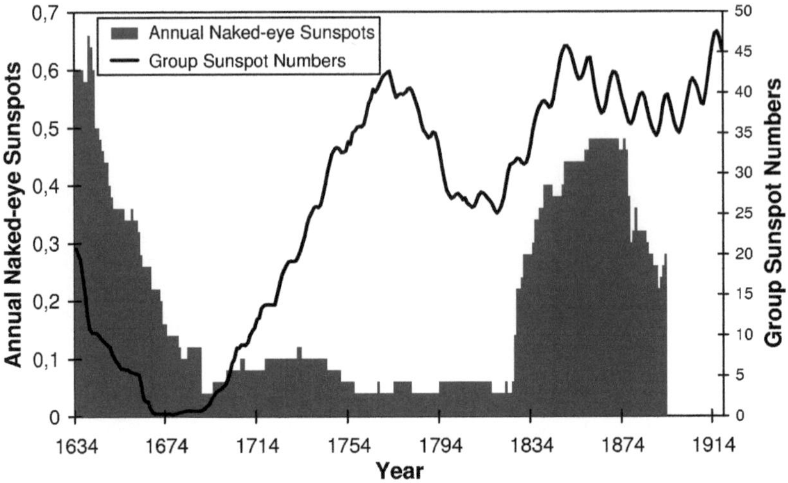

Fig. 2.18. 50-year moving average of annual naked-eye sunspots and Group Sunspot Numbers.

Another problem for the use of the data is that there is a marked upward trend in the annual number of events recorded in catalogues. This could reflect the greater difficulty of finding sunspot records as one goes back in time. Figure 2.19 shows how the secular number of naked-eye sunspots records, from Vaquero et al. (2002), varies from the 2nd century BC to the 19th century AD. The solid line is the best linear fit to the data indicating the increasing trend in the data. This upward trend is equal to 1.1 observed sunspots per century.

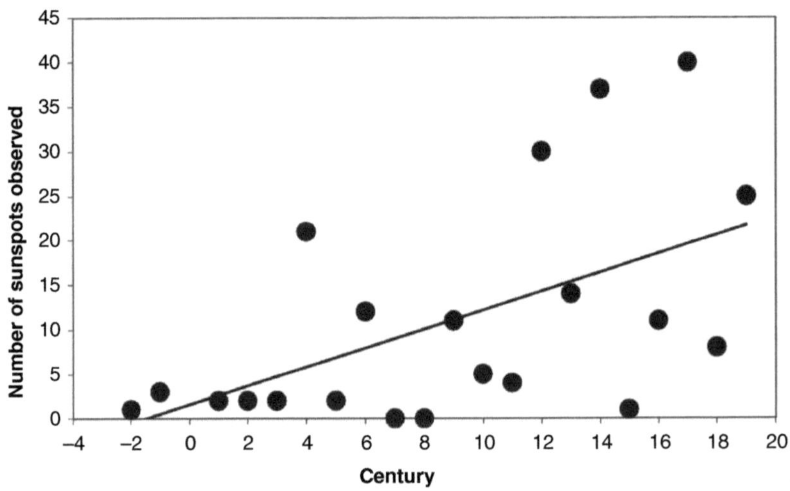

Fig. 2.19. Secular number of naked-eye sunspots recorded, from Vaquero et al. (2002). The line is the best linear fit to the data.

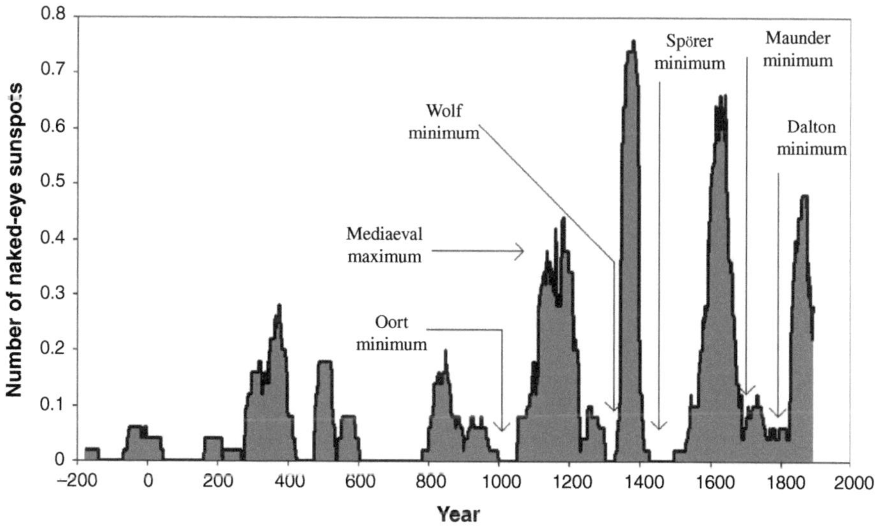

Fig. 2.20. 50-year moving average of the time series of naked-eye sunspots (adapted from Vaquero et al., 2002).

In spite of all the above arguments, one can plot Figure 2.20, constructed by calculating the 50-year moving average of the annual number of naked-eye sunspot observations recorded in the catalogues. The known minima of Oort, Wolf, Spörer, Maunder, and Dalton stand out sharply. The Maunder and minima are somewhat displaced from their original locations. The Mediaeval Maximum is centred on the first half of the 12th century and other maxima are around 1880, 1630, 1380, 880, 630, and 380. Thus, the most important episodes in the history of the Sun during the last millennium, and probably during the last 2000 years, appear qualitatively in the record of naked-eye sunspots. Though climatic, historical, and sociological factors clearly affect the information, these factors are so nonlinear and changeable on such diverse time-scales that the complete record reproduces the solar activity better that one might have expected.

2.4.1 Time Series with Naked-Eye Sunspot Observations

In spite of the evident problems that any time series obtained from the catalogues of naked-eye sunspot observations will contain, several authors have tried to use this information in studies of historical solar activity.

Probably, the most elaborate attempt to construct time series from the historical naked-eye sunspot observation reports was made by Nagovitsyn (2001). He tried to use the data of the Wittmann and Xu (1987) catalogue taking into account not only the mere facts of sunspot observations but also the qualitative description of each event. Thus, he obtained several different annual time series to characterize different aspects of the solar activity:

- Dichotomic series of uniform events (WX_0).
- Quantitative series of uniform events (WX_N).
- Quantitative series of mutually weighted events (WX_C).
- Duration of observations of individual events (WX_T).
- Hypothetical probabilistic parameter of the spatial organization of solar activity and time series of seasonal parameter (WX_Q).

The *dichotomic series of uniform events* (WX_0) was constructed using the value 1 for the cases of "at least one event during a year" and the value 0 for the "no event" case (see Figure 2.21).

Nagovitsyn analysed this series using the self-similarity function (binary statistics of events). This analysis revealed several harmonic components: (1) a 400-year cycle, which had previously been described by Link (1963), (2) a 260-year cycle, the most prominent peak in the amplitude spectrum, (3) a 200-year cycle, very close to the de Vries cycle ($\approx$ 210 years), (4) 60, 90, and 130-year secular cycles, and (5) an 11-year cycle that can be resolved into a multiplet (peaks at 9.7, 10.6, and 11.2 years).

The *quantitative series of uniform events* (WX_N) is a sequence of the numbers of events recorded during each year. Numerous problems exist for the construction of this series such as events that could be the same, problems in the quality of the dates offered in the historical sources, etc. This index is less

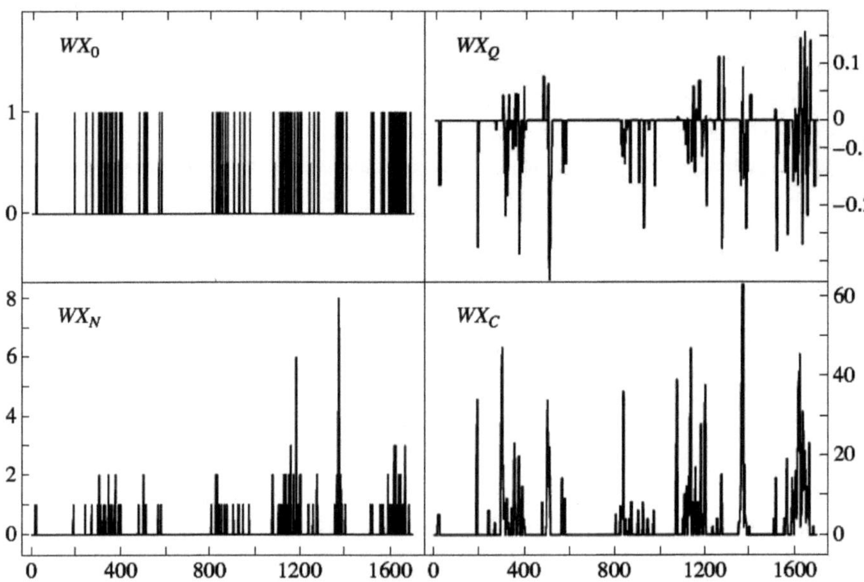

Fig. 2.21. Different time series constructed by Nagovitsyn (2001) based on the Wittmann and Xu (1987) catalogue.

robust than the previous one with respect to the typical problems of historical sources.

The *quantitative series of mutually weighted events* (WX_C) can be constructed assigning to each particular observation a certain score. Thus, one can assume that the observer compares a particular spot with a familiar object according to the especially large size of the feature on the solar disk. The score depended on the characteristics of the report. For example, Nagovitsyn assigned score 10 to the features called "moles", score 15 to those called "spots" or "dots", and score 20 or 30 to events whose descriptions are followed by phrases similar to "as large as a plume" or "... peach" (20) and "melon" (30). Moreover, he took into account the duration of the event observation multiplying the initial score by a factor of 1.2, 1.5 or 2.0 if the event had been observed for 3–6, 7–13, or 14–20 days, respectively.

He also made a histogram of the *duration of observations of individual events* (WX_T) supplementing the cases of events recorded directly for several days T by those where two adjacent events represented as independent catalogue entries (these are often records from different chronicles) are separated by less than 1.5 solar rotations (40 days).

The last time series constructed by him was the *hypothetical probabilistic parameter of the spatial organization of solar activity and time series of seasonal parameter* (WX_Q). He assumes the following hypothesis: "Because of the viewing conditions due to the negative extremum of the Earth's heliographic latitude, a spot recorded in the Wittmann and Xu catalogue observed in the given year within the Gregorian March and seen just for one day is most likely to be located in the southern solar hemisphere. A similar spot observed in the Gregorian September is most likely to be located in the northern hemisphere". He estimates that the probability that an observed naked-eye sunspot is located in the northern hemisphere is equal to

$$p = \frac{1}{\pi}(\arccos \frac{B_0}{R_0} - \frac{B_0}{R_0}\sqrt{R_0^2 - B_0^2}) \tag{2.9}$$

where B_0 is the heliographic latitude of the Earth and R_0 is the radius of a circle (concentric with the visible solar disk) equal to the mean limiting spot visibility radius. Since most of the naked-eye sunspot events were observed for one day only, R_0 is equal approximately to 6° or 7°. Solar N-S asymmetry can be defined as

$$q \equiv \frac{N - S}{N + S} = 2p - 1$$

In Equation (2.9), Nagovitsyn (2001) used 7.5°. He used Equation (2.9) to compute p for various months and to infer the corresponding values of q, which he denoted as WX_Q.

Other authors have used less ambitious and sophisticated indices of naked-eye sunspot records. However, the simplest indices are relatively easy to construct accurately and without ambiguities. Ding et al. (1983) found 11-year, 60-year, and ~250-year periods in the time distribution of naked-eye sunspot records using auto-correlation analysis. Vaquero et al. (2002) constructed the series of the number of naked-eye sunspots from 165 BC to AD 1918. These data were analysed using the multi-taper method (MTM) (Thomson, 1982) and singular spectrum analysis (SSA) (Allen and Smith, 1996; Dettinger et al., 1995; Vautard et al., 1992). The most notable aspect of the MTM spectrum of the sunspot data from 165 BC to AD 1918 is the strong peak at around 250 years. It would be difficult for this period to show up in telescope-based sunspot series because of their relatively short length. There are also other peaks at 115, 85, and 60 years which could be related to the Gleissberg period. Other shorter-period cycles also appear, including the 11-year cycle.

Figure 2.22 shows a wavelet analysis of the annual number of naked-eye sunspots during the period AD 1–1918 (data from Vaquero et al., 2002). There are three intermittent periods in the long-term behaviour. A periodicity of 60 years is found around the early 6th, the 12th, and the 19th centuries. Other periodicities of 100 and 250 years are present in the last half of the record (1200–1918 and 800–1918, respectively). The main overall wavelet peak has a period of 241 years.

As a final comment, we can make a comparison between the solar activity reconstructed using cosmogenic radionuclides, the most reliable proxy to study the long-term behaviour of the Sun, and the series of naked-eye sunspot observations. Figure 2.23 shows the reconstruction of sunspot number made by Solanki et al. (2004) and a 50-year moving average of the annual number of

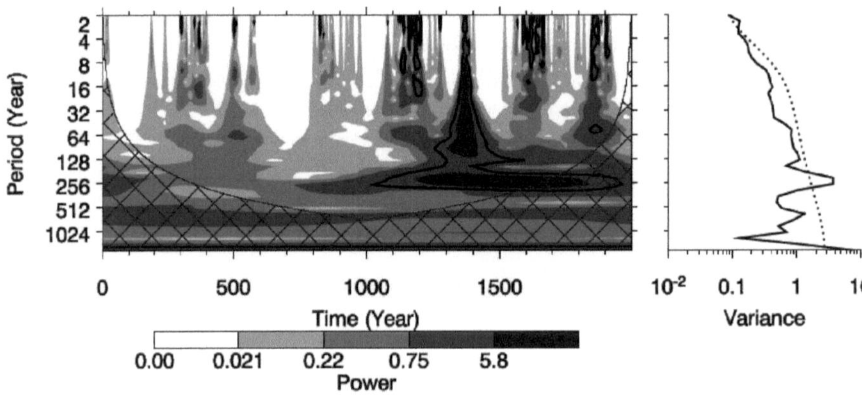

Fig. 2.22. (Left panel) The wavelet power spectrum of the annual number of naked-eye sunspot recorded during the period AD 1–1918 (from Vaquero et al., 2002). Contour levels are chosen so that 75%, 50%, 25%, and 5% of the wavelet power is above each level, respectively. The cross-hatched region is the cone of influence, where zero padding has reduced the variance. The black contour is the 1% significance level, using a red-noise background spectrum. (Right panel) Global wavelet power spectrum (black line). The dashed line is the significance for the global wavelet spectrum, assuming the same significance level and background spectrum as in the left panel.

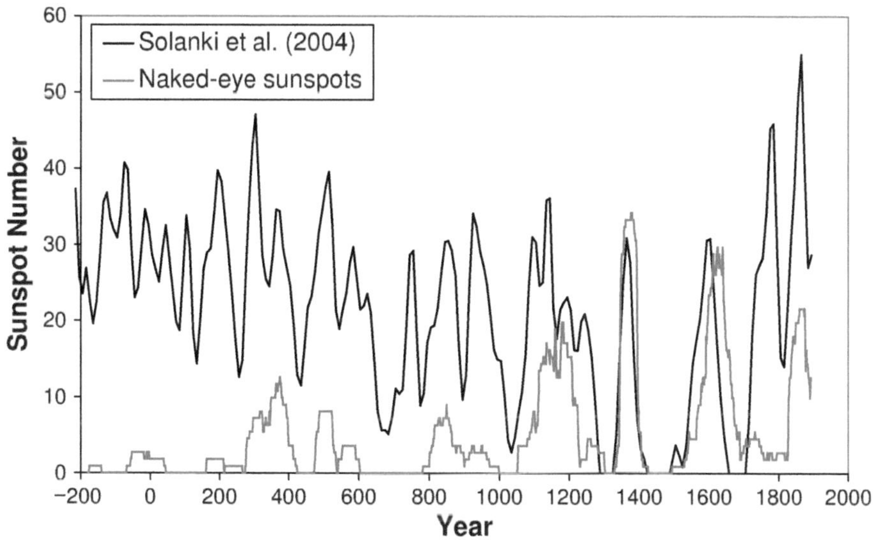

Fig. 2.23. Reconstruction of sunspot number by Solanki et al. (2004) (black line) and using the 50-year moving average of the annual number of naked-eye sunspots by Vaquero et al. (2002) (grey line). Data were provided by I.G. Usoskin.

naked-eye sunspot reconstructed by Vaquero et al. (2002) in arbitrary units. The naked-eye record is not useful for the study of long-term solar activity from 200 BC to AD 800 although some peaks of solar activity also appear in the naked-eye record. For the period 800–1600 approximately, the naked-eye sunspot record shows the most important periods of the solar activity although the amplitudes are variable. The two curves are quite similar during the period AD 1300–1600. Since the year 1600, the use of local sources in the Oriental record has made the correlation worse, and major differences can be observed during the period 1700–1825.

2.4.2 Solar Cycle and Giant Sunspots

Is there any relationship between giant sunspots (suspected as being observable with the naked-eye) and the 11-year solar cycle? Wittmann (1978) concluded that giant sunspots are reliable tracers of the maximum epoch. On some rare occasions, a large sunspot may appear in the time of a minimum, but the majority of giant sunspots appear concentrated around the maximum of the 11-year solar cycle. Figure 2.24 shows a bimodal distribution of large spots (area greater than 1500 millionths of solar hemisphere) with maxima at −1.4 and +0.6 years from the maximum epoch of the sunspot number. Clearly visible is the Gnevyshev gap (also recorded by modern observers of naked-eye sunspots) between the maxima at −1.4 and +0.6 years from epoch 0.

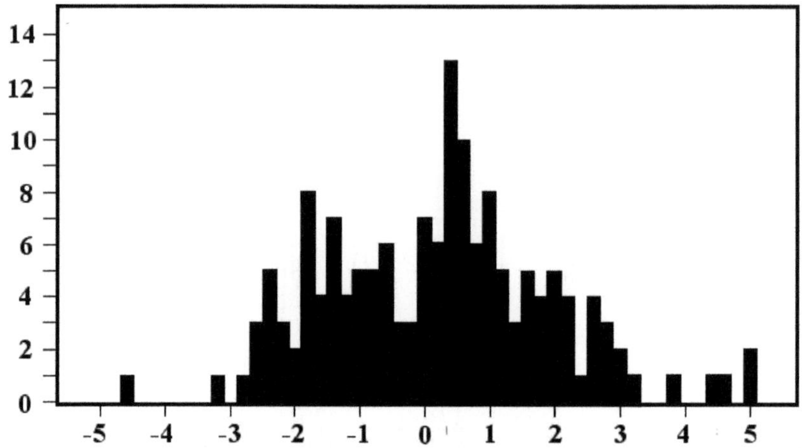

Fig. 2.24. Relative phase of occurrence for the 153 largest spots (areas>1500 millionths of a solar hemisphere) of the period 1872–1985.

Wittmann and Xu (1987) identified 44 maxima of 11-year solar cycles in their data on historical naked-eye sunspot observations. A fit of the whole sample with respect to Julian Day number gives

$$\mathrm{JD(Max.)} = (1722516 \pm 365) + (4060.1 \pm 2.5)\mathrm{N} \tag{2.10}$$

where N is an arbitrary cycle number. Wittmann and Xu (1987) chose N = 178 for the maximum of 1980–83. Also, Equation (2.10) can be written as

$$\mathrm{Year(Max.)} = (4.0 \pm 1.0) + (11.116 \pm 0.007)\mathrm{N} \tag{2.11}$$

if we neglect the differences between the Julian and the Gregorian calendars.

Wittmann and Xu (1987) computed the dates of maxima of the solar cycle using Equation (2.10) and plotted the relative phase for 235 historic sunspots as in Figure 2.24. This figure shows the distribution of historic naked-eye sunspots. A comparison between Figures 2.24 and 2.25 indicates that many historical records either do not represent genuine sunspots or are not representative of the maximum epoch of the solar cycle.

However, these authors went a step further because there are some reports with more reliable information than others. Thus, they selected the 360 most reliable giant spots for the period AD 299–1985. Figure 2.26 shows the distribution of these sunspots. They are concentrated around the epoch 0. With respect to Figure 2.24, the new data set partially fills the Gnevyshev gap. This may be proof that compact groups of small sunspot could also be observed with the naked eye.

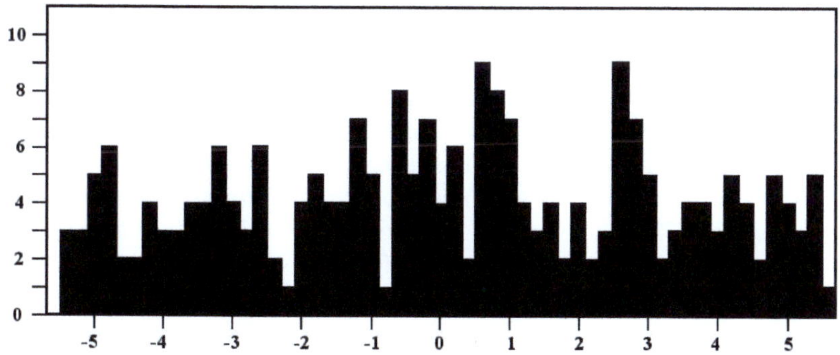

Fig. 2.25. Relative phase of occurrence for the 235 sunspots recorded by Wittmann and Xu (1987) in the period 165 BC–AD 1684.

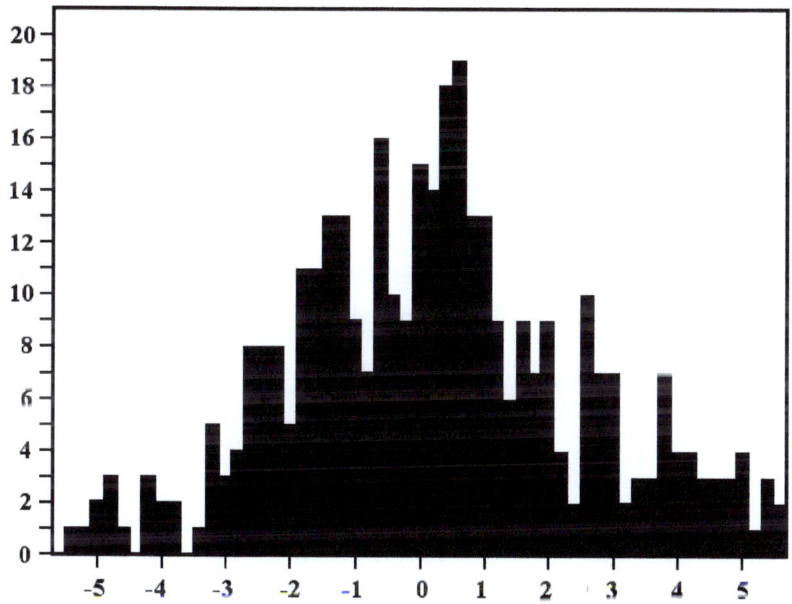

Fig. 2.26. Relative phase of occurrence for the 360 most reliable spots of the period AD 299–1985 (adapted from Wittmann and Xu, 1987).

2.4.3 High-Resolution Record

The most important advantage of the use of indices based on naked eye observations is the high time resolution available. The majority of the observations are dated with a precision of days. The high resolution of these observations as compared with other proxies should be used coherently by space climate researchers.

A very nice example of the use of the high resolution of historical sunspot records is due to Willis et al. (2005). They have recently used a comprehensive collection of catalogues of ancient sunspot and auroral observations from East Asia to identify possible intense historical geomagnetic storms in the interval 210 BC–AD 1918. In previous work, Willis and Stephenson (2001) presented evidence for an intense geomagnetic storm in December AD 1128, using historical auroral and naked-eye sunspot observations in combination. On the one hand, the sunspot described in *The Chronicle of John of Worcester* was seen on 8 December 1128 (see Figure 2.10). And on the other, a red aurora was observed on the night of 13 December 1128 from Songdo, Korea. Moreover, there is additional evidence suggesting recurrent geomagnetic activity around this time. Several Oriental sunspot and auroral records occurred approximately in synodic-solar-rotation periods suggesting two series of recurrent geomagnetic storms.

Willis et al. (2005) compared the East Asian historical sunspot and auroral data-set looking for "approximate coincidences" between pairs of events. They formulated a selection criterion for the automatic identification of such geomagnetic storms. The assumption made in the work was the following: "an historical geomagnetic storm occurred if the time interval, T (measured in days), between the observation of a sunspot and the associated auroral display satisfies the following condition:

$$-8 \leq \mathrm{T} \leq +15$$

This selection criterion was chosen based on three specific assumptions about (1) the duration of sunspot visibility with the unaided eye, (2) the likely range of heliographic longitudes of an energetic solar feature, and (3) the likely range of transit times for ejected solar plasma. Table 2.2 lists the "approximate coincidences" between Oriental sunspot and auroral observations that were found.

Table 2.2. List of "approximate coincidences" between Oriental sunspot and auroral observations, derived from the sunspot and auroral databases by Willis et al. (2005). Dates before 5 October 1582 AD are in the Julian calendar; subsequent dates are in the Gregorian calendar

No.	Sunspot Observation	Auroral Observation
01	1–10 Mar 1137, China	4 Mar 1137, China
02	10, 11 Feb 1185, China & Korea	2 Feb 1185, Japan
03	27 Mar 1185, Korea	26 Mar 1185, Korea
04	3–12 Dec 1193, China	5 Dec 1193, China
05	3–12 Dec 1193, China	6 Dec 1193, China
06	19–31 Dec 1202, China	19 Dec 1202, Japan
07	21 Feb 1204, China	21 Feb 1204, Japan
08	21 Feb 1204, China	22 Feb 1204, Japan
09	21 Feb 1204, China	23 Feb 1204, Japan
10	28 Jan–3 Feb 1370, China	11 Feb 1370, Korea
11	21 Oct 1370, China	27 Oct 1370, Japan
12	17 Apr 1556, Korea	13 Apr 1556, Korea

Table 2.2. Cont.,

No.	Sunspot Observation	Auroral Observation
13	22 May 1618, China	17 May 1618, China
14	15–24 Oct 1620, China	19 Oct 1620, China
15	15–24 Oct 1620, China	20 Oct 1620, China
16	17–20 Mar 1624, China	21 Mar 1624, Korea
17	15, 16 Apr 1624, China	18 Apr 1624, Korea
18	15, 16 Apr 1624, China	19 Apr 1624, Korea
19	15, 16 Apr 1624, China	21 Apr 1624, Korea
20	2 Sep 1625, China	28 Aug 1625, Korea
21	2 Sep 1625, China	16 Sep 1625, Korea
22	29 Jun 1626, China	24 Jun 1626, Korea
23	29 Jun 1626, China	10 Jul 1626, Korea
24	9 Dec 1638, China	23 Dec 1638, China
25	16 Jan 1648, Korea	24 Jan 1648, China

Moreover, a set of appendices discusses the literary and scientific reliabilities of the East Asian sunspot and auroral records. The Scheiner telescopic sunspot drawings from *Rosa Ursina*, that will be reviewed in Chapter 3, were also used to assess the credibility of some of the later historical auroral accounts.

References

Allen, C. W. *Astrophysical Quantities.* Astrophysical Quantities, London: Athlone (3rd edition), 1976.

Allen, M. R. and Smith, L. A. Monte Carlo SSA: Detecting irregular oscillations in the presence of colored noise. *Journal of Climate*, 6:3373–3402, 1996.

Bard, E., Raisbeck, G. M., Yiou, F., and Jouzel, J. Solar modulation of cosmogenic nuclide production over the last millennium: comparison between ^{14}C and ^{10}Be records. *Earth and Planetary Science Letters*, 150:453–462, 1997.

Beer, J. Long-term indirect indices of solar variability. *Space Science Reviews*, 94:53–66, 2000.

Beer, J., Blinov, A., Bonani, G., Hofmann, H. J., and Finkel, R. C. Use of Be^{1}0 in polar ice to trace the 11-year cycle of solar activity. *Nature*, 347:164–166, 1990.

Bicknell, P. J. Did Anaxagoras observe a sunspot in 467 B.C.? *Isis*, 59:87–90, 1968.

Blackwell, H. R. Contrast thresholds of the human eye. *Journal of the Optical Society America*, 36:624–643, 1946.

Bo Shu-ren. Ancient Chinese Observations of Physical Phenomena Attending Solar Eclipses. *Kejishiwenji*, 10:193, 1983.

Boyle, R. An observation of a Spot in the Sun. *Philosophical Transactions*, 6:2216, 1671.

Brody, J. *The Enigma of Sunspot. A Story of Discovery and Scientific Revolution.* Edinburgh, Floris Books, 191 pp, 2002.

Casiri. *Bibliotheca Arabico-Hispana Escurialensis.* volume 1, Madrid, 1760.

Chen, Mei-dong and Dai, Nian-zu, *Studies in the History of Natural Sciences*, 1: 227–236, 1982.

Chu Wen-Hsin. *Li Tai Jih Shih K'ao [A study of the Solar Eclipses Down the Ages].* Shanghai, Comercial Press, 1934.

Clark, D. and Stephenson, F. R. Establishing the case for the Maunder Minimum. *Nature*, 284:592, 1980.

Clark, D. H. and Stephenson, F. R. An interpretation of the pre-telescopic sunspot records from the orient. *Quarterly Journal Royal Astronomical Society*, 19:387–410, 1978.

Collados, M., Martinez Pillet, V., Ruiz Cobo, B., del Toro Iniesta, J. C., and Vázquez, M. Observed differences between large and small sunspots. *Astronomy and Astrophysics*, 291:622–634, 1994.

Copernicus, N. *De revolutionibus orbium coelestium.* Norimbergae, Apud J. Petreium, 1543; [Bruxelles, Culture et Civilisation, 1966], 1543.

Darlington, R. R. and McGurk, P. (eds.). *The Chronicle of John of Worcester. Vol. II.* Clarendon Press, Oxford, 1995.

Dawson, W. Continuation of sun-spot observations, *Astronomical Journal* 10:11, 1890.

Dawson, W. Number of sun-spots, 1884-1886. *Astronomical Journal*, 8:26–29, 1888.

Dawson, W. Sun-spots 1887–1889, *Astronomical Journal*, 9:125, 1889.

Daxecker, F. *The Physicist and Astronomer Christoper Scheiner – Biography, Letters, Works.* Publications at Innsbruck University, Innsbruck, Austria, 2004.

Dettinger, M. D., Ghil, M., Strong, C. M., Weibel, W., and Yiou, P. Software expedites singular-spectrum analysis of noisy time series. *EOS*, 76:12–21, 1995.

Ding, Y.-J., Luo, B.-R., and Feng, Y.-M. Peak years of various solar cycles. *Chinese Astronomy and Astrophysics*, 7:24–30, 1983.

Dodd, W. W. and Schaefer, B. E. An Introduction to astronomical Events Recorded in The Russian Chronicles. *Journal of the Royal Astronomical Society of Canada*, 96:236–249, 2002.

Doggett, L. E. and Schaefer, B. E. Lunar crescent visibility. *Icarus*, 107:388–403, 1994.

Eather, R. H. *Majectic Lights. The Aurora in Science, History, and the Arts.* Washington: American Geophysical Union, 1980.

Eddy, J. A., Stephenson, F. R., and Yau, K. K. C. On pre-telescopic sunspot records. *Quarterly Journal Royal Astronomical Society*, 30:65–73, 1989.

Einhard. *The Life of Charlemagne, translated by Samuel Epes Turner.* New York: Harper & Brothers, 1880.

Emler, J. (ed.). *Fontes Rerum Bohemicarum, II. Cosmae Chronicon Boemorum Cum Continuatoribus.* Praha, 1874.

Eresii, T. *Opera, Quae Supersunt, Omnia Graeca Recensuit, Latina Interpretatus Est.* Paris, Editore Ambrosio Firmin Didot, 1866.

Galileo, G. *Istoria e Dimostrazioni Intorno Alle Macchie Solari.* Roma, 1613.

Galindo Trejo, J. and Allen, C. Maya observations of 13th century transits of Venus? In D. W. Kurtz, (ed.), *IAU Colloq. 196: Transits of Venus: New Views of the Solar System and Galaxy*, pp. 124–137, 2005.

Gnevyshev, M. N. On the 11-Years Cycle of Solar Activity. *Solar Physics*, 1:107–120, 1967.

Gnevyshev, M. N. Essential features of the 11-year solar cycle. *Solar Physics*, 51:175–183, 1977.

Goldstein, B. R. Some medieval reports of Venus and Mercury transits. *Centaurus*, 14:49–59, 1969.

Gualterotti, R. *Discorso di Raffael Gualterotti Gentilhuomo Fiorentino sopra l'apparizione de la nuova stella. E sopra le tre oscurazioni del Sole, e de la Luna nel anno 1605.* Florence, Nella Stamperia di Cosimo Giunti, 1605.

Ham, W. T., Mueller, H. A., and Sliney, D. H. Retinal sensitivity to damage from short wavelength light. *Nature*, 260:153–155, 1976.

Hameed, S. and Gong, G. Influence of atmospheric dust on non-telescopic sunspot observations. *Solar Physics*, 132:409–413, 1991.

Hardy, R. Theophrastus observations of sunspots. *Journal of the British Astronomical Association*, 101:261, 1991.

Heath, A. W. Naked eye sunspots and the solar cycle. *Journal of the British Astronomical Association*, 104:304–307, 1994.

Herschel, W. On the nature and construction of the Sun and fixed Stars. *Philosophical Transactions Series I*, 85:46–72, 1795.

Herschel, W. Additional observations tending to investigate the symptoms of the variable emission of the light and heat of the Sun; with trials to set aside darkening glasses, by transmitting the solar rays through liquids; and a few remarks to remove objections that might be made against some of the arguments contained in the former paper. *Philosophical Transactions Series I*, 91:354–362, 1801a.

Herschel, W. Observations tending to investigate the nature of the Sun, in order to find the causes or symptons of its variable emission of light and heat; with remarks on the use that may possibly be drawn from solar observations. *Philosophical Transactions Royal Society London Series I*, 91:265, 1801b.

Hill, A. R. Note respecting solar spots visible to the naked eye. *Monthly Notices Royal Astronomical Society*, 30:207–208, 1870.

Hoskin, M. A. *William Herschel and the construction of the heavens.* London, Oldbourne, 1963, 1963.

Hosie, A. The First observations of sun-spots, *Nature*, 20:131–132, 1879.

Houston, S. D. *The First Writing: Script Invention as History and Process.* Cambridge, Cambridge University Press, 2004.

Hoyt, D. V. and Schatten, K. H. New information on solar activity, 1779-1818, from Sir William Herschel's unpublished notebooks. *Astrophysical Journal*, 384:361–384, 1992a.

Hoyt, D. V. and Schatten, K. H. Sir William Herschel's notebooks – Abstracts of solar observations. *Astrophysical Journal Supplement Series*, 78:301–340, 1992b.

Hoyt, D. V. and Schatten, K. H. Group sunspot numbers: A new solar activity reconstruction. *Solar Physics*, 179:189–219, 1998.

Hoyt, D. V. and Schatten, K. H. Group sunspot numbers: A new solar activity reconstruction. *Solar Physics*, 181:491–512, 1998.

Hsü, J. C. H. *The Mendes Collection of Shang Dynasty Oracle Bones.* Toronto, Royal Ontario Museum, 1972.

Ibn al-Qifti. *Ta'rikh al-hukama.* edited by J. Lippert, Leipzig, 1903.

Ibn Hayyan. *Al-Muqtabis V.* P. Chalmeta, F. Corriente, and M. Subh (eds.), Instituto Hispano-Árabe de Cultura, Madrid, 1979.

Ibn Hayyan. *Crónica del Califa 'Abdarrahman III an-Nasir entre los años 912 y 942 (al-Muqtabis V).* translated by M. J. Viguera and F. Corriente, Anubar Ediciones, Zaragoza, 1981.

Istock, T. H. Solar retinopathy: A review of the literature and case report. *Journal of the American Optometric Association*, 56:374–382, 1985.

Kanda, S. *Proceedings of the Imperial Academy Japan*, 9:293–296, 1933.

Kane, R. P. Which One is the 'Gnevyshev' Gap? *Solar Physics*, 229:387–407, 2005.

Keller, H. U. and Friedli, T. K. Visibility limit of naked-eye sunspots. *Quarterly Journal Royal Astronomical Society*, 33:83–89, 1992.

Kilgour, F. G. Professor John Winthrop's Notes on Sun Spot observations (1739). *Isis*, 29:355–361, 1938.

Křivský, L. Naked-eye observations of sunspot in Bohemia in the year 1139. *Bulletin of the Astronomical Institutes of Czechoslovakia*, 36:60–61, 1985.

Lee, E. H., Ahn, Y. S., Yang, H. J., and Chen, K. Y. The sunspot and auroral activity cycle derived from Korean historical records of the 11th 18th century. *Solar Physics*, 224:373–386, 2004.

Lewis, R. An account of the same aurora borealis. *Philosophical Transactions Royal Society London*, 37:69–70, 1731.

Link, F. Sur la période de 4 siècles dans l'activité solaire. *Zeitschrift für Astrophysik*, 56:279–281, 1963.

Lynch, D. K. and Livingston, W. *Color and light in Nature*. Cambridge, Cambridge University Press, 2nd edition, 2001.

MacRobert, A. M. Viewing sunspots with just a filter. *Sky & Telescope*, 78:289, 1989.

Maunder, A. and Maunder, E. *The Heavens and their Story*. Boston, Dana Estes and Company, 1909.

Maunder, E. W. A prolongated sunspot minimum. *Knowledge*, 17:173, 1894.

McGurk, P. (ed.). *The Chronicle of John of Worcester. Vol. III.* Clarendon Press, Oxford, 1998.

McKim Malville, J. and Singh, R. P. B. Visual astronomy in the mythology and ritual of india: The Sun temples of Varanasi. *Vistas in Astronomy*, 39:431–449, 1995.

Meeus, J. The transits of Venus 3000 B.C. to A.D. 3000. *Journal of the British Astronomical Association*, 68:98, 1958.

Moore, S. Letter to the Editor: The Revd Thomas William Webb and a transit of Mercury. *Journal of the British Astronomical Association*, 113:305–306, 2003.

Mossman, J. E. A comprehensive search for sunspots without the aid of a telescope, 1981–1982. *Quarterly Journal Royal Astronomical Society*, 30:59–64, 1989.

Nagovitsyn, Yu. A. Solar activity during the last two millennia: Solar patrol in ancient and medieval China. *Geomagnetism and Aeronomy*, 41:680–688, 2001.

Needham, J. H. *Science and Civilisation in China, Vol. III.* Cambridge, Cambridge University Press, 1959.

Rae, P. S. *Portents and Politics in Early Yi Korea: 1392-1519.* Ph.D. thesis, University of Hawaii, U.S.A., 1977.

Robinson, J. and Robinson, M. *Stargazer of Hardwick: Thomas William Webb.* Gracewing, 2006.

Sachs, A. H. and Hunger, H. *Astronomical diaries and related texts from Babylonia.* Vol. I. Wien: Osterreichischen Akademie der Wissenschaften, 377 pp, 1988.

Sarton, G. Early observations of sunspots? *Isis*, 37:69–71, 1947.

A. H. Sayce. *Babylonian Literature.* London, S. Bagster and Sons, 86pp, 1877.

Schaefer, B. E. Telescopic limiting magnitudes. *Publications of the Astronomical Society of the Pacific*, 102:212–229, 1990.

Schaefer, B. E. Sunspot visibility. *Quarterly Journal Royal Astronomical Society*, 32:35–44, 1991.

Schaefer, B. E. Visibility of sunspots. *Astrophysical Journal*, 411:909–919, 1993.

Scheiner, C. *Oculus hoc est: Fundamentum Opticum, in quo ex Accurata Oculi Anatome, Abstrusarum Experientiarum* ... Oeniponti, Apud Danielem Agricolam, 1619.

Schove, D. J. *Sunspot cycles.* Stroudsburg, PA, Hutchinson Ross Pub. Co., New York, 1982.

Scuderi, L. A. Oriental sunspot observations and volcanism. *Quarterly Journal Royal Astronomical Society*, 31:109–120, 1990.

Solanki, S. K., Usoskin, I. G., Kromer, B., Schüssler, M., and Beer, J. Unusual activity of the Sun during recent decades compared to the previous 11,000 years. *Nature*, 431:1084–1087, 2004.

Southall, J. Early pioneers in physiological optics. *Journal Optical Society of America*, 6:827–842, 1922.

Stephenson, F. R. Historical Evidence concerning the Sun: Interpretation of sunspot records during the telescopic and pretelescopic eras. *Royal Society of London Philosophical Transactions, Series A*, 330:499–512, 1990.

Stephenson, F. R. and Willis, D. M. The earliest drawing of sunspots. *Astronomy and Geophysics*, 40:21–22, 1999.

Stuiver, M. and Quay, P. D. Changes in atmospheric Carbon-14 attributed to a variable sun. *Science*, 207:11–19, 1980.

Theophrastus. *Theophrastus's enquiry into Plants and minor works on Odours and Weather Signs.* Translated by A. Hort, London, 1916.

Thomson, D. J. Spectrum estimation and harmonic analysis. *Proceedings of the IEEE*, 70:1055–1096, 1982.

Turnbull, H. W. *The Correspondence of Isaac Newton. Vol III: 1688–1694.* Cambridge University Press, 1961.

Usoskin, I. G. and Kovaltsov, G. A. Long-term solar activity: Direct and indirect study. *Solar Physics*, 224:37–47, 2004.

Usoskin, I. G., Solanki, S. K., and Kovaltsov, G. A. Grand minima and maxima of solar activity: new observational constraints. *Astronomy and Astrophysics*, 471:301–309, 2007.

Vaquero, J. M. A forgotten naked-eye sunspot recorded by Galileo. *Solar Physics*, 223:283–286, 2004.

Vaquero, J. M. and Gallego, M. C. Evidence for a sunspot in A.D. 939 in an Arabian Source. *Solar Physics*, 206:209–211, 2002.

Vaquero, J. M., Gallego, M. C., and García, J. A. A 250-year cycle in naked-eye observations of sunspots. *Geophysical Research Letters*, 29:1997, 2002.

Vautard, R., Yiou, P., and Chil, M. Singular-spectrum analysis: A toolkit for short, noisy chaotic signals. *Physica D*, 58:95–126, 1992.

Vyssotsky, A. N. *Astronomical records in the Russian Cronicles from 1000–1600 A.D.* Med. fran Lunds Astro. Observatorium. Ser. 2. Nr. 126, Historical papers Nr. 22, Lund, 1949.

Wade, P. Naked-eye sunspots, 1980–1992. *Journal of the British Astronomical Association*, 104:86–87, 1994.

Wang, P. K. and Siscoe, G. L. Ancient chinese observations of physical phenomena attending solar eclipses. *Solar Physics*, 66:187–193, 1980.

Weld, A. Observation of a solar spot with the naked eye. *Monthly Notices Royal Astronomical Society*, 11:167, 1851.

Welsh, J. The second voyage to Benin. In R. Hakluyt, (ed.), *The Principal Navigations, Voyages, Traffics, Discoveries of the English Nation.* Glasgow, MacLehose, 1904.

Westfall, R. *Never at Rest.* Cambridge, Cambridge University Press, 1980.

Willis, D. M., Armstrong, G. M., Ault, C. E., and Stephenson, F. R. Identification of possible intense historical geomagnetic storms using combined sunspot and auroral observations from East Asia. *Annales Geophysicae*, 23:945–971, 2005.

Willis, D. M., Davda, V. N., and Stephenson, F. R. Comparison between oriental and occidental sunspot observations. *Quarterly Journal Royal Astronomical Society*, 37:189–229, 1996.

Willis, D. M., Easterbrook, M. G., and Stephenson, F. R. Seasonal variation of oriental sunspot sightings. *Nature*, 287:617–619, 1980.

Willis, D. M. and Stephenson, F. R. Solar and auroral evidence for an intense recurrent geomagnetic storm during December in AD 1128. *Annales Geophysicae*, 19:289–302, 2001.

Wittmann, A. The sunspot cycle before the Maunder minimum. *Astronomy and Astrophysics*, 66:93–97, 1978.

Wittmann, A. D. and Xu, Z. T. A catalogue of sunspot observations from 165 BC to AD 1684. *Astronomy and Astrophysics Supplement Series*, 70:83–94, 1987.

Xu, Z. Solar observations in ancient China and solar variability. *Philosophical Transactions Royal Society London A*, 330:513–515, 1990.

Xu, Z. T. In *Proceedings Kunming Workshop on Solar Physics and Interplanetary Travelling Phenomena*, C. de Jager & C. Biao (eds.), Science Press, Beijing, Vol. 1, p. 108, 1983.

Xu, Z.-T. and Jiang, Y.-T. The solar activity in the seventeenth century re-assessed in the light of sunspot records in the local gazettes of China. *Chinese Astronomy and Astrophysics*, 6:84–90, 1982.

Yau, K. K. C. *An investigation of some contemporary problems in astronomy and astrophysics by way of early astronomical records.* Ph.D. thesis, University of Durham, UK, 1988.

Yau, K. K. C. and Stephenson, F. R. A revised catalogue of Far Eastern observations of sunspots (165 BC to AD 1918). *Quarterly Journal Royal Astronomical Society*, 29:175–197, 1988.

Yunnan Observatory Ancient Sunspot Records Group. A re-compilation of our country's records of sunspots throuhg the ages and an inquiry into possible periodicities intheir activity. *Chinese Astronomy*, 1:347–359, 1977.

3

Solar Drawings

Drawings were the first method used for keeping a record of astronomical observations; for example, drawings of the constellations were being made from very early on. This technique was also applied to other scientific disciplines such as Anatomy, Botany and Zoology. Leonardo da Vinci (1452–1519) was one of the pioneers in this field (Kemp, 2006). In this way, it was possible to fulfil one of the essentials of the scientific method: to tell a colleague, come and look at what I have seen.

The lack of a theory of radiation handicapped the interpretation of the observations much more than the available instrumentation. Moreover, the draftsman was obliged to average rapid changes on the Sun produced intrinsically or induced by the turbulence in the Earth's atmosphere. It is interesting to note how the brain of the astronomer was able to record the best moments and later transfer them to a drawing.

An obvious drawback of visual observations is their subjectivity. Each observer has a personal equation which must be applied to any observation before comparing it with others.

In this chapter we describe approximately 250 years of solar observations using the eye and the hand as the best tools for reproducing the reality on the solar surface. The list is far from complete. Some of the observations may have been destroyed and others are difficult to find and identify (Hoyt, 2000).

3.1 Pretelescopic Instruments

In the previous chapter, we described the naked-eye observations of sunspots. They mainly consist of mere descriptions and only in a few cases can we obtain information about the size or internal structure. One exception was the drawing made by John of Worcester on December 8, 1128, described in the previous chapter. The Worcester drawing was made by the naked eye, but already in the Medieval Ages the techniques had advanced to get sharper and better defined images of astronomical objects.

J.M. Vaquero, M. Vázquez, *The Sun Recorded Through History*, Astrophysics and Space Science Library 361, DOI 10.1007/978-0-387-92790-9_3,

3.1.1 The Camera Obscura

A Florentine manuscript of 1289 written by Sandra di Popozo and entitled *Traite de conduite de la famille* refers to some of these: *glass lenses for spectacles recently invented, of great advantage to old people with weak vision.*[1] This would seem to confirm that the use of pieces of glass, called "vetri lenticchie" or "lenti" (lenses) on account of their lentil (lenticchie) shape, for correcting long-sightedness was known from the end of the 13th century. Roger Bacon (1214–1294) wrote several scientific treatises[2] including *De Iride* (On the Rainbow) in which he said: *This part of optics, when well understood, shows us how we may make things a very long distance off appear as if placed very close, and large near things appear very small, and how we may make small things placed at a distance appear any size we want, so that it may be possible for us to read the smallest letters at incredible distances* (see Crombie, 1971).

The art of producing an image of an object through a pinhole has also been known since antiquity (Lindberg, 1970). The earliest mention of this technique was made by the Chinese philosopher Mo-Ti (5th century BC), who created an inverted image formed by light rays passing through a pinhole into a darkened room. Al Hazen (965–1039) describes eclipse observations in *On the form of the eclipse* commenting that: *The image of the Sun at the time of the eclipse, unless it is total, demonstrated that when its light passes through a narrow, round hole and is cast on a plane opposite to the hole it takes on the form of a moon-sickle.* He also established that the smaller the hole is, the clearer the picture is. The first published picture of a pinhole camera is in fact the drawing of the solar eclipse of January 24, 1544, as observed in Louvain, included in the *De Radio Astronomico et Geometrico* of Gemma Frisius (1508–1555), a mathematician and instrument maker (see Figure 3.1).

Leonardo da Vinci describes pinhole image formation in his *Codex Atlanticus*: *If the facade of a building, or a place, or a landscape is illuminated by the sun and a small hole is drilled in the wall of a room in a building facing this, which is not directly lighted by the sun, then all objects illuminated by the sun will send their images through this aperture and will appear, upside down, on the wall facing the hole.* Figure 3.2 shows a diagram of such a pinhole camera.

Giovanni Battista della Porta (1538–1615) describes a pinhole camera obscura and the combination of concave and convex lenses[3] in his *Magiae naturalis, libri XX*, published in 1589. In this book the following passage is included: *Concave lenticulars will make one see most clearly things that are afar off; but Convexes, things neer hand; so you may use them as your sight*

[1] Based on writings, in 1676, by Franciscus Redi (1626–1697), a professor of the University of Pisa.

[2] He was one of the proposers of the scientific method based on a repeating cycle of observation, hypothesis, experimentation, and the need of independent verification.

[3] It is likely that Girolamo Cardano (1501–1576) in his *De Subtilitate Rerum* (1550) had already described a camera obscura with a lens.

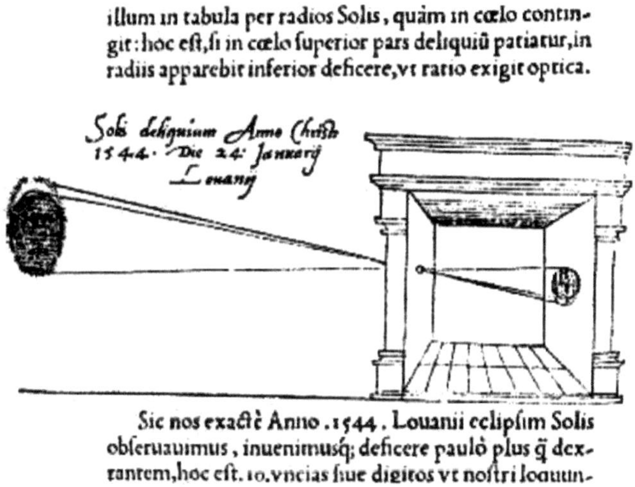

Fig. 3.1. Observation of the total solar eclipse of 24 January 1544 in Louvain with a camera obscura by Gemma Frisius.

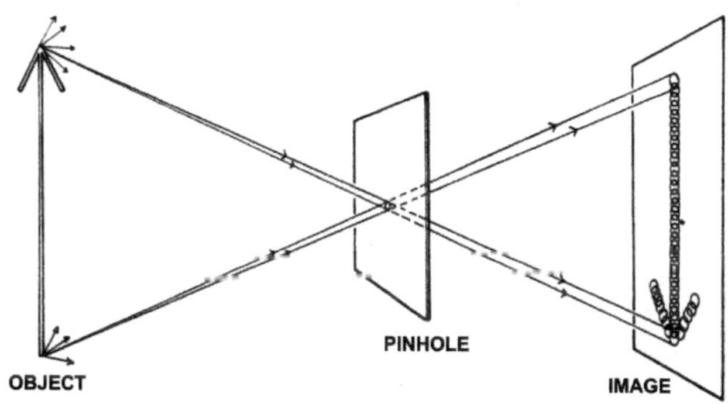

Fig. 3.2. Optical scheme of a pinhole camera.

requires. With a Concave you shall see small things afar off, very clearly; with a Convex, things neerer to be greater, but more obscurely; if you know how to fit them both together, you will see both things afar off, and things neer hand, both greater and clearly (Watson, 2005 p. 44).

Johhannes Kepler first coined the term camera obscura (latin for "dark room") in his book *Ad Vitellionem Paralipomena.* In 28 May 1607 he wished to observe a predicted transit of Mercury across the Sun's disk, and on the appointed day he projected the Sun's image through a small hole in the roof and did indeed observe a black spot (Figure 3.3) that he interpreted to be Mercury. He wrote a small book about this event in 1609: *Phaenomenon singulare, seu Mercurius in Sole.* He describes that this day he was in Prague with Martin Bachazek, when a single Sun ray shone through thin cracks between the shingles. He held a piece of paper against this ray, and there, on

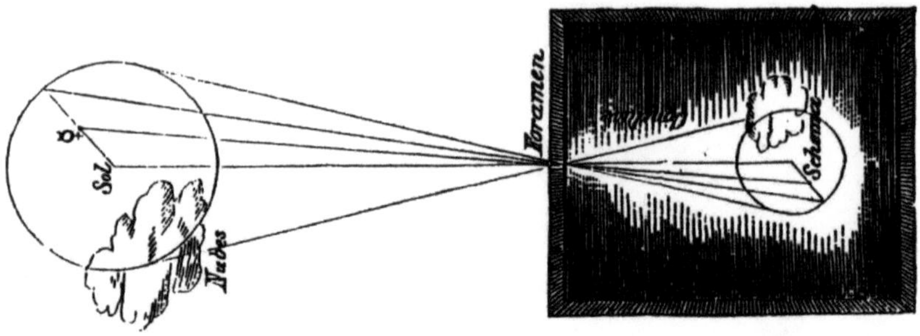

Fig. 3.3. Sunspot observed by Johannes Kepler with a camera obscura and attributed to a transit of Mercury. The image has been rotated 90 degrees to better fit on the page.

the little picture of the Sun thus formed, he espied *a little daub, quite black, approximately like a parched flea.* To ensure against error, he repeated the observation under other conditions: *And to make sure it was not a mark on the paper, we kept moving the paper back and forth so that the light on the paper moved, and everywhere the little black spot appeared together with the light* (Caspar, 1993). In 1620 Kepler showed the English diplomat Henry Wotton (1568–1639) a portable camera obscura of his own design.

It has been suggested that the Dutch artist Jan Vermeer (1632–1675) used a camera obscura to produce his realistic paintings (Seymour, 1964; Steadman, 2001). His canvas was placed beside the screen, and his living models were not seen. Probably, his "optical consultant" was Antony van Leeuwenhoek (1632–1723), the pioneer of microscopy.

Cathedral domes were sometimes used as giant camera obscuras for precise observations of the solar noon. This was the case with the San Petronio cathedral in Bologna, Italy (see Chapter 5 for more details).

The absence of sunspot observations during the 15th and 16th centuries using a camera obscura was possibly related to the existence of prolonged periods of reduced solar activity (Wolf and Spörer minima). Fortunately, the invention of a decisive instrument for astronomy, the telescope, coincided with a period of enhanced solar activity.

3.2 The Invention of the Telescope

The history of the telescope dates from the use of spectacles, which have their own long history, originating in Italy near the end of the 13th century. The first telescope seems to have been invented by Hans Lipperhey (1570–1619), a spectacle-maker of Middelburg (Holland), who on 2 October, 1608 presented a patent of an instrument to the States General in The Hague.[4] King (1955),

[4] The text of the letter was: *The bearer of this, who claims to have a certain device, by means of which all things at a very great distance can be seen as if they*

Bell (1981)[5] and Watson (2005) give an excellent account of the history of this instrument.

In Chapter 1 we established the basic formula for image formation through a telescope. Now, we will provide the historical development of concepts and ideas related to this invention.

William Bourne (1535–1582) described, already in 1580, the formation of images through a convex lens in a report to Lord Burghley.[6] The first design of a telescope was made by Galileo and consisted in a convex and a plano-concave lense. In the *Siderius Nuncius*, Galileo published a ray diagram showing the light bending only at the objective, assigning no obvious role to the eyepiece, an essential tool for the magnification of the image (Figure 3.4).[7]

Johannes Kepler (15761–1630) in his *Dioptrice* suggested the idea of using a convex (positive) eyepiece. However, it was not till the middle of the 17th century that this telescope came into general use, probably because its field of view was much larger than in the Galilean telescope. Figure 3.5 shows a scheme of this design. The objective lens (1) focuses light from a distant object (4) to a focal plane where it forms a real image (5). This image is viewed through an eyepiece (2), which acts like a magnifying glass. The eye (3) sees a magnified inverted image (6) at a large distance.

Once the telescope was invented, its application to solar observations was immediate and rapidly connected to new discoveries.

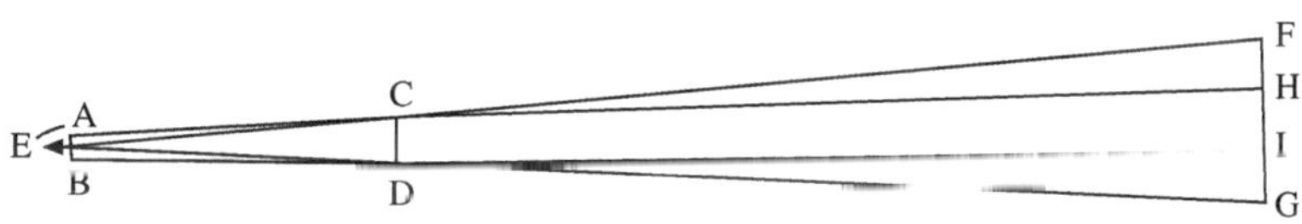

Fig. 3.4. Ray diagram of Galileo telescope in his *Siderius Nuncius*. Looking through a tube (ABCD) the eye (E) sees the full length of the object (FG), but only a fraction of it (HI) when a lens is introduced in the optical path.

---

were nearby, by looking through glasses which he claims all things at a very great distance can be seen as if they were nearby, by looking through glasses which he claims to be a new invention, would like to communicate the same to His Excellency. In fact, Lippershey showed the instrument to Prince Maurice a few days later in the grounds of the Binnenhof, the seat of the States General. However, the application for the patent was refused.

[5] Reprinted version of the 1922 original with an Introduction by J. Passachof.

[6] In its opening pages Bourne explains techniques for making lenses and parabolically curved mirrors then, finally, describes: *The effects of what may bee done with these last two sorts of Glasses: The one concave with a foyle, uppon the hilly side and the other grounde and polished smoothe, the thickest in the myddle, and thinnest towards ye edges, or sydes.*

[7] Antonio Magini (1555–1617) noted at the start of a September 28, 1610 letter to Galileo that if one removed the eyepiece from a Galilean telescope and extended the tube *everything is seen inverted and very clearly, even if quite small.*

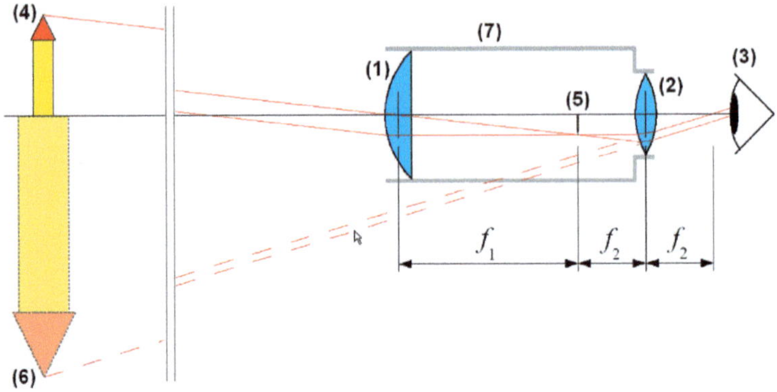

Fig. 3.5. Keplerian telescope. See text for a detailed explanation Courtesy: Wikimedia Foundation Inc.

3.3 First Telescopic Observations of Sunspots

Because of the size and contrast of sunspots, they can be observed by the naked eye under very favourable conditions. As soon as the telescope became available, astronomers began to use it to scrutinize the visible surface of the Sun. This is mainly the history of four scientists, who have the merit of being pioneers in the observations of structures on the apparently perfect solar disk. Parts of the books of Bray and Loughhead (1964), Meadows (1970) and the papers of Mitchell (1916),[8] Coyne (1991) and Casanovas (1997) constitute an adequate background to this topic.

Figure 3.6 shows the number of sunspots during the 17th century, which provide information on the level of solar activity at this time. Curiously, the discovery of the telescope arrived just in time to verify one crucial event: the Maunder minimum.

3.3.1 Thomas Harriot (1560–1621)

Thomas Harriot formed part of a group of scientists under the protection of the ninth Earl of Northhumberland, Henry Percy (1564–1632). He was active in all branches of astronomy but refused to publish his observations and no doubt for this reason his most important discovery remained unknown for a long time (Shirley, 1983; Staiger, 1998 and Chapman, 2006).[9] F.X. von Zach (1754–1832) reported in the Berliner Jahrbuch for 1788 to have found many manuscripts showing that Harriot began his astronomical observations

[8] This paper by W.M. Mitchell was published in volume 24 of the journal *Popular Astronomy* in nine different instalments. In the following, we will refer to the specific page of the reference.

[9] In fact, only two publications can be attributed to Thomas Harriot: the report of his expedition to Virginia and his testament.

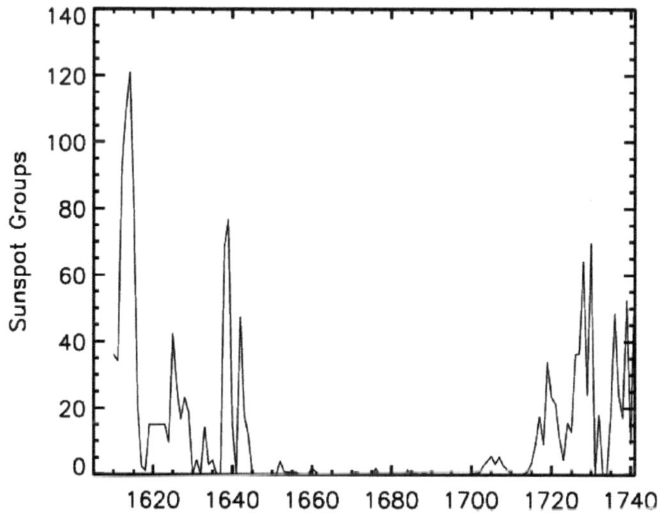

Fig. 3.6. Sunspot Group Number from 1600 to 1740. Source: Hoyt and Schatten (1998).

as early as in the summer of 1609. These manuscripts also contained nearly 200 drawings of sunspots, made from 1610 to 1612, which are preserved on 73 foolscap sheets besides three smaller pieces of paper.

The sun rose over the river Thames from Harriot's home in Syon, and viewing through the morning mist allowed Harriot to use a telescope for solar use. He made the first recorded telescopic observation of the Sun, on 8 December 1610.[10] The short description says *1610, Syon December 8^{d} [Saturday]. The altitude of sonne being 7 or 8 degrees. It being a frost and a mist I saw the sonne in this manner. Instrument 10/1 B. I saw it twise of thrise. Once with the right eye and other time with the left. In the space of a minutes time. After the sonne was too cleare* (Mitchell 1916, p. 150). In the entry of this first observation there is no explicit mention of spots but accompanying it is a sketch showing three spots (see upper part of Figure 3.7).

During 1611 Harriot made only one more effort to repeat the observation, but the solar surface was free of sunspots: *1611 Syon. January 19th. A notable mist. I observed diligently at sondry times when it was fit. I saw nothing but the cleare sonne..*

Later, probably convinced by his friends William Lower (1569–1615) and Christopher Tooke (1572–1630), he turned the telescope on the Sun. Harriot recorded these new observations of December 1, 1611 as follows: *I saw three blacke spots in such order as is here expressed as nere as I could judge, observed by 10/1 [ten-power telescope]. Lower and Tooke also sawe the same. At sundry times all three were seen & observed at once for halfe an houre space, at which*

[10] It corresponds to 18 December in the Gregorian Calendar.

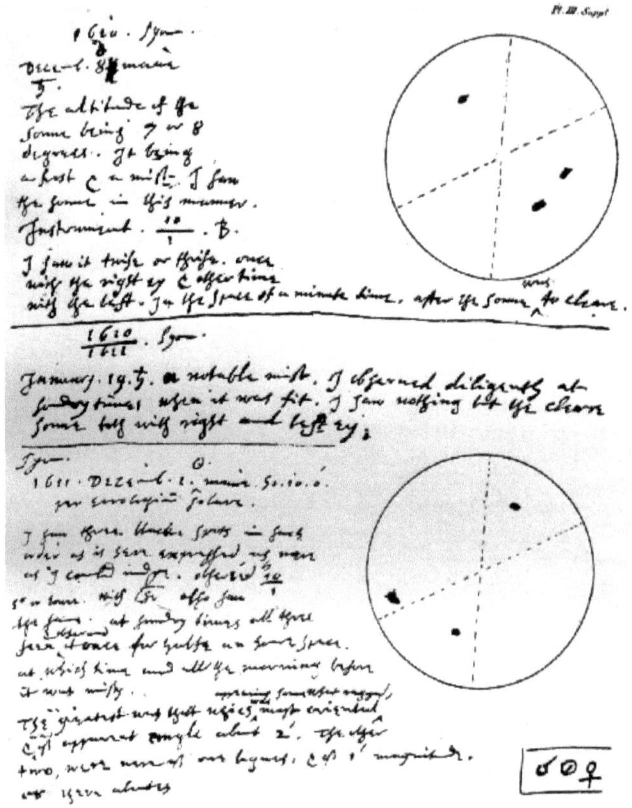

Fig. 3.7. Facsimile of the first sunspot observations by Thomas Harriot.

time and all the morning before it was misty [...] The greatest was that which, appearing somewhat ragged, was most oriental & *was of apparent angle about 2'. The other two, were nere of one bignes:* & *of 1' magnitude, or there aboutes.* The next session was on Tuesday, 3 December, observing eight spots on the disk. Harriot watched how the spots formed and disintegrated noticing that sunspots appeared larger when seen near the centre of the disk, and calculated that the period of solar rotation was over 27 days. He finished with his sunspot observations on January 1613.

The direct method of solar observation did cause Harriot some problems, although he mentioned often having observed in the presence of "thick layer and thin clouds" (Chapman, 1995). He also mentioned once *observing through my coloured glasses.*

Harriot fabricated, collected, distributed, and used many telescopes. All of them were Galilean-type refractors. Shirley (1983) indicates some significant milestones in the development of the telescope attained by Harriot and his lens grinder and observing assistant, Christopher Tooke.

3.3.2 Johannes Fabricius (1587–1616)

David Fabricius (1564–1617) was a Lutheran pastor and an intimate friend of Kepler and Tycho Brahe. His eldest son, Johannes, studied at the University of Leyden, where he became acquainted with the use of the telescope. In early 1611 he returned to his home, in Osteel (East Frisia), where together with his father he started to observe the Sun. They directed the telescope to the edge of the Sun, and when their eyes adjusted to the brightness, they slowly moved toward the centre of the disk.

Johannes wrote a 22-page pamphlet, *De Maculis in Sole Observatis, et Apparente earum cum Sole Conversione Narratio* ("Narration on Spots Observed on the Sun and their Apparent Rotation with the Sun"). The book was printed in Wittenberg (Germany) on 13 June 1611 and constitutes the first known publication on sunspots. In the tract, Johannes related the observations made by him and his father, without giving the times or dates or showing a drawing of the spots, stating the opinion that they were on the Sun and that the Sun therefore probably rotated on its axis. The following is an extract of this book.

> "While observing these things carefully, a blackish spot suddenly presented itself, on one side indeed rather thin and faint, of no little size compared to the disk of the sun. I had at first no little doubt in the reliability of the observation, because a break in the clouds disclosed the rising sun to me, so that I thought that the clouds flying past gave the false impression of a spot on the sun. The observation was repeated perhaps ten times with Batavian telescopes of different sizes, until at last I was satisfied that the spot was not caused by the interposition of clouds.
>
> However, not willing to believe in the manifest testimony of my own eyes, on account of the strange and unusual appearance of the sun, I immediately called my father, at whose house I was then staying, having returned from Batavia, in order that he might be present also to observe this [...] Thus the first day passed, and we left the sun, but not without great longing for its return on the morrow, so that our natural curiosity scarcely bore even the intervention of the night.
>
> Nevertheless we restrained our eagerness by anxious thoughts. For it was not yet certain whether that spot which we had seen would wait for the next observation."[11]

Reading the Fabricius *Prognosticon Astrologicum* for the year 1615, it is possible to state that the date of the observation was 27 February 1611.

After knowing the discovery of Fabricius, J. Kepler exclaimed: *Lucky I, who was the first in this century to have observed the spots*, taking shelter on the uncertain destiny: *How very changeable is the fortune of war in astronomy*

[11] Translated from German by H.L. Crosby, it appeared in Mitchell (1916), pp. 155–159. See also Berthold (1894).

too, since the movable army of conjectures, with vacillating assurance, turns now here now there (Caspar, 1993, p. 167).

3.3.3 Cristoph Scheiner (1575–1650)

Born in Wald, near Mindelheim, in Swabia (Germany), Scheiner became a Jesuit priest at the age of twenty and soon professor of mathematics in the Ingolstadt University and later in the Roman College. One day in March 1611, he and his assistant Johann B. Cysat (1585–1657) were looking through the smoke rising from the university which had caught fire, both saw spots on the Sun (Casanovas, 1997). From October to December the same year, they carried out regular solar observations but making use of coloured filters.[12]

His first sunspot observations were published, in Augsburg, as three letters to Mark Welser (1558–1614)[13] with the title *Tres Epistolae de Maculis Solaribus* (Three Letters on Solar Spots)[14] and cover the period October 21–December 14, 1611. Figure 3.8 shows some of these drawings included in the first letter. At that time, he used first the camera obscura for the measurement of spot positions and then the telescope for a finer examination of the shape. A second observing run covered the period March 16–April 4, 1612. In this case Scheiner rotated the images in such way as if all the observations had been carried out at midday.

When Scheiner communicated his discovery to Theodore Busaeus, the Jesuit provincial, the latter refused to give credit to the observations arguing: *I have read my Aristotle from end to end many times and I can assure you that I have never found in it anything similar to what you mention. Go, my son, calm yourself, and be assured that what you take for spots on the sun are the faults of your glasses or your eyes* (Mitchell, 1916, p. 209). Probably to avoid problems, or requested by his ecclesiastic superiors, Scheiner decided to sign his writings with the expression *Apelles latens post tabulam* (Apelles hiding behind the painting).[15]

Scheiner remained silent but active until 1626, when began the publication (which would ultimately take four years to do) of his monumental work, written in Latin, *Rosa Ursina Sive Sol ex admirando facularum & Macularum suarum Phaenomeno Varius*, a book of 780 pages, dedicated to Paolo Giordano Orsini, Duke of Bracciano, a member of one of the great families in Rome. In total, seventy full-page drawings of the Sun are included, corresponding to the period from December 1624 to June 1627. Figure 3.9 shows one of these drawings made in 1625.

---

[12] He abandoned the construction of this type of telescope due to the increase in prices of the glass.

[13] An Augsburg banker and scholar, who was a friend and patron of Jesuit scholars.

[14] Letters dated 12 November, 19 and 26 December (all 1611) and 16 January 1612.

[15] The story recalls the ancient painter Apelles hiding behind one of his paintings to hear criticisms of his work.

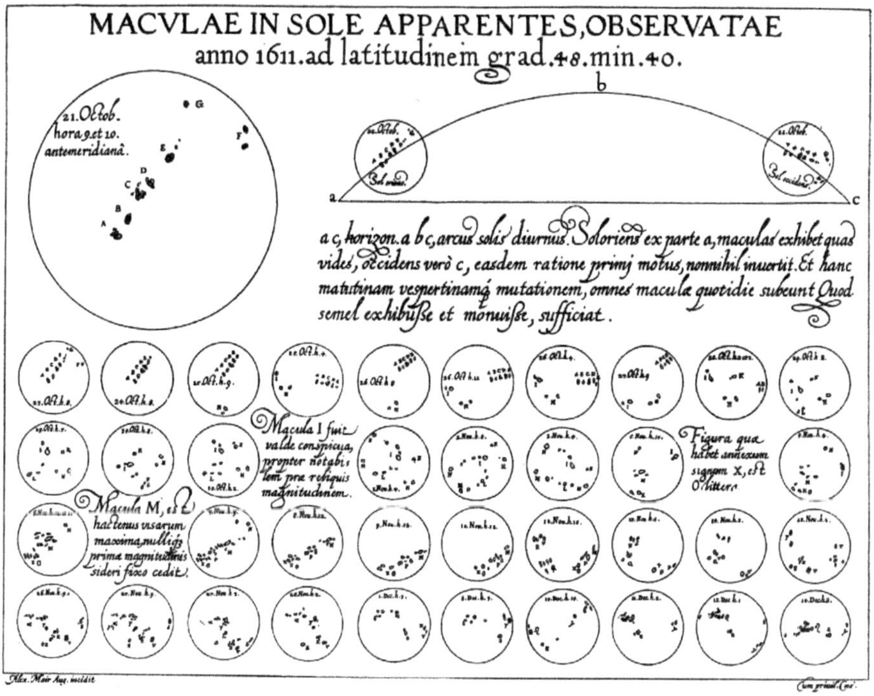

Fig. 3.8. Sunspot drawings from the letters of C. Scheiner to M. Welser describing 39 observing days from 21 October to 14 December, 1611. Sunspots are indicated with the first letters of the alphabet.

One important finding of these early observations was the slight inclination of the Sun's axis of rotation with respect to the ecliptic producing a variable path of sunspots across the disk observed from the Earth (Figure 3.10). Scheiner also found that sunspots occurred only in a narrow belt, "the royal zone", extending 30 degrees on either side of the equator.

For old books on Scheiner, see Von Braunmühl (1891). Daxecker (1996, 2006) are the most recent biographies.

3.3.4 Galileo Galilei (1564–1642)

Galileo Galilei was born in Pisa, the son of a musician, Vincenzio Galilei (1525–1591). In 1581, he was enrolled at the University of Pisa, in the school of medicine. He did not complete this degree, but instead studied mathematics. In 1589, he was appointed to the chair of mathematics in Pisa. Detailed information on Galileo and his scientific works can be found in Favaro (1895),

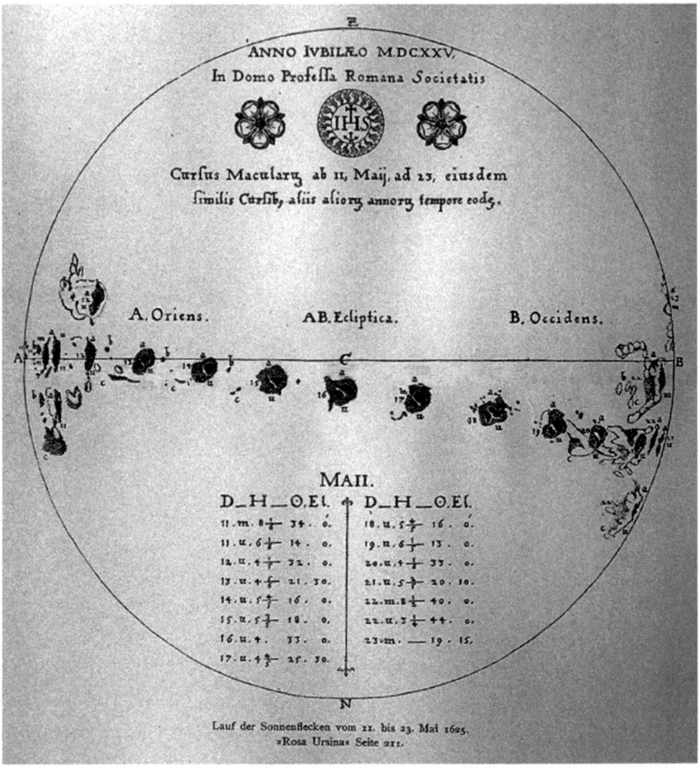

Fig. 3.9. Path of sunspots across the disk. Source: *Rosa Ursina.*

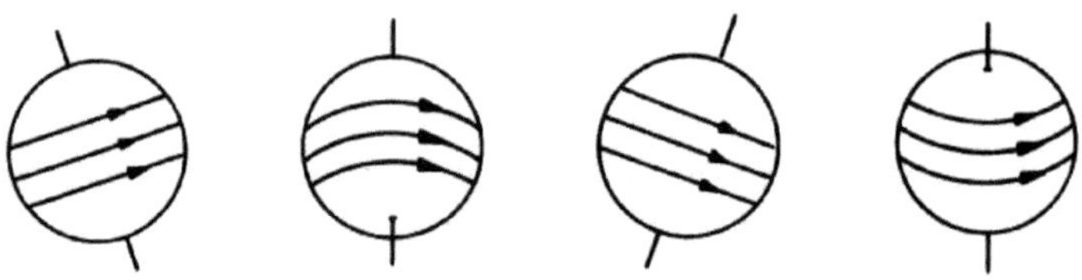

Fig. 3.10. Variation path of sunspots during the year produced by the inclination of the Sun's axis of rotation with respect to the ecliptic. From left to right: December, March, June and September.

Drake (1957),[16] Drake (1978), Machamer (1998), Sobel (1999, 2000), Shea and Artigas (2004) and Bredekamp (2007).

During a visit to Venice, July 26 1609, Galileo had heard about the Dutch instrument that could make distant objects seem near, probably from his friend Paolo Sarpi (1552–1623). Pressed by a tour of Lippershey to Italy to

[16] It includes English translations of *The Starry Messenger* (1610), *Letter to the Grand Duchess Christina* (1615) and excerpts from *Letters on Sunspots* (1613) and *The Assayer* (1623).

announce and sell the project, he decided to build his own instrument. In his book *Il Saggiatore* (The Assayer), published in Rome 1623, he claims to have solved the problem in 24 hours.[17] In fact, he was able to present a telescope, magnifying eight times, to the Venetian doge, Leonardo Doná, with the priest Paolo Sarpi (1552–1623) presenting him to the court. The Venetian Senate was soon impressed by the novel instrument, mainly owing to its military applications as a spyglass. Reeves (2008) provides an excellent account on the lapse between the discovery of the telescope and Galileo's first astronomical observations of the Moon and Jupiter's satellites.

During his life, he built several different telescopes but their exact number is not known. Two of them, made of wood, are preserved at the Science Museum of Florence. A first estimate of their resolution was made by Abetti (1923) giving a value of 20″ for the 14×[18] and 10″ for the 20×. The optical quality of the four lenses has been tested by Greco et al. (1992, 1993). A replica of the larger telescope was made by Cipriani during the visit of G.E. Hale to Arcetri in 1923 and brought by the latter to the United States (see Pettit, 1939).[19] A Galilean telescope used in projection is limited in field by the clear aperture of the eye lens. The 14× would project an image of about one half degree, and the 20× an entire degree. In spring 1611 Galileo visited Rome (from March 29 to June 4), showing his telescopes to the astronomers of the Collegio Romano,[20] among them the astronomers and mathematicians Christopher Clavius (1538–1612), Christoph Grienberger (1561–1636), Paolo Lembo (1570–1618) and Odo van Maelcote (1572–1614), and a large number of cardinals in the residence of the Cardinal Giovanni Battista Bandini (1558–1629). For a history of the Collegio Romano and its relation with Galileo see Wallace (1984).

The first known mention of sunspots by Galileo occurs in a letter dated 1 October, 1611 and addressed to the painter Ludovico Cigoli (Drake, 1957).

---

[17] In *Siderius Nuncius* (Starry Messenger, printed at Venice March 1610) Galileo describes his first experiences observing with a telescope, that already magnified twenty times: *I fitted two glasses, both plane on one side, while on the other side one was spherically convex, and the other concave. The applying my eye to the concave glass I saw objects satisfactorily large and close. Indeed they appeared three times closer and nine times larger than when observed with natural vision only.*

[18] Telescope amplifying the observed object 14 times.

[19] Observing with this telescope, E. Pettit noticed that: *when a solar image with a diameter of about one foot is projected upon a screen in a completely darkened room, the penumbra of only the very largest spots can be seen.*

[20] Institution founded by the Jesuits in 1550. By papal bulls of 1552 and 1556 it received the right to grant doctorates in philosophy and theology as well as the privileges enjoyed by the universities of Paris, Louvain, Salamanca, and Alcala. By 1567 the Collegio Romano had over a thousand students, and Pope Gregory XIII erected a large building to house the students and faculty. Over the years the college gradually became known as the Gregorian University in honour of that pope.

Galileo also enclosed fine drawings made from 3 to 11 May in a letter dated 2 June, 1612 to Maffeo Barberini (1568–1644), the future pope Urban VIII, claiming to have started his sunspot observations in December 1610. These drawings are preserved in the Vatican Museum. In any case, his first continuous observing period was from 12 February to 3 May, 1612 summarized in 13 drawings of the solar disk in four sheets of paper. He used numbers to represent the different sunspots.

The observations during summer 1612 (2 June–8 July) were published in *Istoria e dimostrazione intorno alle macchie solari* (History and demonstrations concerning sunspots) published in 1613 (Figure 3.11). This publication contains the letters exchanged with Mark Welser responding to the claims of Scheiner.[21]

In his second letter to Mark Welser (14 August 1612), Galileo describes his observing method, whose invention he attributed to Benedetto Castelli (1578–1643): *Direct the telescope upon the sun [...] Having focused and steadied it, expose a flat white sheet of paper about a foot from the concave lens; upon this will fall a circular image of the sun's disc, with all the spots that are*

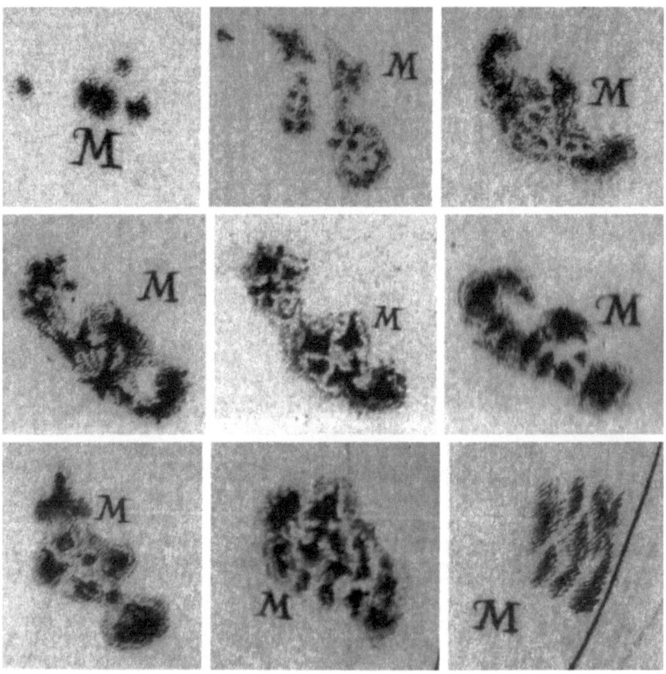

Fig. 3.11. Evolution of a sunspot group from 19 to 27 June, 1612 (from top to bottom and from right to left) observed by Galileo.

[21] The brief letters from Welser to Galilei are dated: 6 January, 1 June, 28 September and 5 October (all 1612). The long replies of Galilei to Welser are dated 4 May, 14 August and 1 December (all 1612).

on it arranged and disposed with exactly the same geometry as in the sun. The more the paper is moved away from the tube, the larger this image will become, and the better the sunspots will be depicted. Thus they will all be seen without damage to the eye [...] In order to picture them accurately, I first describe on the paper a circle of the size that best suits to me, and then by moving the paper towards or away from the tube I find the exact place where the image of the sun is enlarged to the measure of the circle I have drawn (Drake, 1957, p. 115).

Also in this letter, Galileo described a method, based on a modest camera obscura, to observe sunspots easily: *For without any instruments, from any little hole through which sunlight passes, there emerges an image of the sun with its spots, and at a distance this becomes stamped upon any surface opposite the hole [...] If in church someday your Excellency sees the light of the Sun falling upon the pavement at a distance from some broken windowpane, you may catch this light upon a flat white sheet of paper and there you will perceive the spots.* (Drake, 1957, p. 116).

Galileo became blind at the age of 72, from a combination of cataracts and glaucoma (see Sobel, 1999, p. 354). This had nothing to do with his telescopic observations of the Sun a quarter of a century earlier. At the beginning his observations were made near sunset, but soon he applied the projection method, which allowed him to make observations also at midday. In neither case could he have damaged his eyes.

Galileo was an active member of the Lyncean Academy (Freedberg, 2002), probably the first scientific society, founded by his friend Federico Cesi (1585–1630). He was also a master of perspective drawing, for example chiaroscuro, a pure exercise in the art of representing three dimensions in two through the use of light and shadow on complex geometric forms. Galileo's drawing ability improved further when he was admitted, on 18 October 1613, to the Accademia del Disegno (Academy of Drawing), founded in 1561 by G. Vassari (1511–1574). However, his artistic ability was actually more important for the lunar drawings, and not so much for the solar ones. See Kemp (1990) for a book on the relationship between visual arts and science in Europe.

3.3.5 The Scheiner–Galileo Debate

The Galileo–Scheiner debate was mainly a dispute about who had been first to observe sunspots. However, the two scientists also disputed over other matters.

In 1612 Galileo wrote to the Grand Duke of Florence, Cosimo II (1590–1621) : *Repeated observations have finally convinced me that these spots are substances on the surface of the solar body where they are continuously produced and where are also dissolved, some in shorter and others in longer period. And by the rotation of the Sun, which completes its period in about a lunar month, they are carried round the Sun.*[22] Galileo also argued that the sunspots showed irregular shapes, whereas a planet should give a circular shadow.

On the other hand, Scheiner defended the purity of the Sun, attempting to free it from any defect or spot. He attributed the observed dark structures to opaque clouds orbiting very close to the Sun. However, his hypothesis was not only based on philosophical concepts but also on his own experience. In his 1612 letters he argued: *if they were on the Sun their motion would imply that the Sun rotates, and we should see the spots return in the same order and in the same position they had among themselves and with respect to the Sun. So far they have failed to reappear although other spots have followed the first ones across the solar disc. This is a clear argument that they are not on the Sun.* In the third of his letters to Mark Welser, Galileo commented on the solar imperfections: *Well, if alteration were annihilation, the peripatetics would have some reason for concern; but since it is nothing but mutation, there is no reason for such hostility to it*[23] (Drake, 1957, p. 142).

Scheiner's interpretation was supported by another Jesuit, Jean Tarde (1561–1636), a genuine defender of the planetary hypothesis presented in his book *Borbonica sideria.* Baumgartner (1987) describes his theory in detail. A first problem arose with the excessive number of planets necessary to explain the dozens of sunspots which Tarde had observed on 25 August, 1615. Figure 3.12 shows some of his sunspot drawings.

Galileo saw also the significance of the sunspot observations to the new heliocentric theory. In his *Dialogue concerning the two chief world systems*, he

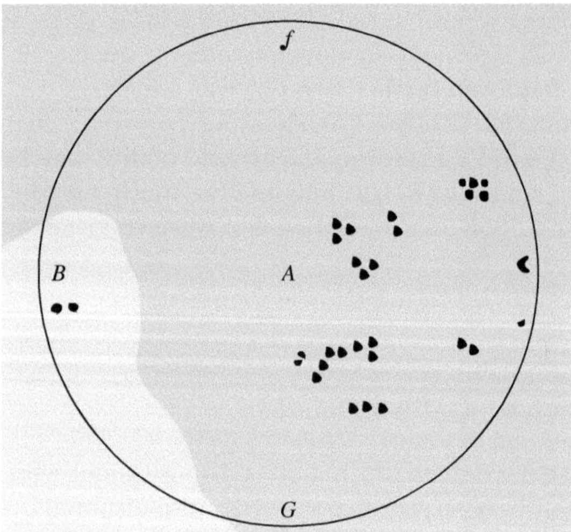

Fig. 3.12. Sunspot drawings by Jean Tarde in August 1615 contained in his book *Borbonica sideria* conserved at the library of the Paris Observatory.

[22] Cited in Newton, H.W., 1958, *The Face of the Sun*, p.31.

[23] Peripatetics, the ones walking about, were the followers of Aristotle.

shows that it is possible to accommodate the observed motions of sunspots in both the Copernican or Ptolemaic systems. However, the planetary motions required in the heliocentric theory are simpler and more plausible.[24] A Sun rotating with a period of 27 days and with its rotation axis inclined seven degrees with respect to the ecliptic, fits the sunspot motions (Figure 3.10). See also Mueller (2000) for more details.[25] After his famous trial before the Holy Office, Galileo indirectly defended the Copernican theory by supporting publications of disciples such as *Discourse of Comets*[26] by Mario Guiducci (1585–1646), printed in 1619.

Scheiner dedicated the last part of his life to debunking the Copernican theory. In 1650 his book entitled *Introductory treatise in favor of a moving Sun and a stable Earth against Galileo Galilei* was published posthumously.

3.3.6 L. Cigoli (1559–1613)

Ludovico Cardi, called Cigoli, was a painter and architect, and a close friend of Galileo. In mid-September 1611 he wrote to Galileo about the observations made by the painter Domenico Cresti, Il Passignano (1559–1636), who believed the spots to originate near the centre of the Sun and to make their way to its surface along spiral lines.[27] In 23 March 1612 Cigoli sent to Galileo 26 daily drawings of the Sun, made by him in Rome during the period 18 February–23 March, by placing a thick green glass over the eyepiece. The drawings were accompanied by a date and a time.[28] Cigoli commented: *I do not believe I have told you that I have a telescope, and it is sufficiently good to enable me to see the clock of St. Peter's and its hand from St. Maria Maggiore, but not the numbers of the hours as distinctly as I saw it with yours.*

In his treatise *Prospettiva Pratica* (1613), Cigoli studied the perspective projection on a spherical surface, decisively influencing the observations of Galileo. The spots were progressing on the solar surface in such a way that they became foreshortened to his sight as they moved towards the edge of the sphere. Cigoli was also very helpful in making arrangements with Mathias

24 In the words of Galileo: *Whatever can be accomplished through fewer things is done in vain through more.* A nice description of Occam's razor.

25 Smith (1985) concluded that actually the apparent movement of sunspots was a crucial test for the Copernican theory. However, the arguments presented by Galileo were confusing and poorly expressed. He supported the idea that Galileo developed his proof only months before committing the *Dialogue* to press.

26 There is an English translation by S. Drake, published by University of Pennsylvania Press, 1960.

27 This proved to be erroneous. It was based on the assumption that the Sun rotates very rapidly, changing its aspect during the course of the day.

28 Much of what is known of his friendship with Galileo derives from their correspondence in the period 1609–1613. Twenty-nine letters from Cigoli to Galileo have survived and, unfortunately, only two from Florence to Rome (published in italian by Matteoli, 1959). The originals are preserved at the National Library of Florence.

Greuter (1564–1638) to engrave the observations of Galileo (Chappell, 1975). For more details on the life of Cigoli and his relation to astronomy, see Reeves (1997).

3.3.7 Other Observers

Francesco Sizzi (1585–1618), Florentian astronomer and astrologer, criticized the astronomical discoveries of Galileo, especially those related with the Jupiter satellites, in his *Dianoia Astronomica*, published in 1611.[29] However, Galileo was kind to him and in a letter to Filippo Salviati (1582–1614) commented: *I had much rather gain the friendship of Sig. Sizzi by forgiving him all insults that have him as an enemy though conquest. And for that reason I have managed also to apologize for him among the Jesuit fathers, who read his puerelities with vast amusement* (Drake 1958). In 1613 Sizzi wrote a letter to Orazio Morandi (1570–1630) in Rome, in which he reported, though rather cryptically, the annual variation of the paths of the sunspots (Figure 3.10) based on some French observations.

Petrus Saxonius (1591–1625) observed the Sun in February–March 1616 in Nuremberg (Germany). Figure 3.13 reproduces one of his drawings included in the book *Maculae solares ex selectic observationibus*. Most of his writings are preserved at the library of Schweinfurt (Germany).

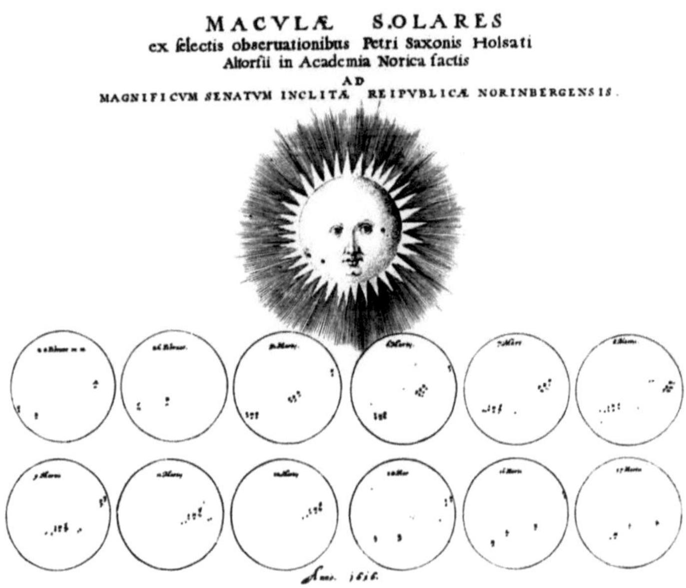

Fig. 3.13. Sunspot drawings by P. Saxonius (24 Feb–17 March 1616), printed at Altdorf (Nuremberg, Germany).

[29] He argued that the satellites are invisible to the naked eye and therefore can have no influence on the Earth, and therefore would be useless, and therefore do not exist.

Athanasius Kircher (1602–1680) was a German jesuit, who published around forty works in different fields.[30] In April 1625 he looked at sunspots through a telescope and was so impressed to comment: *from that day onwards astronomy became one of my chief studies.* Ten years later he studied the Sun jointly with Scheiner. In the manuscript entitled *The Cristian Philosopher*, Cotton Mather (1663–1728)[31] commented on the ideas of Kircher about the Sun presented in in his *Mundus Subterraneus*[32]: *Kircher supposed the Sun to be a body of wondrous fire, unequal in surface, composed of parts which are of different nature, some fluid, some solid: The disque of it, a sea of fire wherein waves of astonishing flame have a perpetual agitation* (Figure 3.14). Mather also remarked that instead this gleeful view, Hooke and Newton preferred to

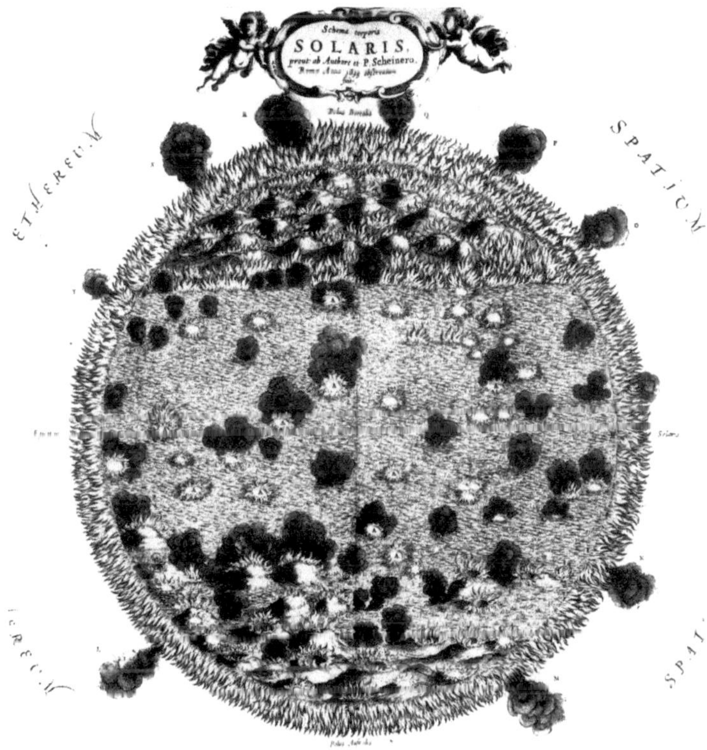

Fig. 3.14. A view of the solar surface by A. Kircher based on Scheiner's observations (*Mundus Subterraneus*, p. 64).

[30] In his *Catoptric Stegranography*, he describes methods for projecting texts using both sunlight and candles, with the help of both flat and concave mirrors, and a convex lens.

[31] He was a socially and politically influential New England Puritan minister, often remembered for his connection to the Salem witch trials.

[32] Mainly dedicated to volcanic eruptions, the Latin version was published in two volumes (1665 and 1678) in Amsterdam. A Dutch translation appeared in 1682.

describe the Sun as *a solid and opaque body.* For more details of his biography see Fletcher (1970), Godwin (1979) and Findlen (2004).

After his frustrated attempt with the camera obscura, Kepler continued nevertheless to be interested in the Sun. He rapidly commented on the letters of Scheiner, supporting the view that sunspots did not belong to the Sun. To be able to produce such opacity, Kepler imagined them like the stains we observe on a red hot iron, or like slag or dross on the surface of molten metal. He tried to do his own observations by covering the lens with a little hole and a blue filter in the eyepiece, but he says[33]: *nevertheless it is an hour and I still see my writing red and I have a painful itching in the eyes.*

Pierre Gassendi (1592–1655) was a persevering and intelligent observer, as is evident from his notebook carefully kept from 1618 until 1652 and filling over 400 pages. His main work *Opera Omnia* was published in Lyon in 1658. Figure 3.15 reproduces one of his sunspot drawings. Guerrini (2001) presents an unpublished letter from Pierre Gassendi to his friend Nicolas Fabricius (1580–1637), Lord of Peiresk, commenting on the planetary sunspot theory of Malapert and about the Scheiner–Galileo polemic.

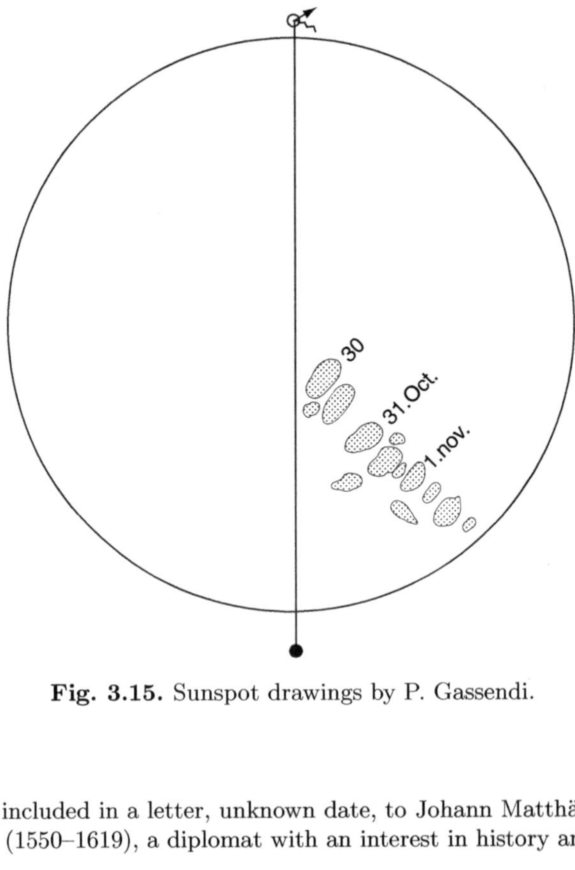

Fig. 3.15. Sunspot drawings by P. Gassendi.

[33] Comments included in a letter, unknown date, to Johann Matthäus Wacker von Wackenfels (1550–1619), a diplomat with an interest in history and philosophy.

Simon Marius (1573–1624), or Mayr, had a major debate with Galileo on the discovery of the satellites of Jupiter. He was also an observer of sunspots in 1611. Around 1602, he met David Fabricius during a short stay in Prague, and later pointed out the priority of his sunspot publication in the book *Mundus Jovialis.*

3.3.8 Instrumental Development

Harriot and Fabricius were not aware of the method of observing the Sun by projecting the image on a sheet of paper. As commented previously, this method was first used by Galileo but the merit of its discovery is usually given to Benedetto Castelli (1578–1643), a friend of Galileo's residing at Padua (see Figure 3.16). Probably a disciple of Cigoli, Sisgismondi Coccapani (1583–1642), also used this technique for observing the Sun and the Moon (Reeves, 1997, p. 6).

The first observations were done with an azimuthal mounting. Therefore, we may reasonably assume that the projected image not only had drift, due to the absence of a tracking system, but also rotated throughout the day. In his *Rosa Ursina*, Scheiner describes the first equatorial mounting, called "Heliotelescopium", suggested by his companion C. Grienberger, professor of mathematics at the Roman College (Figure 3.17). This instrument eliminated the need to measure the solar altitude and the determination of the vertical, necessary parameters for the orientation of the ecliptic on the projected solar image.[34]

The use of filters was already described by Petrus Apianus (1495–1552) in his *Astronomicum Caesareum*. Scheiner mentions that a blue or green glass

Fig. 3.16. Scheiner and an assistant observing the Sun by the projection method.

[34] With the equatorial mounting, the angle between the meridian and the ecliptic depends only on the obliquity and the solar longitude.

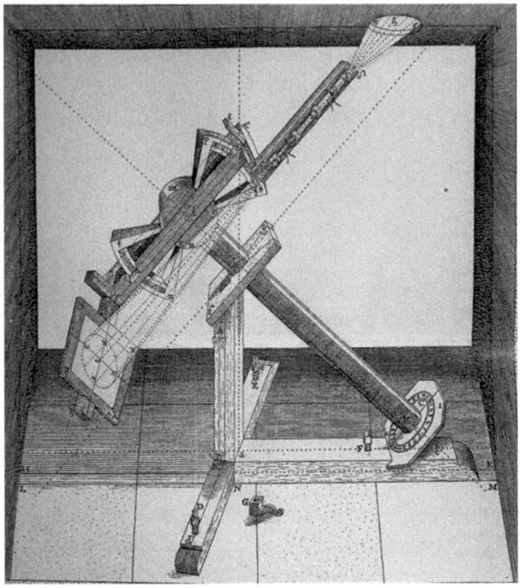

Fig. 3.17. Helioscopium of C. Scheiner, described in detail in the third volume of the *Rosa Ursina.*

keeps the eye uninjured even at midday. In the section dedicated to William Herschel we will speak in more detail on this topic.

Anton Maria Schyrleus de Rheita (1597–1660) was an astronomer and optician. In his *Oculus Enoch et Eliae*, besides describing one of his inventions, an eyepiece for a Keplerian telescope which left the image inverted, it also contained a long section on binocular telescopes, which greatly influenced other telescope-makers and opticians in the next century.

Robert Hooke (1635–1703) published, in 1676, a booklet entitled *A description of helioscopes and some other instruments*, where he summarized the development of solar telescopes in the period immediately following Lippershey's invention. He built a portable camera obscura around 1694. In anticipation of the heliostat, the use of a plane mirror to direct light into a fixed telescope for astronomical purposes seems to have been first proposed by Hooke, at a meeting of the Royal Society in 1668, in discussing the very long focus telescopes of Huygens, but the problem was how to make acceptably flat mirrors with the existing technology.

The design of the lens by Galileo was soon superseded by Johannes Kepler's system of two convex lenses, although Scheiner was probably the first to build and use such a telescope.

The refracting telescope had the problem of chromatic aberration, which reduced the contrast of the images, covering up the small details. This can be avoided by combining two different types of glass. However, I. Newton suggested a different approach, using a curved mirror to reflect and focus the light. He used a second, flat, smaller mirror in front of the curved mirror to catch the light rays and reflect them out the side of the tube to an eyepiece, which would magnify the image. At that time, the lenses of the refracting telescopes were limited to about 5 cms in diameter. A mirror could be made much larger than this. Variations of the Newtonian design were soon made by James Gregory (1638–1675) in 1663 and L. Cassegrain (1629–1683) in 1672. See Figure 3.18 for a scheme of the telescopes.

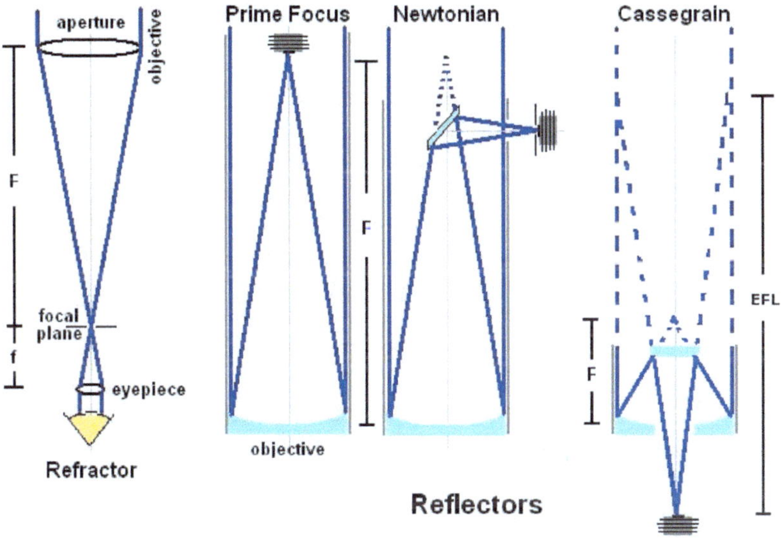

Fig. 3.18. Schemes of the different types of telescopes.

3.4 The Maunder Minimum

Eddy (1976) verified a previous claim of Maunder (1894) on the practical absence of sunspots during the period 1650–1715. During this period the appearance of a sunspot was an occasion to write a paper on it. Hoyt and Schatten (1996) examined in detail how well the Sun was observed during this period. They identify 88 observers making 7170 observations. Taking together the days with recorded sunspot observations and reports on days without sunspots on the disk, they obtain an upper estimate of a 98% of observing coverage during this period (see also Kovaltsov et al., 2004; Usoskin et al., 2006).

Historical observations are an essential tool for knowing about the details of this important event of solar variability (Beckman and Mahoney, 1998; Vaquero, 2007). Table 3.1 summarizes the main solar observers during the Maunder minimum.

Obviously, it could always be claimed that, during this period, the solar observations were not continuous or that some important observations have

Table 3.1. Astronomers during the Maunder minimum with more than 1500 observations. Source: Hoyt and Schatten (1996)

Period	Observer	Period	Observer
1662–1664	Weigel (Jena)	1676–1683	J. Flamsteed (Greenwich)
1660–1665	J. Picard (Paris)	1687–1694	J. Flamsteed
1642–1684	Hevelius (Gdansk)	1675–1690	H. Siverus (Hamburg)
1661–1671	M. Fogel (Hamburg)	1677–1703	G.C. Eimmart (Nuremberg)
1682–1718	Ph. La Hire (Paris)	1703–1715	W. Derham (Upminster)

been lost or simply not published. In this context, any piece of information on the solar behaviour during the Maunder minimum is essential to reconstructing the solar activity and to understanding the long-term variability of the magnetism of our star.

In fact, the existence of such discontinuities in the solar activity has been verified with the measurement of ^{14}C and ^{10}Be in tree rings and ice cores, respectively (Beer et al., 2006; Muscheler et al., 2007; Usoskin et al., 2007).

3.4.1 J. Hevelius (1611–1687)

Johan (or Jan, or Johannes) Hevel (or Hevelke, Hewel, Hewelcke; latinized Hevelius) was born in Gdansk (Poland), studied Law in Leiden and then travelled to London and Paris where he came into contact with other astronomers.

We have previously commented (Chapter 1) on the effects of chromatic aberration on observations. This effect can be minimized by building telescopes of extremely long focal length. Hevelius was a pioneer in this field and the largest classical refractor at that timer, 47 m long, was used by him in the Sternenburg Observatory in Dantzing.

Figure 3.19 illustrates how Hevelius observed the Sun by projection. An outer wall was pierced and a socket was mounted to hold a sphere. The ball was fitted with a telescope that could swivel and thereby track an object across the sky. The room was darkened and the telescope image was projected onto a moveable easel.

In his *Selenographia* there are 26 solar drawings (pp. 500–524, Chapter 5, and the Appendix) which accurately show the penumbra of sunspots, and their changes in shape with rotation of the Sun (see Figure 3.20). In the Appendix, Hevelius gives a value for the mean period of solar rotation of 27 days. In his *Cometography* he included also drawings of sunspots showing their evolution (Figure 3.21).

3.4.2 The Paris Observers

The French king Louis XIV is known as the Sun King, but more for his power than for his interest in solar observations. However, his name is linked forever to the construction of the Paris Observatory. In 1665 the physicist

Fig. 3.19. J. Hevelius observing the Sun by projection. Reprinted from his book *Machina celestis*.

and astronomer Adrien Auzout (1622–1691) convinced Jean Baptiste Colbert (1619–1683), a powerful minister of finances, and Louis XIV to construct "l'Observatoire Royal". Inaugurated officially on 1 May 1672 by the king, its first director was Giovanni Domenico Cassini (1625–1712). The two men had a decisive influence on starting a programme of regular solar observations.

The first astronomical experience of Jean Picard (1620–1682) was to observe the solar eclipse of August 1645 with P. Gassendi. In March 1666 he started recording his solar observations in notebooks, a habit that continued when he moved to the Paris Observatory in 1673. After his death the solar programme was conducted by Phillipe La Hire (1640–1719) for at least 35 more years.

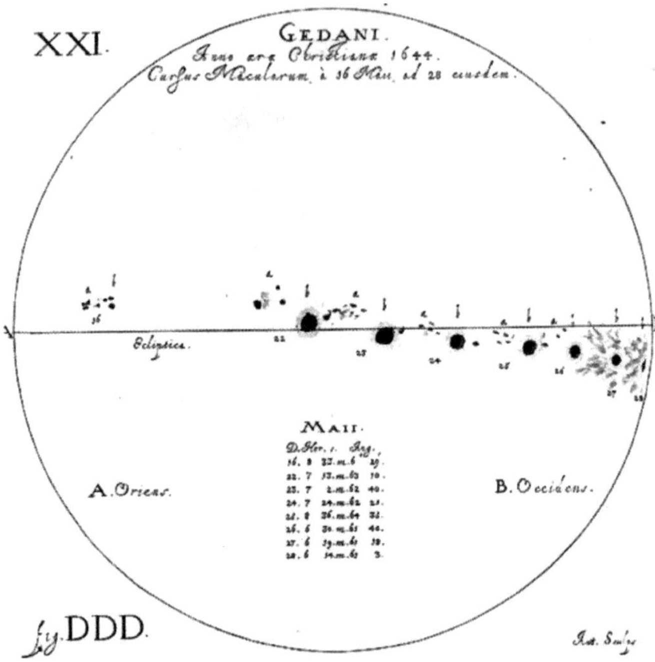

Fig. 3.20. Sunspot drawing by J. Hevelius in the *Selenography*.

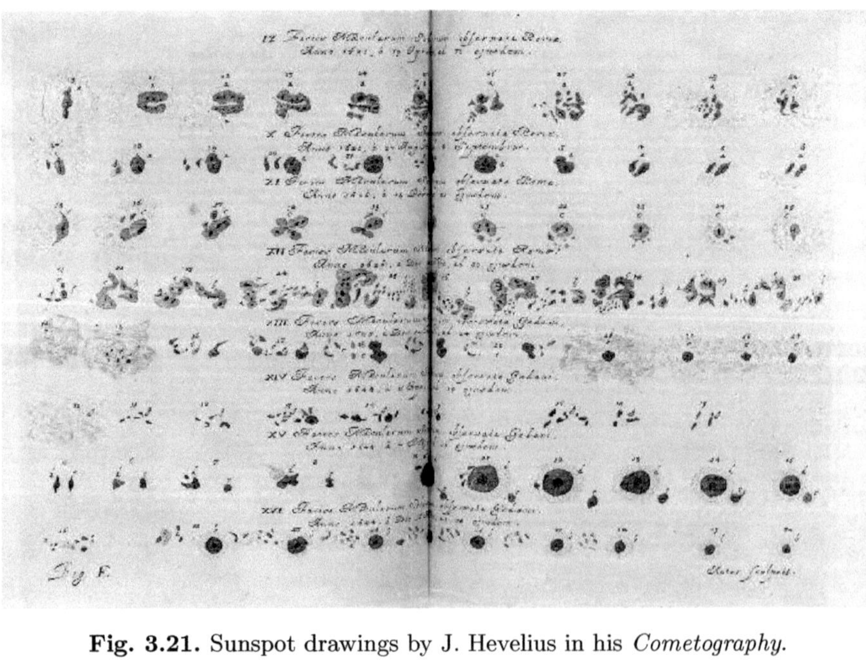

Fig. 3.21. Sunspot drawings by J. Hevelius in his *Cometography*.

To detect sunspots, Picard and his colleagues used telescopes with long focal lengths (> 16 feet). It is likely that the size of the diffraction disc ranged between 2 and 3 arcseconds (Ribes and Nesme-Ribes, 1993), as can also be guessed from some of the drawings.

In a letter to J. Hevelius, dated 18 August 1671, he commented after the appearance of a large sunspot crossing the solar disk in the period 3–13 August 1671: *was so much the better pleased at discovering it since it was ten whole years since he had last seen one, no matter how great the care he had taken from time to time to watch for them.*[35] Picard remarked that the sunspot was composed of small spots disposed in the shape of an arc, as the tail of a scorpion (Picolet, 1978).

3.4.3 William Derham (1657–1735)

William Derham was born in Worcester (England), graduated from Trinity College, Oxford at the early age of 18 and was ordained a priest in 1681 (Atkinson, 1952). He observed sunspots from 1703 to 1715 and some of his unpublished observations remain in the Cambridge University Library.

He reckoned that sunspots were the results of volcanic eruptions and fancied seeing smoke rising from them. He commented in 1711 that *there are doubtless great intervals when the sun is free, as between the years 1660 and 1671, 1676 and 1684*, (Derham, 1711). He also drew specific attention to the variability of sunspot numbers.

3.4.4 Nicholas Bion (1652–1733)

Nicholas Bion was a maker of astronomical instruments and a globe and sphere merchant. His places of birth and death are unknown but we know that his workshop was located at Paris, Quai de l'Horloge, and also that Louis XIV (1638–1715) appointed him royal maker of mathematical instruments. He wrote a profusely illustrated treatise (1709) on the construction and use of mathematical instruments.

In another work (Bion, 1699), he explained the use of celestial and terrestrial globes. As for sunspot activity, in four pages of this book Bion describes some sunspot observations obtained at the Paris Observatory "in the absence of J. Cassini" the head of the Observatory. Figure 3.22 reproduces the plate with the drawings of the observations.

3.4.5 John Flamsteed (1646–1719)

John Flamsteed was the son of a prosperous merchant in Denby near Derby, Derbyshire, England. He studied astronomy between 1662 and 1669 on his

[35] In another letter to M. Fogel, dated 21 August, Piccard affirmed that his previous observation of a sunspot had been 13–14 August 1661.

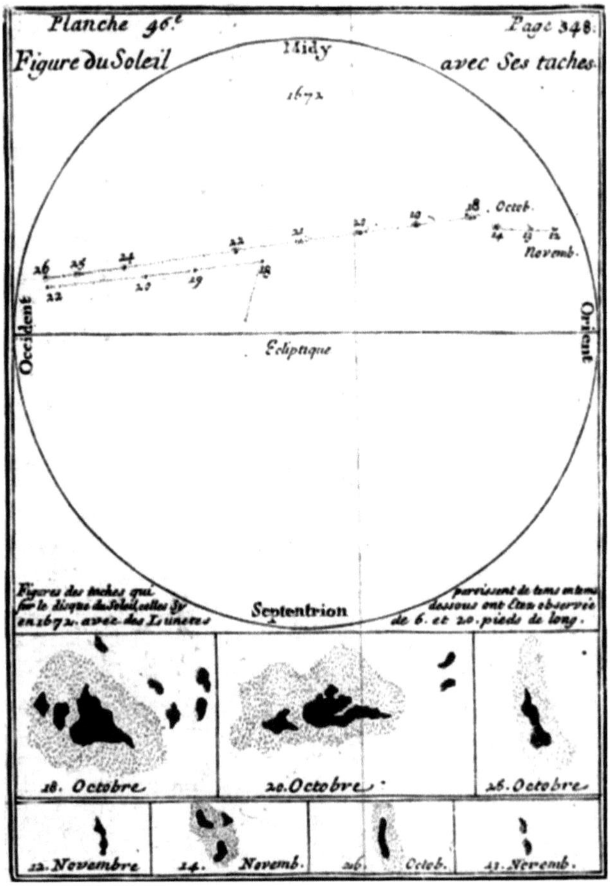

Fig. 3.22. Observations of three different sunspots during the year 1672 by N. Bion. Image available from GALLICA (Bibliothèque Nationale de France).

own and against his father's wishes. He was employed by King Charles II as Britain's first Royal Astronomer on 4 March 1675, on the recommendation of Jonas Moore.

Founder of the Royal Greenwich Observatory, his most famous achievement was to produce a 3000-star British Catalogue. In addition to this work, he made observations of the Sun's motion, measured the latitude of Greenwich, calculated the inclination of the ecliptic and the position of the equinoxes, created tables of atmospheric refraction and tidal patterns, and devised a method of observing absolute right ascension.

Based on his *Historia Coelestis Brittanica* and in his letters to W. Derham, Hoyt and Schatten (1995b) reconstructed the daily chronology of his solar observations from 1676 to 1700. The reduced number of sunspots during the Maunder minimum is verified in a note sent to the Royal Society informing them of a sunspot in 1684: *On the 25th of April past, as I was measuring the distance of a planet from the Sun about an hour before noon, I discovered a large spot entered within his disk a little distant from his following limb. These appearances however frequent in the days of Scheiner and Galileo have been so rare of late that this is the first only one I have seen in his face since December 1676.*[36]

3.4.6 Charles Malapert (1581–1630)

Charles Malapert was Jesuit who worked together with his companion Alexius S. Polonus (1593–1653). They used telescopes obtained from Scheiner in Mons (Belgium) to observe sunspots. In 1630 Malapert was called to Madrid to occupy a newly created chair in the Jesuit "Colegio Imperial". On his way, he fell ill and died shortly after crossing the Spanish border. Polonus reached Madrid alone and stayed for several years. In 1634 he constructed an ingenious planetarium for the University (Birkenmajer, 1967).

In 1633, the book *Austrica sidera heliocyclia astronomicis hypothesibus illigata* was posthumously published, based on Malapert's own sunspot observations and on others communicated to him by Father Wely in Coimbra, J.C. Cysat in Ingolstadt, Simon Perovius (1586–1656) in Kalisz (Poland) and C. Scheiner in Rome.

3.4.7 G. Kirch and G. Schultz

As was the case with the letters from Galileo and Scheiner to Mark Welser, this means of communication was very useful during the Maunder minimum. This was also true for the correspondence between J. Flamsteed and W. Derham, but their sources of information remain unknown.

Between 1681 and 1692, 43 letters to Gottfried Kirch (1639–1710)[37] were written by Gottfried Schultz (1643–1698), all of which have now been lost, and 32 were written by Kirch to Schultz (Herbst, 2005). It appears that between 19 May 1684 and 3 October 1686 Schultz observed the Sun for 77 days, during which time he reported only three sunspots (May 1684, July 1684 and April 1686).

3.5 The Rise of Solar Activity and the Dalton Minimum: 18th and 19th Centuries

After the Maunder minimum, solar activity started to increase, bringing about a renewed interest on the part of astronomers for observing sunspots. This

[36] H.W. Newton, 1955, *The Face of the Sun*, page 34.

[37] German astronomer, studied under J. Hevelius in Danzing. He was the first to discover a comet using a telescope.

trend was momentarily interrupted with the onset of the Dalton minimum elapsing from 1790 to 1820.

3.5.1 Louis Feuillée (1660–1732)

Louis Éconches Feuillée (sometimes spelled Feuillet) was a French member of the Order of the Minims, explorer, astronomer, geographer, and botanist. The Academie of Sciences sent him on several expeditions to South America.

His sunspot drawings were published in two volumes of the *Journal des observations physiques, matematiques et botaniques*, in 1714. Figure 3.23 shows an example of these drawings.

3.5.2 Christian Horrebow (1718–1776)

Christian Horrebow was an active astronomer in Copenhagen (Denmark) with major interests in stellar parallaxes, meteorology and the Sun. He started his sunspot observations in 1761 and continued doing so until his death. His

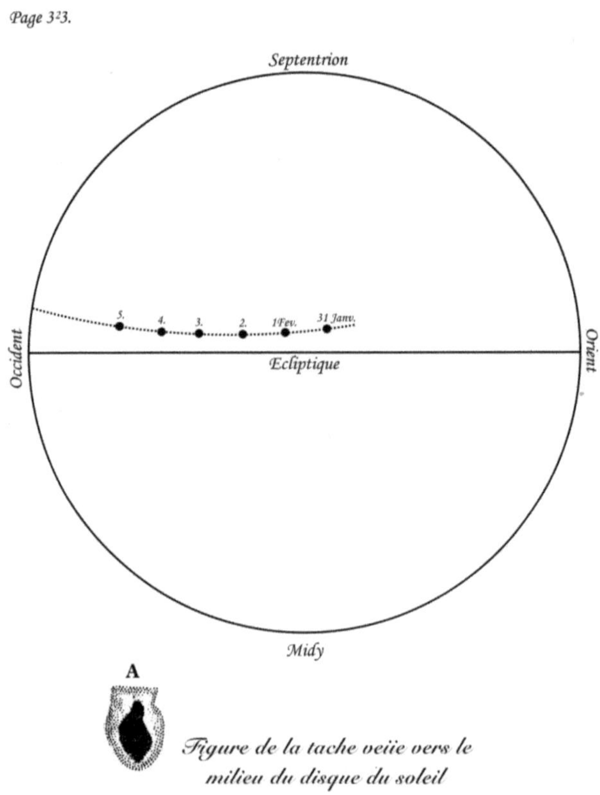

Fig. 3.23. A sunspot crossing the solar disk. Drawings by L. Feuillee.

observations are contained in 15 volumes of notebooks, presently located at the University of Aarhus. Hoyt and Schatten (1995a) critically studied these notebooks, revising former estimations of the level of solar activity at that time by Thiele (1859).

He was probably the first to suggest a periodicity in the number of sunspots. This is evidenced in one unpublished manuscript of Horrebow, dated 1776, discovered by R. Wolf. He remarked: *Even though it follows from observations that changes and variations in sunspots are frequent, one cannot find any definite rule by which the order and duration of this variation is completed. The main reason for this is that astronomers have made very few attempts so far at frequent observations of sunspots, undoubtedly because they felt that nothing of any interest for astronomy or physics would come of it. However, it is hoped that by means of frequent observations one will also find a period here similar to the movements of other heavenly bodies* (Izenman, 1985).

After Horrebow's death, Thomas Bugge (1740–1815) and Erasmus Lievog continued the observations for a few more years, but the scientific priorities of the observatory changed and the systematic observation of sunspots was cancelled.

3.5.3 Johann Hyeronimus Schroter (1745–1816)

Schroter was an amateur astronomer who set up in Lileinthal (near Bremen) his own observatory. In 1779 he acquired a 91cm long achromatic refractor with a 50 mm lens to observe the Sun, Moon, and Venus. His solar observations are mainly contained in his book *Beobachtungen über die Sonnenfackeln und Sonnenflecken*, printed in Erfurt in 1789. Figures 3.24 and 3.25 illustrate the fine detail of sunspots resolved in his observations.

3.5.4 Johann Caspar Staudacher (1731–ca. 1796)

Staudacher was observing the Sun and making detailed drawings of sunspots from 15 February 1749 to 31 January 1796, which are now preserved at the Astrophysikalisches Institut Potsdam (Germany). Arlt (2008) describes in detail the characteristics of this material, evaluating its usefulness for the determination of sunspot positions. He drew his solar images with black ink with the solar disk painted in yellow until 1768. Concentric circles, obviously representing 15 degrees steps, were added from January 1764 (see Figure 3.26 for an example).

3.5.5 William Herschel (1738–1822)

Born in Hanover (Germany), William Herschel arrived in England in 1757 to begin a career in music. He was the first to take advantage of large reflecting telescopes and to make important discoveries (Uranus) with them.

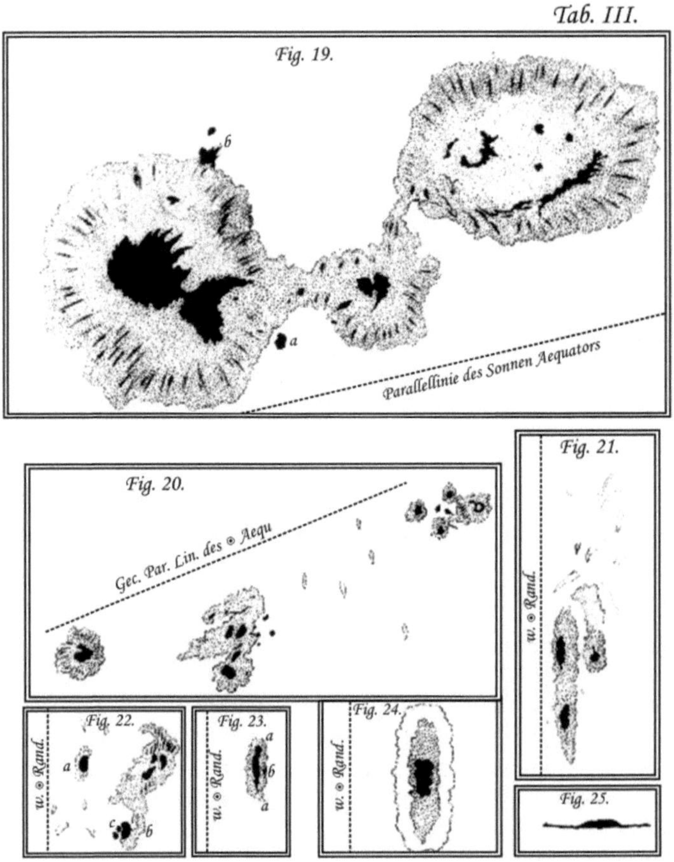

Fig. 3.24. Sunspot drawings by J.H. Schröter.

Although primarily interested in the night sky, Herschel studied the Sun motivated by the possibility of finding a planet closer to the Sun than Mercury. Soon he became interested in sunspots and other solar structures. He observed sunspots from 1779 to 1818, with most of his observations made from 1799 to 1806. Hoyt and Schatten (1992) provide a summary of the contents of his two solar notebooks preserved at Churchill College in Cambridge, England.[38]

He was worried about the way of filtering out the solar light: *The instrument I wished to adapt for solar inspection, was a Newtonian reflector, with 9 inches aperture; and my aim was, to use the whole of it open. I began with a red glass; and [...] took two of them together [...] the eye could*

[38] These notebooks have been microfilmed and are available from the American Philosophical Society in Philadelphia, the University of Notre Dame, Harvard University, and the National Research Council of Canada.

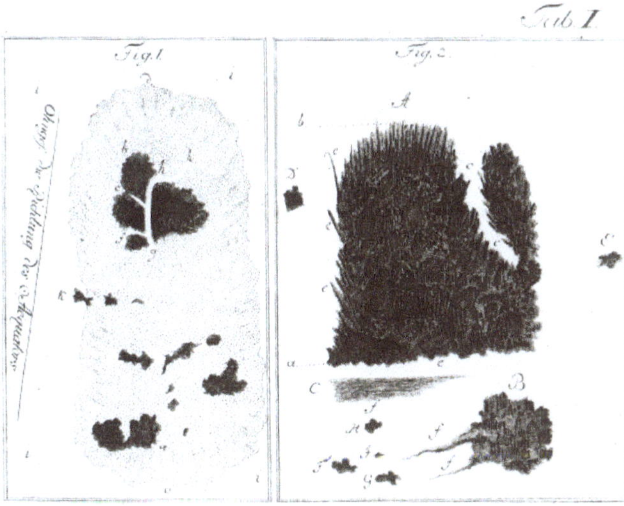

Fig. 3.25. Detailed sunspot drawing by J.H. Schröter.

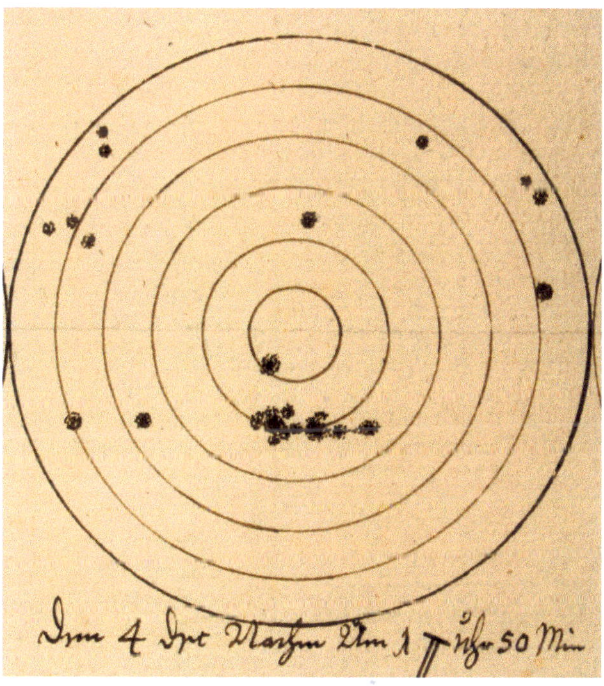

Fig. 3.26. Drawing of the solar disk and different sunspots of 4 December 1768 by J.C. Staudacher. Figure 5 by Arlt (2008).

not bear the irritation, from a sensation of heat [...] I now took two green glasses; but found they did not intercept light enough. I therefore smoked one of them....they still transmitted considerably more light than the red glasses, they remedied the former inconvenience of an irritation arising from heat [...] nothing remained but to find such materials as would give us the colour...of a pale green light, sufficiently tempered for the eye to bear its lustre. To find a suitable filter material, Herschel says, "I placed a prism in the upper part of a window [...] then I intercepted the colours [...] successively, by the glasses." He noticed that with "combinations of differently colored darkening glasses [...] some of them, I felt a sensation of heat, though I had but little light; while others gave me much light, with scarce any sensation of heat." (Herschel, 1800).

Herschel also experimented with filters made of different fluids. He reported on this matter: *I viewed the sun with a skeleton eyepiece, into the vacancy of which may be placed a moveable trough, shut up at the ends with well-polished plain glasses, so that the sun's rays may be made to pass through any liquid contained in the trough, before they come to the eye-glass. Through spirit of wine, I saw the sun very distinctly [...] March 8, I viewed the sun through water. It keeps the heat off so well, that we may look for any length of time [...] April 26, I viewed the sun through Port wine, and without smoke in the darkening glasses* (Herschel, 1801a).

Herschel believed sunspots were openings in the solar atmosphere leading to a cool habitable interior (Herschel, 1795; Kawaler and Veverka, 1981). Owing to the fact that Herschel was the one who had expressed this idea, it remained prevalent in the solar community for several decades. The work of Jonathan Lane (1819–1879) was the first step leading to the view of a solar interior with temperatures of millions of degrees (Lane, 1870).

3.5.6 J.A. Alzate (1737–1799)

Alzate was a Mexican astronomer who observed the Sun almost every day from 1769 to 1793 (Vaquero, 2004; Vaquero et al. 2007). We will comment more on his work in Chapter 5.

3.5.7 J.W. Pastorff (1767–1838)

His sunspot observations took place from 1819 to 1833. He observed very large sunspots on 24 May and 21 June, 1828 in Drossen (Germany) (Pastorff, 1828). The notebooks of his observations are now housed in the archives of the Royal Astronomical Society, including 1477 sunspots drawings. Hoyt and Schatten (1995c) have revised his sunspot observations.

3.5.8 John Herschel (1792–1871)

John Herschel was the only son of William Herschel. In 1834 he travelled to the Cape of Good Hope, mainly to map the sky of the southern hemisphere. For his solar observations at the Cape, he used projection through refractors, and filters of two sheets of green glass and one sheet of cobalt blue glass.

He proposed the use of a simple unsilvered glass surface, the Herschel wedge, shaped to prevent reflections from the rear surface from reaching the surface. This system was improved by Ignazio Porro (1801–1875), using a similar glass but placed at an angle so that the reflected light was polarized. A Nicol prism between the eyepiece and the eye could be rotated to extinguish the light.

In his book *Outlines of Astronomy*, he presented in the first plate some solar drawings that we reproduce here (Figure 3.27).

3.5.9 Temple Chevallier (1794–1873)

A British clergyman, Chevallier was Professor of Astronomy at the University of Durham and founder of its Observatory. He made regular and continuous sunspot observations from 1847 to 1849.

3.5.10 Frederick Howlett (1821–1908)

Howlett's drawings of sunspots extending over a period of 35 years are included in eight volumes preserved at the Royal Astronomical Society of

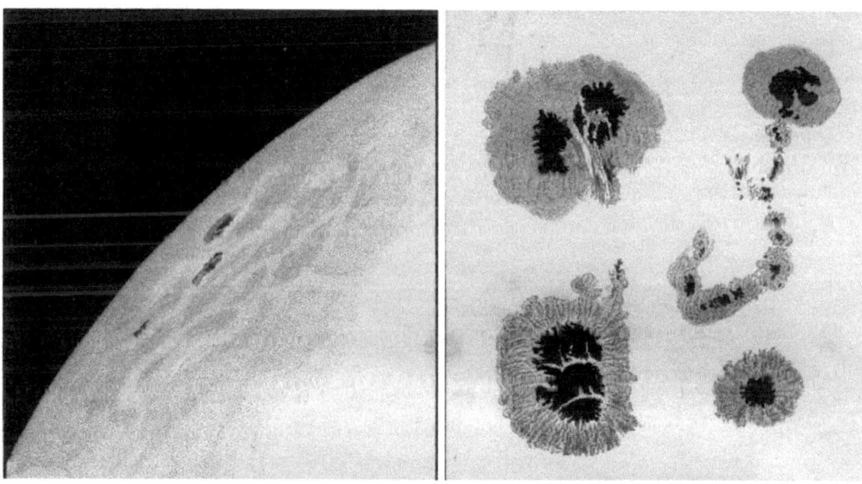

Fig. 3.27. (Left) Drawing of a sunspot group and surrounding bright faculae close to the solar limb. (Right) Drawing of a group of sunspots. Source: J. Herschel *Outlines of Astronomy*.

London. The method of observing was by projecting the solar image on a screen, placing the image at a distance from the eyepiece of the 3-inche aperture refractor, with a magnifying power of 80 linear, to produce an image of 32 inches in diameter (Howlett 1877, 1894) .

John Herschel explained that the colours Howlett described were evidently owing to chromatism in the eyepiece and what he needed was a good achromatic object-glass, preferably of the type he himself had constructed, which had the concave lens turned towards the light.

3.5.11 S.H. Schwabe (1789–1875)

Born in Dessau (Germnany), we mentioned Schwabe in Chapter 1 as the discoverer of the 11-year solar cycle in 1843 (Schwabe, 1844). His observational work was originally aimed at discovering possible intramercurial planets. Starting on 11 October 1825, he observed the Sun virtually every day that the weather allowed, and did so continuously for 42 years. His sunspot drawings are preserved in the Library of the Royal Astronomical Society.

He was also one of the first to comment on a possible relation between sunspots and Earth temperatures: *I register the height of the barometer and thermometer three times in the course of each day, but the annual mean numbers deduced from these observations have not hitherto indicated any appreciable connection between the temperature and the number of spots.*[39]

3.6 Sunspot Drawings in the Photography Era

The first daguerreotype of an astronomical object was of the Moon and was obtained by John William Drape (1837–1882) in 1840 with a 20-minute exposure. On 2 April 1845, Louis Fizeau (1819–1896) and Leon Foucault (1819–1868) obtained the first solar photograph, with an exposure time of 1/60 second. A new "objective" detector was available to the detectors with the possibility of storing and preserving the observations for a long time. However, drawings continue to be a useful technique today.

3.6.1 G. Spörer (1822–1895)

Spörer was one of the first to note a prolonged period of absence of sunspots, the Maunder minimum (Spörer, 1889). An astronomer of the Astrophysical Observatory at Potsdam (Germany), since 1860 he devoted his leisure hours to the observations of sunspots until 1893. He published four comprehensive memoirs describing his solar work, that were also regularly published in the journal *Astronomische Nachrichten.* Most of his records are photographic but he also made sunspot drawings.

---

[39] Cited in *Cosmos*, by A. Von Humboldt (1769–1859) and Sawyer (1948).

3.6.2 Samuel P. Langley (1834–1905)

In 1887, Langley became secretary of the Smithsonian Institution and established the Astrophysical Observatory there. He continued his study of the solar spectrum and made new determinations of the solar constant of radiation and, in 1904, announced his conclusion that this parameter was variable.

Until 1876 sunspots constituted the main focus of his previous research at the Allegheny Observatory. Figure 3.28 shows an extraordinary sunspot drawing, where the granulation and a light-bridge crossing the sunspot umbra are clearly visible (Langley, 1889).

3.6.3 S. Chevalier (1852–1930)

Born in St. Laurent des Autels (France), Chevalier joined the Jesuits in 1871, traveling to China in 1883, where he became director for many years of the Observatory Zo-Sé (China). This observing station was built in 1900 some 20 km from the Zikawei Observatory (Shanghai) and was equipped with twin 40 cm refractors of 7 m focal length carried on the same equatorial mounting. One telescope was used for visual observations (see Figure 3.29). His observations were published in the 19 volumes of the *Annals of the Observatory of Zo-Sé* between 1907 and 1922.

3.6.4 E.L. Trouvelot (1827–1895)

A French emigrant to the United States in 1852, Trouvelot first worked in agriculture. He progressively began to shift his interest toward astronomy. Joseph Winlock (1826–1875) invited him to join the staff of the Harvard College in 1872, where he could take advantage of the existing telescopes to produce

Fig. 3.28. Sunspot drawing by S.L. Langley on 21 September 1870 using the 13 inch refractor at Allegheny Observatory.

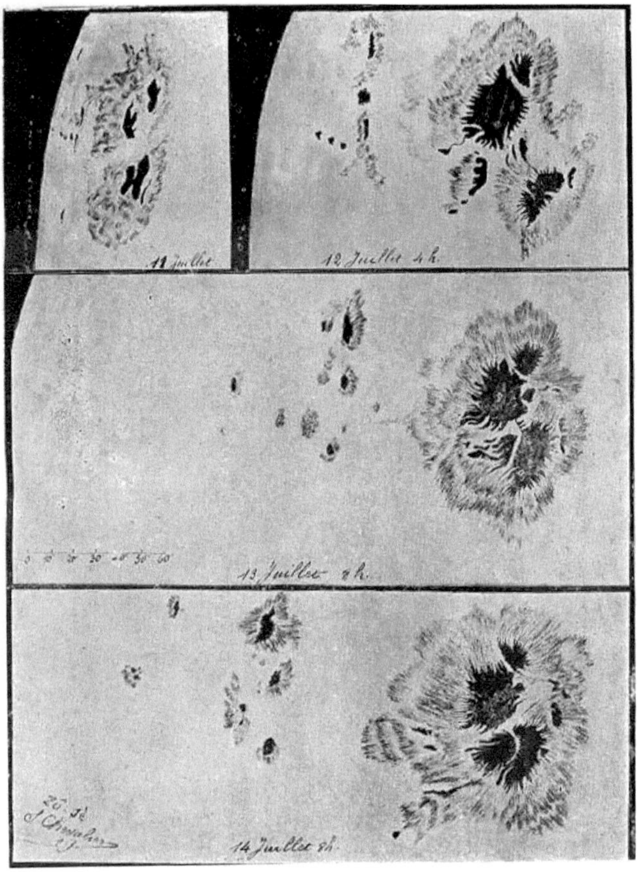

Fig. 3.29. Sunspot drawings by S. Chevalier made on 12, 13 and 14 June of 1905.

excellent astronomical drawings. By 1882, Trouvelot had returned to France and joined the Meudon Observatory.

He explained clearly the idea behind his drawings: *While my aim in this work has been to combine scrupulous fidelity and accuracy in the details, I have also endeavored to preserve the natural elegance and the delicate outlines peculiar to the objects depicted.* The best of his drawings were reproduced using chromolitography. Figure 3.30 reproduces a large group of sunspots. We can clearly see the influence of the personal profile on the drawing, producing an artistic view of the sunspot (Trouvelot, 1882).

3.6.5 Stonyhurst Observatory

Stonyhurst Observatory was founded in 1838 in Lancashire (England) by the Jesuits and mainly conducted research on the relationships between solar activity and terrestrial magnetism. Solar observations started in 1846 under

Fig. 3.30. Drawing of "veiled" sunspots by E.L. Trouvelot (June 17, 1875). Reproduced with permission from New York Public Library.

the direction of Alfred Weld (1823–1890). After joining the Society of Jesus in 1855, Walter Sidgreaves (1837–1919) taught chemistry and mathematics at Stonyhurst and became acting director of the observatory in the period 1863–68 during the time that his superior, Father Stephen Joseph Perry (1833–1889), another of the Stonyhurst priest-astronomers, was engaged in his own theological studies. It was Perry who in 1881 began a series of observations of sunspots by means of drawings of the solar disk on a large scale (27 cm in diameter), of which 3800 in all were made (Udías, 2003). Figure 3.31 shows a set of daily drawings made by Perry in 1887.

The observatory was officially closed in 1947 with a short renaissance from 1957 to 1976.

3.6.6 Gyula Fényi (1845–1927)

A similar history had another Jesuit observatory founded under the auspices of the cardinal Lajos Haynald (1816–1891). Carl Braun (1831–1907) was a disciple of A. Secchi and used an 18 cm telescope to observe the Sun. After his retirement, Gyula Fényi, born in Sopron, was appointed director of the observatory, where he had already been working between 1871 and 1874. Between 1885 and 1917 he made solar observations, completing about 40 000 (see Tóth et al., 2002). The collection of sunspot drawings is now preserved at the library of the Heliophysical Observatory of the Hungarian Academy of Sciences (see Figure 3.32 for an example).[40]

[40] Digitized images from 1880 to 1919 are at ftp://fenyi.solarobs.unideb.hu/pub/HSID/

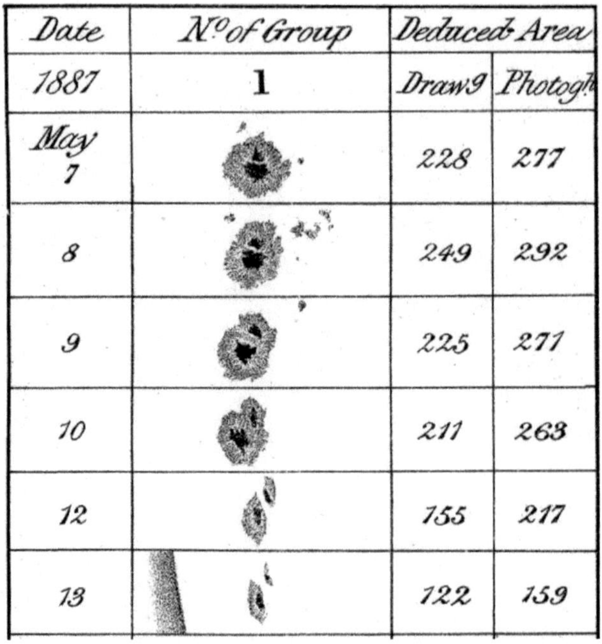

Date	Nº of Group	Deduced Area	
1887	1	Draw9	Photogh
May 7		228	277
8		249	292
9		225	271
10		211	263
12		155	217
13		122	159

Fig. 3.31. Drawings showing the evolution of a sunspot made by S. Perry at the Stonyhurst Observatory in 1887.

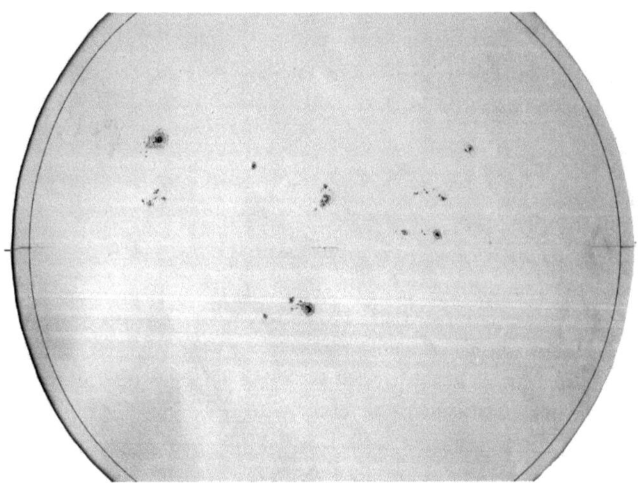

Fig. 3.32. Sunspot drawings made by J. Fenyi at the Kalocsa Observatory, 1 July 1906.

3.7 The First Granulation Drawings

In a letter to G.B. Baliani (1582–1666), dated 12 March 1614, Galileo stated: *It seems to me that I observe that all the face of the Sun is, so to speak, of heterogeneous light, that is, as it were surrounded by a delicate cloud of unequal transparency.* Years later, Schiener described in his *Rosa Ursina* the appearance of the solar surface: *If one uses a helioscope of excellent quality, one sees in addition to spots and facula, the following unknown and hitherto unpublished phenomena: 1. That the entire surface of the visible hemisphere is varied, composed of shaded regions and points of light. 2. That the entire surface resembles a ruffled sea, and that this appearance continuously changes.* (Mitchell, 1916, p. 349).

In 1801, William Herschel observed with a reflector of three metres focal length, the presence of small structures on the solar surface which he called corrugations: *I call that very particular and remarkable unevenness, ruggedness, or asperity, which is peculiar to the luminous solar clouds, and extends all over the surface of the globe of the Sun. As the depressed parts of the corrugations are less luminous than the elevated ones, the disc of the Sun has an appearance which may be called mottled* (Herschel, 1801b, p. 265).

James Nasmyth (1808–1890) was also a notable constructor of telescopes and assiduous observer of the Sun. A successful industrialist and engineer, Nasmyth made extensive solar observations (see Figure 3.33) after he retired from business in Manchester and moved to a new home near Hawkshurst in Kent. At a meeting of the Manchester Literary and Philosophical Society, held in 1861 March 5, he announced that the entire solar surface was made up of a multitude of filaments, which he called "willow-leaf" (Nasmyth, 1865). This finding was based on exceptional observations obtained on 20 July 1860, with a Cassegrain–Newtonian telescope of 50 cm of aperture on an altazimuthal

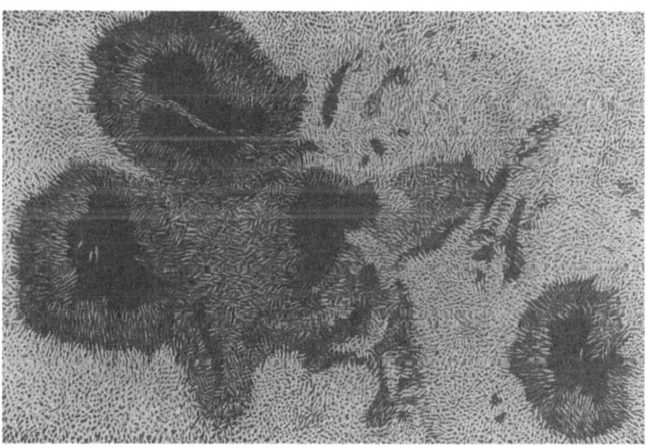

Fig. 3.33. Drawing of a large sunspot group and the surrounding photosphere made by J. Nasmyth on 5 June 1864.

mounting. The surface structures seemed to have a fixed size crossing each other in all directions. Following the interpretation by W. Herschel of a habitable Sun with a cool interior, he commented: *the thickness of the layer does not appear to be very deep, as I can see down through the interstices which are left here and there between them, and through which the dark or the dark stratum (intergranular lanes) is rendered visible.* It is the occurrence of the infinite number of these interstices, and the consequent visibility of a part of the dark stratum, that gives to the general solar surface that peculiar and well-known mottled appearance which has for a long time been familiar to observers of the Sun. In 1862 he announced that the Sun's surface was covered by a compact pattern of thin bright filaments shaped much like "willow-leaves". A historical analysis of the debate is made by Bartholomew (1976).

Nasmyth felt that his discovery was of great importance, as is expressed in two letters to three prestigious astronomers of that time: George Airy, Warren de la Rue and John Herschel. The latter responded enthusiastically, supporting his familiar consideration of the Sun as a habitable place: *What can they be? Are they huge phosphorised fishes? If so, what monsters! Or are they crystals? a kind of igneous snow-flakes? floating in a fluid of their own specific gravity? Some kind of solidity or coherence they must have, or they would not retain their shape in the violent movements of the atmosphere which the change of the spots indicate.* In a lecture delivered at the end of 1861 he described the solid flakes as being not only the immediate sources of solar light and heat, but as, perhaps *organisms of some peculiar and amazing kind* (Herschel, J. 1867, *Familiar Lectures on Scientific Subjects*, Alexander Strahan, pp. 47–90).

The initial enthusiasm with which this claim was greeted turned to controversy when the Nasmyth view was challenged by William Rutter Dawes (1799–1868). He had been observing these structures with the 5-foot refractor at Ormskirk since 1830. For him to see the mottled appearance of the solar surface required telescope diameters of only 8 cm and a magnification of 60. With a large aperture ($\sim$ 20 cm), the surface is made up of luminous masses imperfectly separated from each other by rows of minute dark spots. He was probably the first to use the word granulation, although Bartholomew (1976) suggested that this term had been used earlier by Warren de la Rue (1815–1889) in private correspondence with J. Herschel.

Dawes (1864) describes some interesting geometrical ratios: *the proportion of the area of the less luminous spaces to that of the most luminous masses, is subject to very considerable change.* E.J. Stone also observed the granulation with the 12.8-inch equatorial at the Royal Greenwich Observatory, giving the name of "rice grains" to these structures (Stone, 1864).

W. Huggins (1824–1910) was also actively involved at that time in the ongoing debate in the Royal Astronomical Society on the physical interpretation of the mottled appearance of the Sun's visible surface (see in Figure 3.34 one of his drawings). In 1866, he established some observing facts easily subscribed by modern astronomers (Huggins, 1866). Essentially, he found that the term rice-grain was appropriate for telescopic powers less than 100, whereas higher powers showed them to be more oval or round with ragged outlines. This is

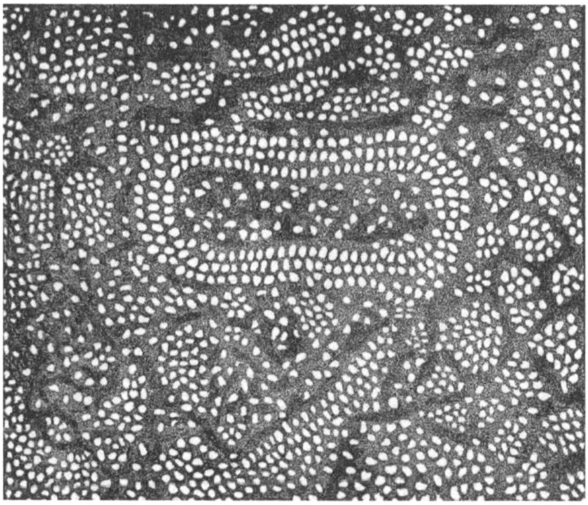

Fig. 3.34. Drawing of the solar granules by W. Huggins. Reproduced from *The Sun* of C. Young (1894) and Huggins (1866).

to be interpreted in terms of the rounding produced by degrading the spatial resolution of the atmosphere-telescope system.[41] Figure 1.20 (Chapter 1) illustrates this effect very well.

The most realistic drawing of solar granulation (Figure 3.35) was made by A. Secchi (1818–1878), director of the Observatory of the Roman College, and

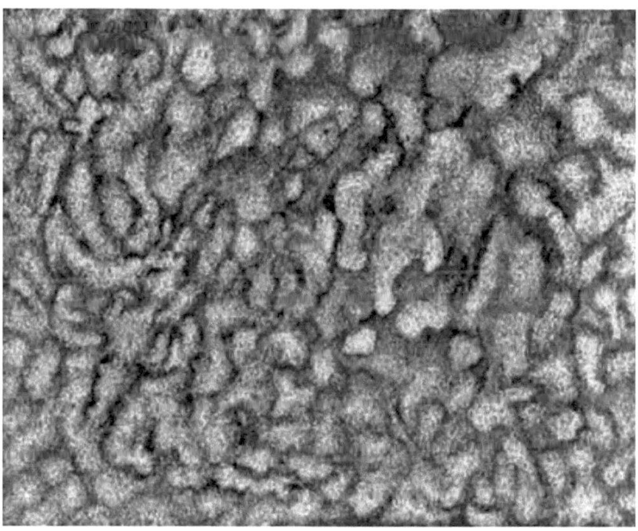

Fig. 3.35. Drawing of the solar granules by A. Secchi. Reproduced from his book *Le Soleil.*

[41] Huggins was aware of these effects when he wrote in 1885: *We live at the bottom of a deep ocean of air, and therefore every object outside the Earth can be seen by*

a pioneer of stellar spectroscopy and solar–terrestrial relations. His observations of our star were made on the roof of St. Ignatius Church in Rome with a small telescope.

3.8 Sunspot Fine Structures

Different observers tend to give new names to structures, which correspond to the same physical phenomena. Table 3.2 summarizes the different terminology used by William Herschel.[42]

3.8.1 Penumbra

We have seen how the early drawings were already able to distinguish between the umbra and penumbra. The latter is mainly characterized by its filamentary structure.

In the letter of 16 January to Mark Welser under the title *Accuratior Disquisitio*, C. Scheiner describes the filamentary structure of the penumbra: *the boundary of all the spots is rough, as if surrounded with little fibers of white and black, and the majority of the sunspots are brighter about their borders than in the centre* (Mitchell, 1916, p. 431).

One of the first detailed descriptions was made by Dawes (1852): *The interior edge of the penumbra appears extremely jagged; the bright ridges on its surface, which are directed nearly towards the centre of the spot, being seen projected to irregular distances on to the umbra.*

Langley (1874) seemed to be worried over the modern debate about the height of the penumbral filaments, as he commented: *these filaments are not*

Table 3.2. Glossary of William Herschel's sunspot terminology. From Hoyt and Schatten (1992)

Before 1800 Jan 31	1800 Jan 31 to 1801 May 19	1801 May 19 to 1818	Modern-day Terminology
Spots	Openings	Openings	Sunspots
Penumbra or shelving sides	Flats	Shallows	Penumbrae
Nodules	Nodules	Nodules	Bright Points
Punctures	Punctures	Pores	Pores

us only as it looks when viewed through this great depth of air. Professor Langley has shown recently that the air mars, colours, distorts, and therefore misleads and cheats us to an extent much greater than was supposed. Cited in Eddy and Ise (1979), p. 39.

42 However, the new nomenclature never caught on, partly because it reflected the heterodox view of a cool Sun and partly because adequate names for the various features already existed.

only more brilliant at the umbral edge (as if their extremities were curved upward, and less obscured, where partly elevated above a darker supervening medium), but this tendency may be traced in good definition, all over the penumbra, in which they have a certain tendency to unite in narrow sheets or plates, which superposed, form the fascicles called "thatch-straws" by Mr. Dawes.

3.8.2 Umbral Structures

Dawes (1864) remarked that the umbra seems to be perforated near its centre by a perfectly "black hole", which he considered to be the real nucleus. It corresponds to our dark nucleus described in Chapter 1.

Langley (1874) reported that the whole umbra is seen at times to be nearly or wholly made up of sunken banks of the filaments. For him, the umbra has the appearance of a submerged penumbra, with the dark nuclei being only deeper portions of its shade.

3.8.3 Light-Bridges

Light-bridges are bright structures associated with sunspots and seem to be related with the process of disintegration. Scheiner already commented in the *Rosa Ursina* (p. 535): *Many spots appear to be gradually inundated by luminous material from all sides, the luminous material spreads itself out, the spot decreases in size until no trace of a shadow remains, and only a bright faculae is seen..*

Its granular structure was noticed by Nasmyth, who commented: *The filaments (granules) in question are seen and appear well defined at the edges of the luminous surface where it overhangs "the penumbra", as also in the details of the penumbra itself, and most especially are they seen clearly in the details of the bridges as I term those bright streaks which are so frequently seen stretching across from side to side over the dark part of the spot* (Nasmyth, 1862).

Figure 3.36 shows sunspot observations by E. Trouvelot in 1872, where very bright light–bridges are clearly visible.

3.9 Faculae

In the last letter from Welser to Galileo (5 October 1612) the name facula is mentioned for the first time, probably referring to observations by C. Scheiner: *Since you like to hear on the discoveries of my friend, I copy what I have recently received from him concerning some new observations [...] are "little torches", that is regions which are brighter than those surrounding them, so that they shine more brilliantly than these.*

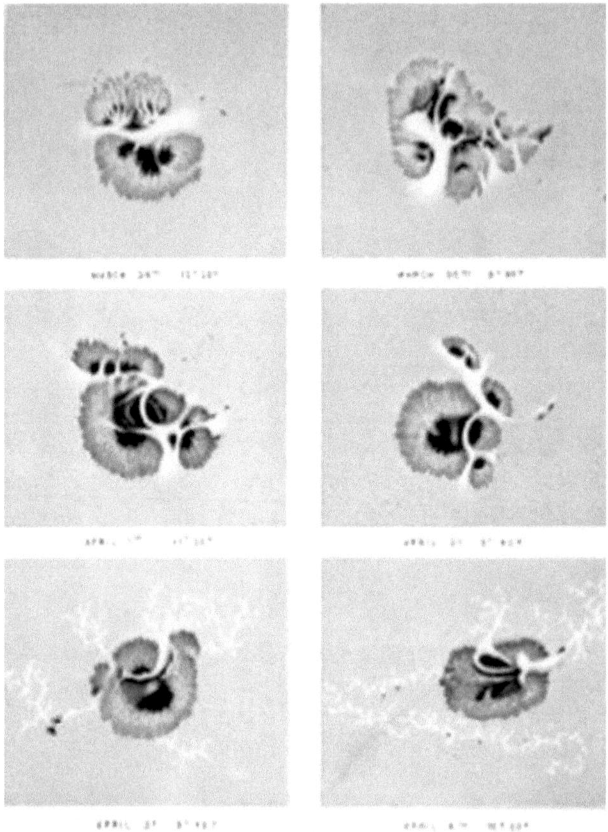

Fig. 3.36. Drawing of bright light-bridges by E.L. Trouvelot. Litographic print issued by Harvard College Observatory in 1876 Credit: Science Museum London, Science and Society Picture Library.

In his *Rosa Ursina*, Scheiner described in more detail these structures: *I and R.P. Ioannes Baptista Cysat, the sky being most clear, examining the sun most carefully by moving the telescope, found it free from spots, but dotted with brighter regions which I call "faculae"*. Scheiner also mentioned that faculae were frequently seen at the spot position after the latter had disappeared and in other cases the faculae precede the spot.

C. Huygens questioned the existence of these structures in his *Kosmotheoros*, published in 1698: *Nor could ever have the luck to discern those bright spots they brag so much of in the Sun as well as of his dark ones, tho the latter I have very often saw; so that with very good reason I can doubt whether there's any such thing. For, in all the exact observations, I could never find any such pretended to be seen any where but just about his dark spots, and it is no great wonder than those parts which are so near the darker, should appear somewhat brighter than the rest.*[43]

Stephen Gray (1666–1736) noticed that faculae were brighter at the solar limb: *The faculae doe not appear, as was said, but when near the limb of the sun and the nearer thereto the brighter, the reason whereof I take to be that property of flame to appear brighter when most contrasted* (Clark and Murdin, 1979; Soon and Yaskell, 2003, p. 58).

3.10 White-Light Flares

We have described in Chapter 1 the main characteristics of a flare. This is a phenomenon easily observed using special filters, but only the most energetic leave a signature in the visible light of the photosphere. This was the case that allowed Richard Carrington (1826 1875) to observe this phenomenon. In his 1860 paper in *Monthly Notices of the Royal Astronomical Society* he describes the observation:

> "While engaged in the forenoon of Thursday, September 1, in taking my customary observation of the forms and positions of the solar spots, an appearance was witnessed which I believe to be exceedingly rare. The image of the sun's disk was, as usual with me, projected on to a plate of glass coated with distemper of a pale straw color, and at a distance and under a power which presented a picture of about 11 inches diameter. I had secured diagrams of all the groups and detached spots, and was engaged at the time in counting from the chronometer and recording the contacts of the spots with the cross-wires used in the observation, when within the area of the great north group (the size of which had previously excited great remark), two patches of intensely bright and white light broke out, in the positions indicated in Figure 3.37. My first impression was that by some chance a ray of light had penetrated a hole in the screen attached to the object glass, for the brilliancy was fully equal to that of direct sun light; but by at once interrupting the current observation, and causing the image to move [...] I saw I was an unprepared witness of a very different affair. I therefore noted down the time by the chronometer, and seeing the outburst to be very rapidly on the increase, and being somewhat flurried by the surprise, I hastily ran to call some one to witness the exhibition with me, and on returning within 60 seconds, was mortified to find that it was already much changed and enfeebled. Very shortly afterwards the last trace was gone. In this lapse of 5 minutes, the two patches of light traversed a space of about 35,000 miles." For more details on the life of Carrington see Clark (2007).

[43] An English translation is available at:http://www.phys.uu.nl/~huygens/cosmotheoros_en.htm

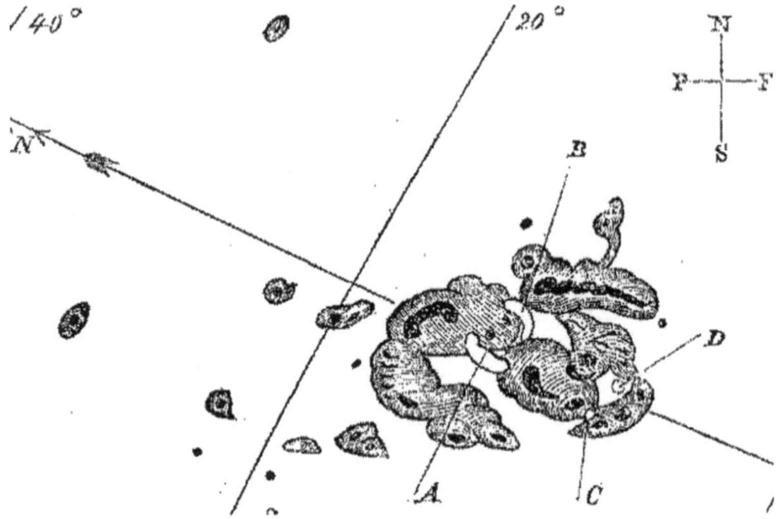

Fig. 3.37. Drawing of the flare observed by R. Carrington on 1 September 1859. "A" and "B" represent the positions in which the flare appeared. "C" and "D" the place where it disappeared.

The event was independently observed by R. Hodgson at his home at Highhate with a refractor of six inches and a pale neutral tint sunglass: *While observing a group of solar spots on the 1st September I was suddenly surprised at the appearance of a very brilliant star of light, much brighter than the sun's surface* (see Hodgson, 1860). One day later, auroras were visible in the two hemispheres reaching low latitude sites (Cliver, 2006).

An earlier, plausible report of a white-light flare has been found in the notebook of Stephen Gray, who on 27 December 1705 noticed a "flash of lightning" near a sunspot. The document is now preserved at the Cambridge University Library (Hoyt and Schatten, 1996).

3.11 The Outer Layers of the Sun

Visual observations of the outer layers outside the totality of solar eclipses was an impossible task for many centuries. Only after the application of the spectrograph was this possible with the slit partially open and placed tangent to the solar limb. Jules Janssen, William Huggins, Warren de la Rue and Norman Lockyer were pioneers in this field.

Prominences were the favourite target of these pioneering studies. Lorenzo Respighi (1824–1889) made some nice drawings at the Campidoglio Observatory, Rome. Carl Frederic Fearnley of University of Christiania in Oslo made detailed drawings of prominences in 1872 and 1873 using a 15 cm refractor equipped with a spectroscope.

The Observatory of Madrid also collected numerous drawings of solar structures, including prominences. Figure 3.38 shows one example.

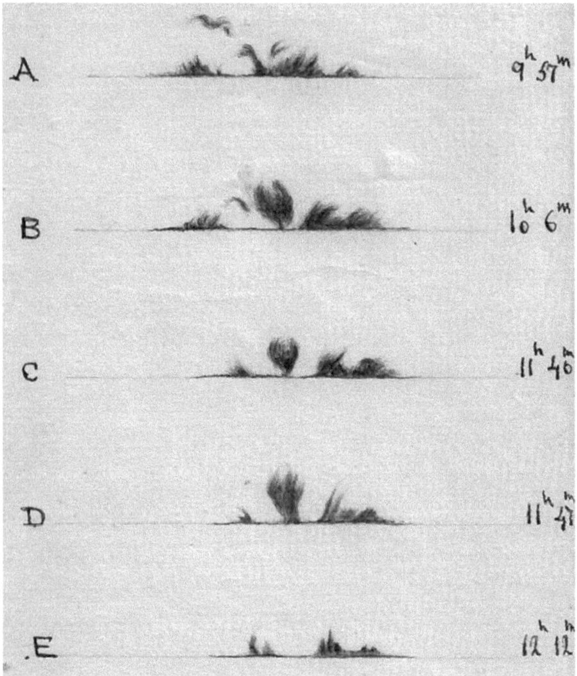

Fig. 3.38. Evolution of a solar prominence at the Observatory of Madrid (6 July 1917).

Figure 3.39 shows coronal loops drawn by Angelo Secchi from observations in the H_α line on 5 October, 1871. The similarity with space observations displayed in Chapter 1 is remarkable.

Fig. 3.39. Drawing of coronal loops by A. Secchi. Reproduced from C. Young *The Sun*.

From 1886 to 1917 Gyula Fényi collected a long series of prominence observations using a spectroscope of six prisms.

We will come back to the observations of the solar outer layers in the next chapter, dedicated to eclipses.

3.12 The Influence of the Eye in Solar Drawings

In the previous chapter we have described the main properties of our available detector, the human eye, and its relation with naked-eye observations of sunspots. Here, we will describe its influence in the subjectivity inherent in solar drawings.

3.12.1 Eye Aberrations

Similar to the lenses of a telescope, the eye is affected by aberrations that degrade the retinal image and ultimately limit spatial vision. The lower order aberrations (defocus and astigmatism) can be corrected with spectacles, but not the higher orders. However, Artal et al. (2006) have shown that the neuralvisual system is adapted to the eye's particular aberrations. They found that the distribution of aberrations between the cornea and lens appears to allow the optical properties of the eye to be relatively insensitive to variations arising from eye growth or exact centration and alignment of the eye's optics relative to the fovea. These results may suggest the presence of an auto-compensation mechanism that renders the eye's optics robust despite large variation in the ocular shape and geometry.

The history of the Mars channels reflect the effect of another eye condition, Daltonism or colour-blindness. Colour-blind people perceive any modification of the intensity of light as a change of colour, and they are more sensitive to contrast effects than those with normal colour vision. Schiaparelli's earlier training in draftsmanship had given him the ability *to transcribe quickly onto paper the almost cinematic impressions of the figures observed in the field of the telescope.* However, his eye was strongly affected by colour-blindness; thus, as he himself admitted, he *failed to distinguish gradations of red and green*, and he once described the general appearance of the major markings as *almost like that of a chiaroscuro made with Chinese ink upon a general bright background.* On the other hand, his colour-blindness seems to have made him more sensitive to delicate markings at the threshold of visibility; as a record of fleeting impressions, his observations are unrivaled (Sheehan, 1996). How these or related illnesses have affected astronomers drawing solar structures is a matter of future studies.

3.12.2 The Influence of the Brain

In a letter to his friend Piero Aliotti, bishop of Forli, the great Michelangelo Buonarroti (1475–1564) recommended painting more with the brain than with the hands. This makes us pause to consider how a preexisting idea about a certain observed pattern can influence the way an astronomer draws it. Again the channels of Mars are a good reference. Apart from instrumental problems and the influence of atmospheric turbulence, Percival Lowell (1855–1916) drew the "canali" surely because he was previously convinced of their existence. This can also be applied to the interpretation of visual solar observations.

3.13 Physics from Drawings

3.13.1 The Wilson Effect

C. Scheiner was interested in the depth of sunspots. However, his writings are ambiguous in this respect. Sometimes he considered the spots as elevations and other times as depressions. It was necessary to wait a century for a detailed study of this subject.

Alexander Wilson (1714–1786) was a professor of practical astronomy at the University of Glasgow. He used a Gregorian telescope, with a 26 inch focus, which magnified 112 times. On 23 Nov. 1769 he observed that *the umbra appeared much contracted on that part which lay towards the centre of the disc, while the other parts of it remained nearly of their former dimensions* (Wilson, 1774). He noticed that it was the opposite of what was expected from a pure effect of perspective. Fortunately the sunspot was long-lived and reappeared with nearly the same size on the eastern limb, observing the same effect but now on the eastern side of the penumbra.

He then suggested that the central part or nucleus was beneath the level of the Sun's spherical surface. In his own words, it was *a vast excavation in the luminous matter of the Sun.*

He proudly remarked that except for a few conclusions concerning the rotation of the Sun around its axis, and the inclination of its axis to the plane of the ecliptic, all that had been investigated before him on sunspots was a matter of conjecture. His sunspot model consisted of a vast cavernous opening, having the nucleus (umbra) at the bottom, and the penumbra forming its sloping sides.

In the French Memoires of 1770, Joseph Lalande (1732–1807) criticized the hypothesis of Wilson arguing that only a few spots, the roundish ones, behave the way Wilson proposed. He presented an alternative sunspot model: *that the spots us phenomena arise from dark bodies like rocks, which by an alternate flux and reflux of the liquid igneous matter of the Sun, sometimes raise their heads above the general surface. That part of the opaque rock, which at any time thus stands above, gives the appearance of the umbra, whilst those parts, which in each lie only a little under the igneous matter, appear to us as the surrounding penumbra.*[44] Wilson (1783) reacted rather vigorously criticizing

[44] Both Wilson and Lalande used the word "nucleus" for the umbra and "umbra" for the penumbra.

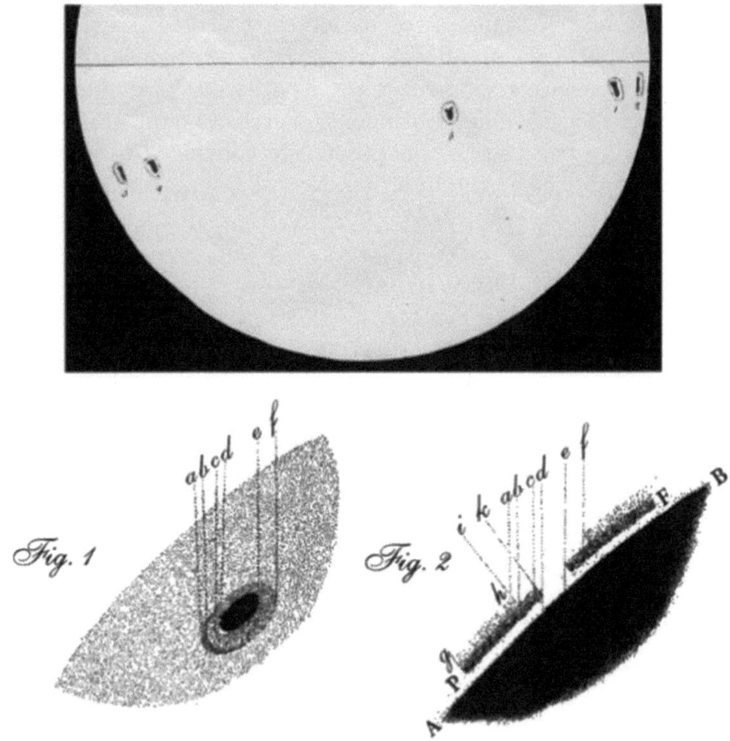

Fig. 3.40. Sunspot drawings included in Wilson (1774).

the adequacy of the old sunspot observations of J. Picard and P. La Hire used by Lalande for this specific purpose and stressing the need to use optical arguments. He remarked: *No doubts ought to arise of the spots being themselves what direct observation declares them, namely, excavations in the sun, as actually demonstrated by competent observations.*

With this aim in mind Wilson modelled a globe and the spots upon it according their proper dimensions. Then put it in a wooden frame and *viewed it afar off when set upon a stand, whilst the globe was turned slowly round, and subtended an angle at the telescope equal to the apparent diameter of the Sun. By an object glass micrometer I then took the distances from the limb when the farthest penumbra of different spots vanished, as also the distances of the nuclei just when disappearing.* In summary, a modern experimental procedure of simulation of a phenomenon under study.

F. Howlett selected a set of sunspots from the Stonyhurst collection, that in his opinion did not follow the Wilson behaviour. He claimed that the statements and observations of Wilson were extremely circumstantial (Howlett, 1886), although he remarked that they were different types of sunspots manifesting different degrees of profundity, as well as of general form, he also mentioned the possible influence of the refraction on the observations close to the solar limb. Cortie (1898a,b) was in this case the defender of the Wilson theory, based on sunspot drawings by Ricco in Palermo and Catania. He concludes that the sunspot darkness is due to absorption.

Nowadays we know that it is not the case of a real geometrical depression, rather an effect related with the transparency of the solar material to the radiation (see Chapter 1). The visible light originates at different heights in the umbra, penumbra and photosphere. The dependence of the opacity on temperature implies that the umbra is more transparent than the umbra. In other words, we see through to deeper layers in the umbra.

We can now consider a sunspot located close to the limb (Figure 3.41). Since the penumbral radiation originates at a lower height than the photospheric radiation, the photosphere will obscure part of the penumbra on the side of the spot remote from the limb. The penumbra closer to the limb, on the other hand, will not be obscured. Since the umbral radiation originates in a layer located deeper than the layers contributing to the penumbral radiation, part of the umbra will be obscured by the penumbra. Hence, the size of the umbra will decrease (increase) faster than expected from perspective foreshortening as the sunspot is moved by solar rotation towards (away from) the solar limb (Collados et al. 1987; Maltby 2000).

The height difference between the umbra and the photosphere, called the Wilson depression, is of the order of 500 km.

3.13.2 Solar Rotation

Sunspot drawings provide an excellent tool for applying the tracer technique to determine solar rotation. For that purpose we need full-disk images of the

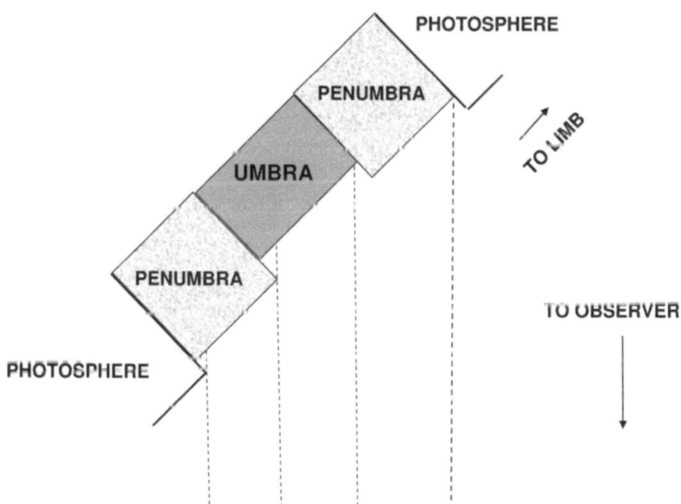

Fig. 3.41. Sketch of a roundish sunspot located close to the solar limb.

Sun and not detailed drawings of the sunspots. The time should be included, a fact which was not included in many of the historical documents.

Determination of Sunspot Positions

The location of the feature on the Sun is the first step to be carried out. Given a reference point (the centre of the image of radius R), this can be done using two different types of coordinates: cartesian (x,y) and polar (r, θ)

$$r = (x^2 + y^2)^{1/2};\ \theta = \arctan y/x$$
$$x = r\cos\theta;\ y = r\sin\theta$$

The angular distance from the disk centre is given by: $\sin\rho = r/R$.

Heliographic Coordinates

Generally, it is convenient to convert these values to heliographic coordinates fixed on the Sun, namely the latitude B and the distance from the central meridian, l. For this purpose we need to define first the reference planes (Figure 3.42):

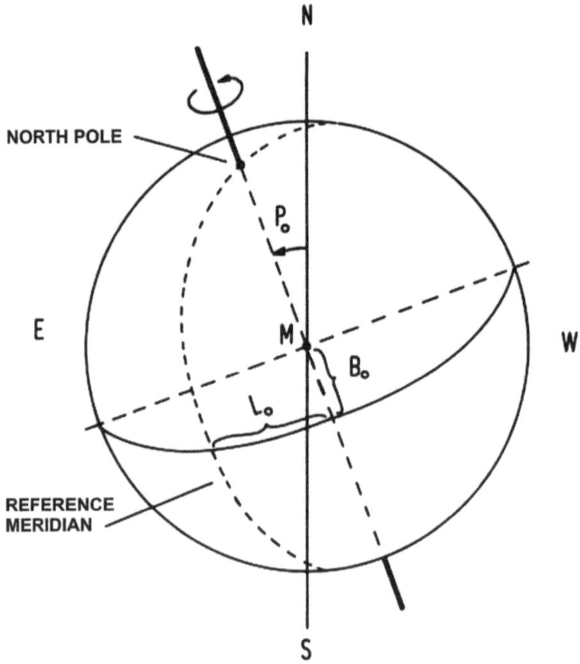

Fig. 3.42. Heliographic coordinates.

$$\sin \mathrm{B} = \cos \rho \sin \mathrm{B}_0 + \sin \rho \cos \mathrm{B}_0 \sin \theta$$

$$\sin \mathrm{l} = \frac{\cos \theta \sin \rho}{\cos \mathrm{B}}$$

To determine the heliographic longitude L from the measured l we need a simple addition: $\mathrm{L} = \mathrm{L}_0 + \mathrm{l}$.

The "zero" meridian ($L_0 = 0$) is defined as the central meridian which passed through the apparent centre of the disk on 1 January 1854 at Greenwhich local noon. The Carrington rotation number (CRN) is given by

$$\mathrm{CRN} = \mathrm{int}[\mathrm{R}_0 + \frac{(\mathrm{JD} - \mathrm{JD}_0)}{27.2753}]$$

where int (X) is given to the nearest integer $\leq$ X. Here R_0 is the Carrington number for any known day and can be obtained from the almanacs and JD_0, the Julian day number for that day.

Carrington rotation refers to the Earth-related synodic period, $\mathrm{P}_{\mathrm{syn}}$. The sidereal period, P_{sd}, is the solar rotation period as seen from a fixed point in the sky. Both are related by

$$\frac{1}{\mathrm{P}_{\mathrm{sd}}} = \frac{1}{\mathrm{P}_{\mathrm{syn}}} + \frac{1}{\mathrm{P}_{\mathrm{E}}}$$

where $\mathrm{P}_{\mathrm{E}} = 365.25$ days.

A set of eight disks, called Stonyhurst disks, enables the calculation of the latitude and longitude of individual sunspots or sunspot groups with the aid of simple arithmetic. Each disk is designed to be used with a 6-inch drawing which is usually placed over the top of the disk so that it shows through the disk drawing.

Essential for these calculations is an accurate determination of the rotation elements (i, Ω), where i is the inclination of the Sun's rotation axis and Ω the angle between the crosspoint of the solar equator with the ecliptic and the equinox point (Wöhl, 1978).

Calculation of Rotation Rates

The obtained rotation values were fitted to a law of differential rotation: $\omega = a + b \sin^2 B$. Table 3.3 shows the results for different analyses of historical data.

Rotation vs. Activity Level

Variations in the solar differential rotation, using sunspots as tracers, with the phase of the solar activity cycle have been measured (Balthasar and Wöhl, 1980; Gilman and Howard, 1984; Gupta et al., 1999) although the results fail to show high statistical significance. Javaraiah et al. (2005) found

Table 3.3. Determinations of the solar rotation from different sources during the 17th century

Author(s)	Observer	*a*	*b*
Eddy et al. (1976)	Hevelius	13,99 ± 0,14	6.99 ± 2,73
Eddy et al. (1977)	Scheiner	13.38 ± 0.04	0.90 ± 0.62
Eddy et al. (1977)	Hevelius	13.95 ± 0.12	4.75 ± 3.33
Herr (1980)	Harriot	13.59 ± 0.12	2.22 ± 1.22
Abarbanell and Wöhl (1981)	Hevelius	13.59 ± 0.07	3.71 ± 1.43
Yallop et al. (1982)	Scheiner	13.25 ± 0,03	2.42 ± 0.32
Yallop et al. (1982)	Hevelius	13.48 ± 0.03	2.27 ± 0.95
Ribes et al. (1987)	Picard	13.23 ± 0.07	8.34 ± 0.89
Casas et al. (2006)	Galileo	13.42 ± 0.11	4.96 ± 1.40

that the solar equatorial rotation rate, determined from sunspot group data during the period 1879–2004, decreased over the last century, whereas the level of activity has increased considerably. Therefore, a change in the rotation rates during the Maunder minimum can be expected. This could proceed in an abrupt or gradual way. To study this behavior Casas et al. (2006) represent (Figures 3.43 and 3.44) the coefficients of the differential rotation fits with respect to the Sunspot Group Number averaged for the observing period. As a reference, we have included the mean values given by Balthasar et al. (1986) corresponding to the solar cycles 12–20 that cover the period 1874–1976.

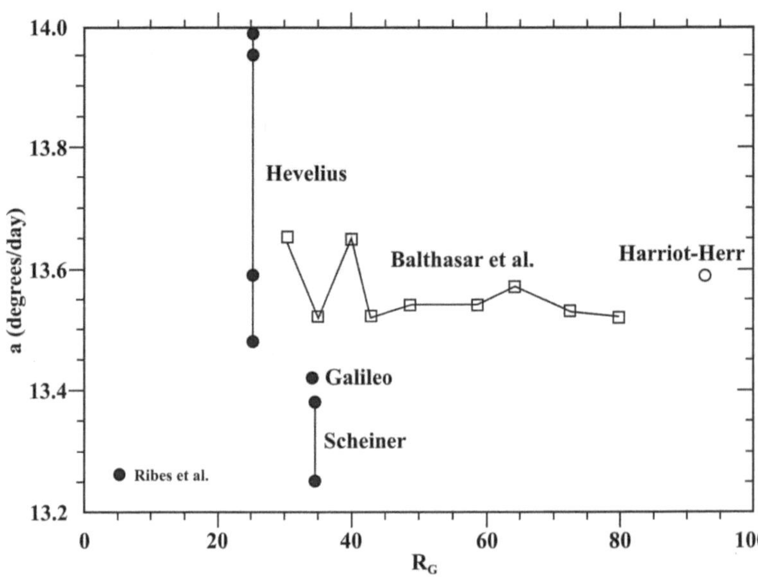

Fig. 3.43. The equatorial rotation, coefficient, *a*, vs. the Sunspot Group Number. Adapted from Casas et al. (2006).

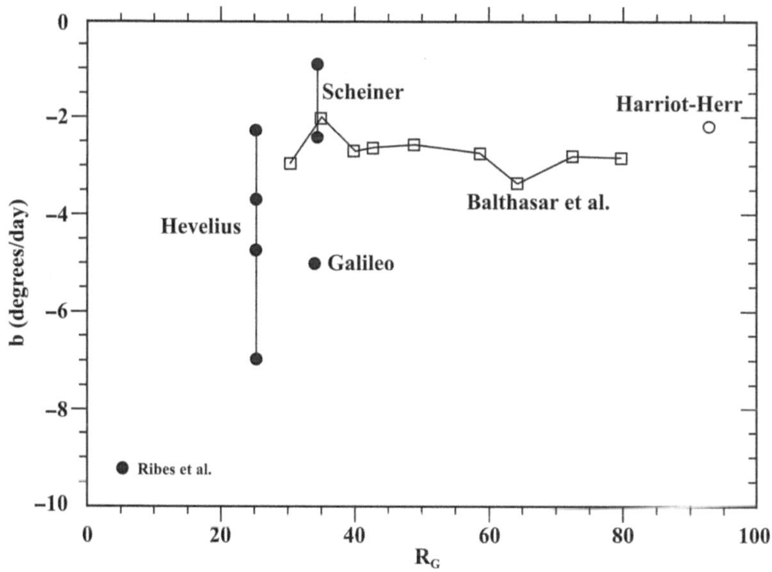

Fig. 3.44. Relation between the coefficient, b, expressing the level of differential rotation, and the Sunspot Group Number. Adapted from Casas et al. (2006).

The historical observations are quite scattered over time and in a second approach we have represented the derived rotation values as a function of the maximum of R_G during the cycle where the observations were made. The result for the equatorial rotation is shown in Figure 3.45. A similar trend is found in the values of b.

Excluding the data of Eddy et al. (1976, 1977), Casas et al. (2006) found the following relation between the equatorial rotation with the maximum value of the cycle, R_{Gmax}:

$$\omega = (13.61 \pm 0.03) - \frac{(7.12 \pm 1.76)}{R_{Gmax}}$$

In 1874 the Greenwich Photoheliographical Results started a set of full disk photographs of the Sun, from which it was possible to determine the sunspot positions. The programme was cancelled in 1976, just the year when the paper of J. Eddy on the Maunder minimum showed the importance of long-term monitoring of our star. Balthasar et al. (1986) were the first to fully digitize these data, analysing in detail different aspects related to sunspot rotation.

3.13.3 Sunspot Areas

We have a full disk image of the Sun containing sunspot drawings. Given the radius of the solar disk, R, in millimetres and knowing the scale, we can calculate the apparent area of a sunspot, A_M in square millimetres. The

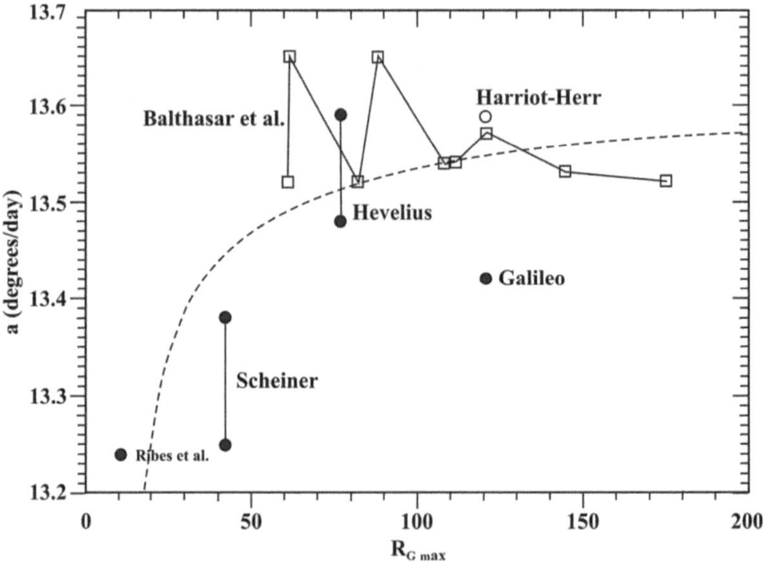

Fig. 3.45. Variation of the equatorial rotation, expressed by the coefficient, a, with the maximum value of $\mathrm{R_G}$ corresponding to the observing period. Adapted from Casas et al. (2006).

correction for perspective effects can proceed **by** applying the formula

$$\mathrm{A_S} = \frac{\mathrm{A_M} 10^6}{2\pi \mathrm{R}^2 \cos\rho} \simeq \frac{\mathrm{A_M} 10^6}{2\pi \mathrm{R}^2 \cos \mathrm{B} \cos(\mathrm{L} - \mathrm{L_0})}$$

where ρ is the angular distance from the centre of the disk to the sunspot and the area, A_S, is expressed in millionths of the visible hemisphere.

Warren de la Rue (1815–1889) compiled sunspot areas during the period 1832–1868 at the Kew Observatory, partially from Stonyhurst drawings, data which were analysed by Vaquero et al. (2002, 2004). Baranyi et al. (2001) have found systematic differences in the determination of sunspot areas between photographic and drawings data bases.

3.14 Modern Solar Drawings

3.14.1 The Fraunhofer Institut "Maps of the Sun"

The "Maps of the Sun" were produced by the Fraunhofer Institut[45] (Freiburg, Germany) under the direction of K.O. Kiepenheuer (1910–1975) and the coordination of K. Brunnckow. They contain detailed drawings of sunspot groups,

[45] Now, Kiepenheuer Institut für Sonnenphysik.

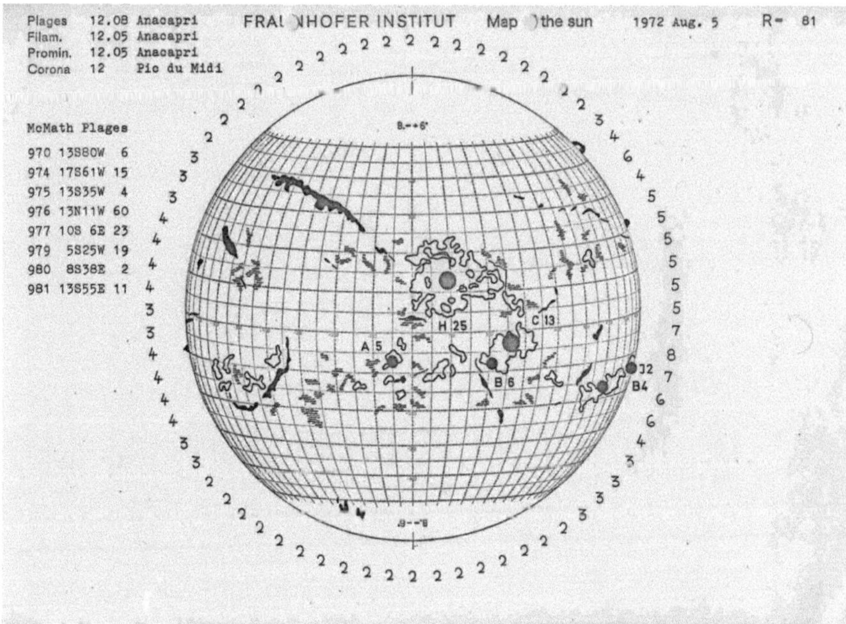

Fig. 3.46. One of the Fraunhofer maps of the Sun (5 August 1972). Courtesy: H. Wöhl (KIS, Freiburg).

calcium plages, flare positions, corona, prominences and filaments for the years 1956–1973. They were elaborated in collaboration with more than ten solar observatories. The maximum number of distributed copies was about 250 and the maps were mailed in general 14 days after the observations.[46] The originals are now preserved at the Institute. Figure 3.46 shows one of these maps.

3.14.2 Potsdam

The current Astrophysical Institut Potsdam is the successor of the Berlin Observatory,[47] founded in 1700 with Gottfriend Kirch as the first director, and the Astrophysical Observatory Potsdam, founded 1874.

Full disk drawings are made regularly as an auxiliary tool for the interpretation of solar radio observations. They started in 1942 and have continued up to the present. Figure 3.47 shows one of these drawings.

[46] One of the contributing observatories (Wandelstein Observatory, Munich) continued with the drawings until about 1987. This material is available at ftp://ftp.ngdc.noaa.gov/STP/SOLAR_DATA/SOLAR_IMAGES/Wendelstein/

[47] The Berlin Observatory became known world-wide when Johann Gottfried Galle discovered the planet Neptune in 1846.

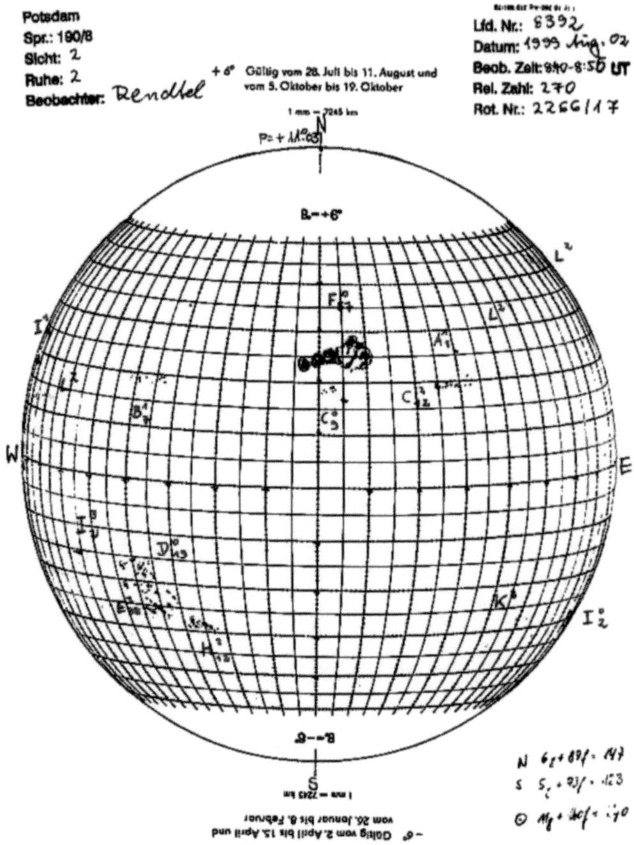

Fig. 3.47. Sunspot drawings on a Stonyhurst disk made at the Potsdam Observatory on 2 August 1999.

3.14.3 The Mt. Wilson Sunspot Drawings

The daily sunspot drawings at the 150 foot solar tower of the Mount Wilson Observatory were begun on 4 January, 1917. They are archived digitally and include a value of the magnetic field strengths of sunspots (Figure 3.48). The sunspot drawing is done in pencil, on a 25.5 by 50 cm pad of paper. Solar north is at the top and east at the right-hand side.

3.14.4 Kanzelhöhe

As it was known since the 1930s that the Sun affects the ionosphere and therefore the propagation of radio frequencies, the former Deutsche

Fig. 3.48. Sunspot drawing at the Mount Wilson Observatory on 2 August 1999.

Luftwaffe founded some solar observatories during the Second World War.[48] Because of the favourable climate and the convenient possibility of coming up to the site all through the year with the cable railway Kanzelbahn, the largest observatory was built at Kanzelhöhe near Villach (Carinthia). In 1943 operation was started with very modern equipment. Soon after the end of the war, the Royal Air Force, which was responsible then for the observatory, transferred supervision to the University of Graz. The official transfer to the re-born Republic of Austria took place in 1949. In the meantime, the engineer-corps of the Royal Army built one more dome at the top of the Gerlitzen mountain for the observation of the solar corona.

Since 1947 at least one daily drawing of the sunspot groups has been produced at the 25 cm projection-image of a telescope with d/f = 110/1650 mm. Figure 3.49 shows one of these full-disk drawings. They are digitized and available at http://cesar.kso.ac.at. Due to the folding of the light path the drawing is side reversed (i.e. the E-limb is to the right). The sheet is rotated so that azimuth = 0 points to sky North. The alignment is checked by tracing a spot's movement across the sheet due to the Earth's rotation.

Lustig and Dvorak (1984) and Hanslmeier and Lusting (1986) used these sunspot positions to study solar rotation and meridional motions, respectively.

3.14.5 Specola Solare Ticinese

Daily sunspot drawings have been carried out at the Specola Solare Ticinese since October 1957. Between 1957 and 1980 the Specola belonged to the Edigenösssiche Sternwarte of ETH Zurich and the corresponding drawings were

[48] They included the observatories of Zugspitze, Wendelstein, Schauinsland on German territory, Kanzelhöhe in Austria and Syracuse in Italy.

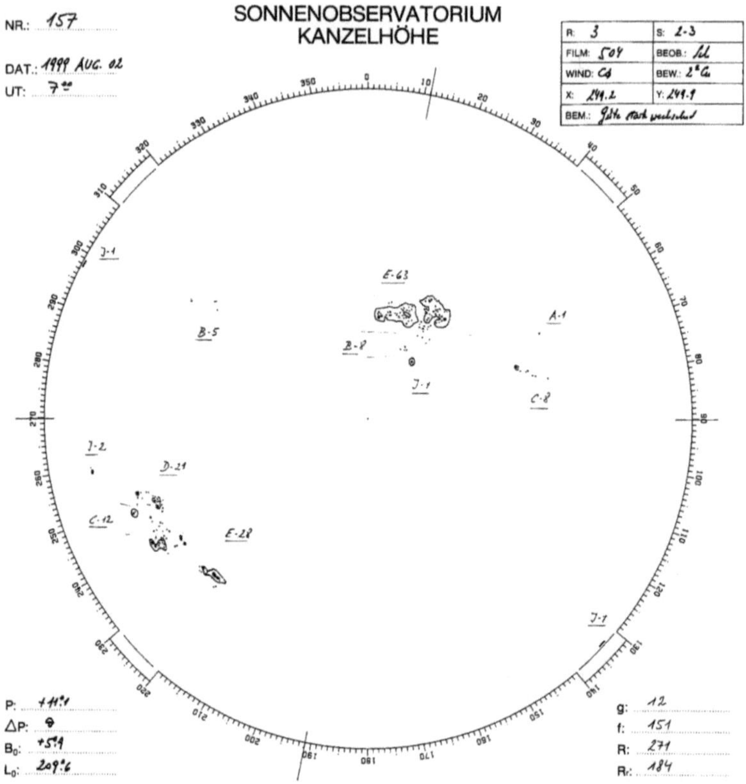

Fig. 3.49. Sunspot drawing made at Kanzelhöhe Observatory on 2 August 1999.

archived by that institute. Starting from 1981 the Specola has been independent and the original drawings are archived locally. Since January 1981 the drawings have been scanned and made available on the Internet (http://www.specola.ch/e/drawings.html).

The drawings are always performed with the white light projection method (diameter of the Sun image 250 mm) according to the Zürich standard. A Zeiss Coudé refractor is used (D/f 150/2250 mm). The drawings are oriented with a precision of 0.1°. Figure 3.50 presents one example.

3.14.6 Rome Solar Phenomena

The Bulletin *Solar Phenomena* was published, from January 1958 to June 1998, by the Rome Astronomical Observatory (Monte Mario) on a monthly basis, and together with the sunspot drawings, measurements of the sunspot magnetic fields were also included. Massimo Cimino (1908–1991) and Maria Torelli were directors of the observatory during this period. Fig. 3.51 shows one of these records.

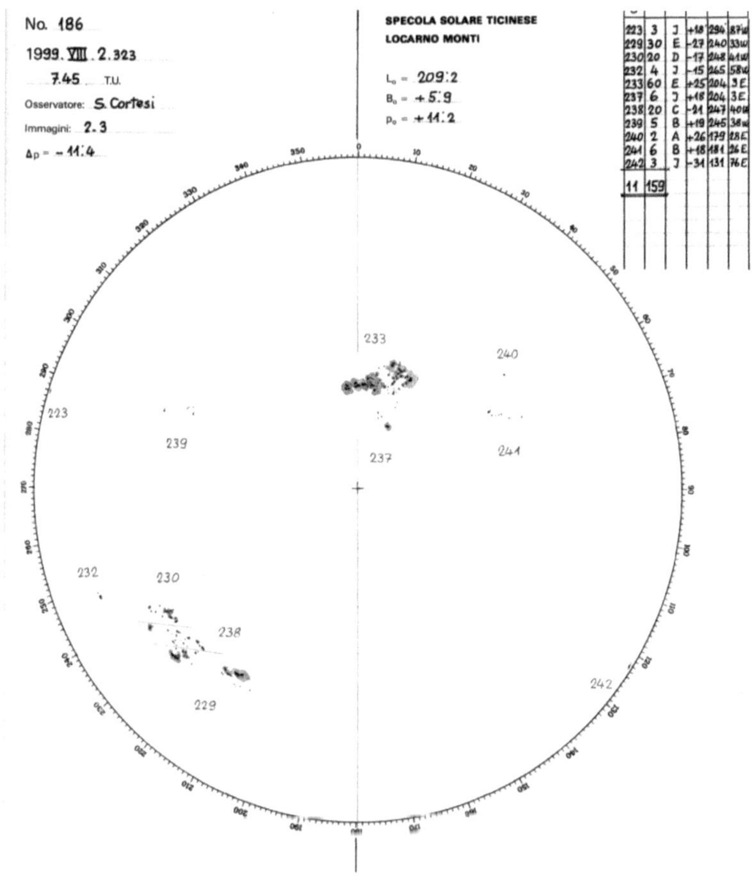

Fig. 3.50. Sunspot drawings made at the Specola Solare Ticinese (2 August 1999).

3.14.7 Cartes Synoptiques and Catalogues of Filaments and Active Regions

Published since 1919 by the Paris Observatory, these consist of maps and tables organized by solar rotation. Lucien d'Azambuja (1884–1970) and M. Roumens prepared the first series of "Cartes Synoptiques de la Chromosphère Solaire" . They include filaments, facular plage contours and the position of the sunspots. Up until 1945 the drawings were only based on images of the K line; later H_α images were also included (Mouradian, 1998). In 1960 M.J. Martres assumed responsibility for the publication and maintained the continuing database until 1991 (Martres, 1998). The World Data Center A for Solar–Terrestrial Physics has digitized the Carte Synoptiques (Coffey & Hanchett, 1998).

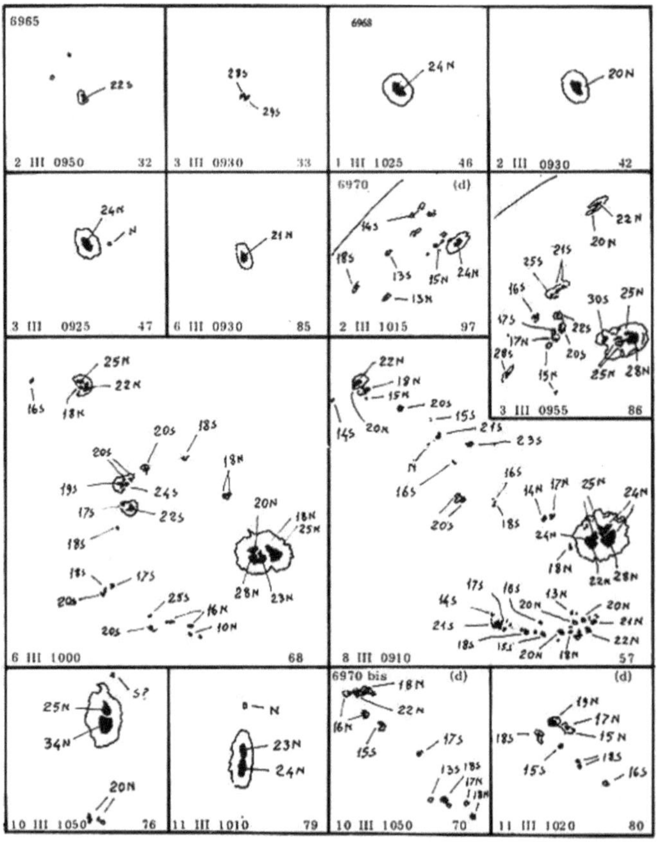

Fig. 3.51. Sunspot drawings with values of the longitudinal magnetic fields made in March 1978 by V. Croce and F. Casamassima.

References

Abarbanell, C. and Wöhl, H. Solar rotation velocity as determined from sunspot drawings of Hevelius, J. in the 17th-century. *Solar Physics*, 70:197–203, 1981.

Abetti, G. I cannocchiali di Galileo e dei suoi discepoli. *L' Universo*, 4:685–692, 1923.

Arlt, R. Digitization of sunspot drawings by Staudacher in 1749–1796. *Solar Physics*, 247:399–410, 2008.

Artal, P., Benito, A., and Tabernero, J. The human eye is an example of robust optical design. *Journal of Vision*, 6:1–7, 2006.

Atkinson, A. D. William Derham, F.R.S. (1657–1735). *Annals of Science*, 8:368–392, 1952.

Balthasar, H., Vazquez, M., and Wöhl, H. Differential rotation of sunspot groups in the period from 1874 through 1976 and changes of the rotation velocity within the solar cycle. *Astronomy and Astrophysics*, 155:87–98, 1986.

Balthasar, H. and Wöhl, H. Differential rotation and meridional motions of sunspots in the years 1940–1968. *Astronomy and Astrophysics*, 92:111–116, 1980.

Baranyi, T., Gyori, L., Ludmány, A., and Coffey, H. E. Comparison of sunspot area data bases. *Monthly Notices Royal Astronomical Society*, 323:223–230, 2001.

Bartholomew, C. F. Discovery of the solar granulation. *Quartely Journal of the Royal Astronomical Society*, 17:263–289, 1976.

Baumgartner, F. J. Sunspots or Sun's planets – Jean tarde and the sunspot controversy of the early 17th-century. *Journal for the History of Astronomy*, 18:44–54, 1987.

Beckman, J. E. and Mahoney, T. J. The Maunder minimum and climate change: Have historical records aided current research? In U. Grothkopf, H. Andernach, S. Stevens-Rayburn, and M. Gomez, (ed), *Library and Information Services in Astronomy III*, volume 153 of *Astronomical Society of the Pacific Conference Series*, p. 212–217, 1998.

Beer, J., Vonmoos, M., and Muscheler, R. Solar variability over the past several millennia. *Space Science Reviews*, 125:67–79, 2006.

Bell, L. *The Telescope.* Dover Publications, 1981.

Berthold, G. *Der Magister Johann Fabricius und die Sonnenflecken: nebst einem Excurse über David Fabricius: eine Studie.* Leipzig, Veit & Comp., 1894.

Birkenmajer, A. Alexius Sylvius Polonus (1593 ca. 1653), a little-known maker of astronomical instruments. *Vistas in Astronomy*, 9:11–12, 1967.

Bion, N. *L'usage des globes celeste et terrestre et des sphéres suivant les different systèmes du Monde*, Paris, 1699.

Bray, R. J. and Loughhead, R. E. *Sunspots.* The International Astrophysics Series, London: Chapman and Hall, 1964.

Bredekamp, H. *Galilei, der Künstler.* Berlin, Akademie Verlag, 2007.

Casanovas, J. Early observations of sunspots: Scheiner and Galileo. In B. Schmieder, J. C. del Toro Iniesta, and M. Vàzquez, (ed), *ASP Conf. Ser. 118: 1st Advances in Solar Physics Euroconference. Advances in Physics of Sunspots*, pp. 3–20, 1997.

Casas, R., Vaquero, J. M., and Vazquez, M. Solar Rotation in the 17th century. *Solar Physics*, 234:379–392, 2006.

Caspar, M. *Kepler.* New York, Dover Publications, 1993.

Chapman, A. The Astronomical Work of Thomas Harriot (1560–1621). *Quartely Journal Royal Astronomical Society*, 36:97–107, 1995.

Chapman, A. Thomas Harriot: the Englishman who beat Galileo. *Astronomy Now*, 20:68–70, 2006.

Chappell, M. Cigoli, Galileo and Invidia. *The Art Bulletin*, 57:91–98, 1975.

Clark, D. H. and L. Murdin. The enigma of Stephen Gray astronomer and scientist (1666–1736). *Vistas in Astronomy*, 23:351–404, 1979.

Clark, S. *The Sun Kings.* Princeton University Press, 2007.

Cliver, E. W. The 1859 space weather event: Then and now. *Advances in Space Research*, 38:119–129, 2006.

Coffey, H. E. and Hanchett, C. D. Digital "Cartes Synoptiques de la Chromosphere Solaire et Catalogues des Filaments et des Centres d'Activite". In D. F. Webb, B. Schmieder, and D. M. Rust, (eds.), *IAU Colloq. 167: New Perspectives on Solar Prominences*, volume 150 of *Astronomical Society of the Pacific Conference Series*, pp. 488–492, 1998.

Collados, M., del Toro Iniesta, J. C., and Vazquez, M. A statistical study of the geometrical Wilson effect. *Solar Physics*, 112:281–293, 1987.

Cortie, A. L. On the level of sun-spots and the cause of their darkness. *Astrophysical Journal*, 7:239–248, 1898a.

Cortie, A. L. The Wilsonian theory and Mr. Howlett's drawings of sun-spots. *Monthly Notices Royal Astronomical Society*, 58:91–95, 1898b.

Coyne, G. V. Sunspots: the Historical Background. In C. P. Sonnett, M. S. Giampapa, and M. S. Matthews, (eds.), *The Sun in Time*, pp. 3–10, 1991.

Crombie, A. C. *Robert Grosseteste and the Origins of Experimental Science 1100–1700.* Oxford, Clarendon Press, 1971.

Dawes, W. R. Description of a new solar eye-piece, *Memoirs of the Royal Astronomical Society*, 21:161, 1852.

Dawes, W. R. Results of some recent Observations of the Solar Surface, with Remarks. *Monthly Notices Royal Astronomical Society*, 24:161–165, 1864.

Daxecker, F. *Das Hauptwerk des Astronomen P. Christoph Scheiner Rosa Ursina sive Sol: eine Zusammenfassung.* Innsbruck, Universitätsverlag, 1996.

Daxecker, F. *Der Physiker und Astronom Christoph Scheiner.* Innsbruck, Universitätsverlag Wagner, 2006.

Derham, W. Observations upon spots in the Sun, 1703 to 1711, with a letter of Crabtree in 1640 upon the same subject. *Philosophical Transactions*, 27: 270–278, 1711.

Drake, S. *Discoveries and opinions of Galileo.* Anchor Books, 1957.

Drake, S. Galileo Gleaning III: A kind work for Sizzi. *Isis*, 49:155–165, 1958.

Drake, S. *Galileo at Work: His scientific biography.* University of Chicago Press, 1978.

Eddy, J. A. The Maunder minimum. *Science*, 192:1189–1202, 1976.

Eddy, J. A., Gilman, P. A., and Trotter, D. E. Solar rotation during the Maunder Minimum. *Solar Physics*, 46:3–14, 1976.

Eddy, J. A., Gilman, P. A., and Trotter, D. E. Anomalous solar rotation in the early 17th century. *Science*, 198:824–829, 1977.

Eddy, J. A. and Ise, R. New Sun: The Solar Results From Skylab. NASA SP-402. *NASA Special Publication 402*, 1979.

Favaro, A. *Le Opere di Galileo Galilei, Vol. V.* Ministero della Istruzione Pubblica, 1895.

Findlen, P. (ed.). *Athanasius Kircher: The Last Man Who knew Everything.* Routledge, 2004.

Fletcher, J. E. Astronomy in the life and correspondence of Athanasius Kircher. *Isis*, 61:52–67, 1970.

Freedberg, D. *The Eye of the Lynx: Galileo, his Friends, and the Beginnings of Modern Natural History.* University of Chicago Press, 2002.

Gilman, P. A. and Howard, R. Variations in solar rotation with the sunspot cycle. *Astrophysical Journal*, 283:385–391, 1984.

Godwin, J. *Athanasius Kircher, A Renaissance Man and the Quest for Lost Knowledge.* London, Thames and Hudson, 1979.

Greco, V., Molesini, G., and Quercioli, F. Optical tests of Galileo's lenses. *Nature*, 358:101, 1992.

Greco, V., Molesini, G., and Quercioli, F. Telescopes of Galileo. *Applied Optics*, 32:6219–6226, 1993.

Guerrini, L. Nel dedalo del cielo. Gassendi e le macchie solari in un inedito del 1633. *Nuncius*, 16:643–672, 2001.

Gupta, S. S., Sivaraman, K. R., and Howard, R. F. Measurement of Kodaikanal white-light images – III. Rotation rates and activity cycle variations. *Solar Physics*, 188:225–236, 1999.

Hanslmeier, A. and Lustig, G. Meridional motions of sunspots from 1947.9-1985.0. I – Latitude drift at the different solar-cycles. *Astronomy and Astrophysics*, 154:227–230, 1986.

Herbst, K.-P. Briefe als Quelle für Beobachtungsdaten von Sonnenflecken: dargestellt am Beispiel des Briefwechsels zwischen Gottfried Kirch und Gottfried Schultz von 1681 bis 1692. *Acta Historica Astronomiae*, 25:92–102, 2005.

Herr, R. B. Solar Rotation in 1612: Galileo vs. Harriot. In *Bulletin of the American Astronomical Society*, volume 12 of *Bulletin of the American Astronomical Society*, pp. 504, 1980.

Herschel, W. On the nature and construction of the Sun and Fixed Stars. *Philosophical Transactions Royal Society London Series I*, 85:46–72, 1795.

Herschel, W. Investigation of the powers of the prismatic colours to heat and illuminate objects; with remarks, that prove the different refrangibility of radiant heat. To which is added, an inquiry into the method of viewing the Sun advantageously, with telescopes of large apertures and high magnifying powers. *Philosophical Transactions Royal Society London Series I*, 90:255–283, 1800.

Herschel, W. Additional observations tending to investigate the symptoms of the variable emission of the light and heat of the Sun; with trials to set aside darkening glasses, by transmitting the solar rays through liquids; and a few remarks to remove objections that might be made against some of the arguments contained in the former paper. *Philosophical Transactions Royal Society London Series I*, 91:354–362, 1801a.

Herschel, W. Observations Tending to Investigate the nature of the Sun, in order to find the causes or symptons of its variable emission of light and heat; with remarks on the use that may possibly be drawn from solar observations. *Philosophical Transactions Royal Society London Series I*, 91:265, 1801b.

Hodgson, R. On a curious appearance seen in the Sun. *Monthly Notices Royal Astronomical Society*, 20:15–16, 1860.

Howlett, F. Rev. F. Howlett's drawings of Sun-spots. *Monthly Notices Royal Astronomical Society*, 37:364, 1877.

Howlett, F. On the asserted foreshortening of the inner side of the penumbra of spots when near the Sun's limb. *Monthly Notices Royal Astronomical Society*, 46:447–451, 1886.

Howlett, F. Sunspot drawings, remarks on three volumes of, presented to the Society. *Monthly Notices Royal Astronomical Society*, 55:73–76, 1894.

Hoyt, D. V. Early telescopic sunspot records. *Applied Historical Astronomy, 24th meeting of the IAU, Joint Discussion 6, Manchester, England, meeting abstract*, 6, 2000.

Hoyt, D. V. and Schatten, K. Group sunspot numbers: a new solar activity reconstruction, *Solar Physics*, 178:206–219, 1998.

Hoyt, D. V. and Schatten, K. H. Sir William Herschel's notebooks – Abstracts of solar observations. *Astrophysical Journal Supplement Series*, 78:301–340, 1992.

Hoyt, D. V. and Schatten, K. H. A new interpretation of Christian Horrebow's sunspot observations from 1761 to 1777. *Solar Physics*, 160:387–392, 1995a.

Hoyt, D. V. and Schatten, K. H. A revised listing of the number of sunspot groups made by Pastorff, 1819 to 1833. *Solar Physics*, 160:393–399, 1995b.

Hoyt, D. V. and Schatten, K. H. Observations of sunspots by Flamsteed during the Maunder Minimum. *Solar Physics*, 160:379–385, 1995c.

Hoyt, D. V. and Schatten, K. H. How well was the Sun observed during the Maunder minimum? *Solar Physics*, 165:181–192, 1996.

Huggins, W. Results of some Observations on the Bright Granules of the Solar Surface, with Remarks on the Nature of these Bodies. *Monthly Notices Royal Astronomical Society*, 26:260–265, 1866.

Izenman, A. J. J. R. Wolf and the Zurich sunspot relative numbers. *The Mathematical Intelligencer*, 7:27–33, 1985.

Javaraiah, J., Bertello, L., and Ulrich, R. K. Long-term variations in solar differential rotation and sunspot activity. *Solar Physics*, 232:25–40, 2005.

Kawaler, S. and Veverka, J. The Habitable Sun – One of Herschel, William's Stranger Ideas. *Journal Royal Astronomical Society Canada*, 75:46–55, 1981.

Kemp, M. *The Science of Art.* Yale University Press, 1990.

Kemp, M. *Leonardo da Vinci: The Marvellous Works of Nature and Man.* Oxford University Press, 2006.

King, H. C. *The History of the Telescope.* New York, Dover Publication Inc., 1955.

Kovaltsov, G. A., Usoskin, I. G., and Mursula, K. An upper limit on sunspot activity during the Maunder minimum. *Solar Physics*, 224:95–101, 2004.

Lane, J. H. On the theoretical temperature of the Sun, under the hypothesis of a gaseous mass maintaining its volume by its internal heat, and depending on the law of gases known to terrestrial experiment. *American Journal of Science, 2n Ser.*, 1:57, 1870.

Langley, S. P. On the structure of the solar photosphere. *Monthly Notices Royal Astronomical Society*, 34:255–261, 1874.

Langley, S. P. *The New Astronomy.* Boston & New York, Houghton, Mifflin Co., 1889.

Lindberg, D. C. The theory of the pinhole images from antiquity to the thirdteenth century. *Archive for History of Earth Sciences*, 36:299–325, 1970.

Lustig, G. and Dvorak, R. Solar rotation rate from stable recurrent sunspot tracings. *Astronomy and Astrophysics*, 141:105–107, 1984.

Machamer, P. (ed.). *The Cambridge Companion to Galileo.* Cambridge University Press, 1998.

Maltby, P. Sunspots: Wilson effect. In P. Murdin (ed.), *Encyclopedia of Astronomy and Astrophysics*, 2000.

Martres, M. J. Lucien d'Azambuja. In D. F. Webb, B. Schmieder, and D. M. Rust, (eds.), *IAU Colloq. 167: New Perspectives on Solar Prominences*, volume 150 of *Astronomical Society of the Pacific Conference Series*, pp. 3–10, 1998.

Matteoli, A. *Macchie di Sole e Pittura, Carteggio L. Cigoli-G. Galilei (1609-1613).* Revista di Storia N. 32, Academia degli Euteleti, San Miniato, 1959.

Meadows, A. J. *Early Solar Physics*, Pergamon Press, 1970.

Mitchell, W. M. The history of the discovery of the solar spots. *Popular Astronomy*, 24, 1916.

Mouradian, Z. The new solar activity synoptic maps of observatoire de Paris-Meudon. In K. S. Balasubramaniam, J. Harvey, and D. Rabin, (ed.), *Synoptic Solar Physics*, volume 140 of *Astronomical Society of the Pacific Conference Series*, pp. 197–204, 1998.

Mueller, P. R. An Unblemished success: Galileo's sunspot arguement in the dialogue. *Journal for the History of Astronomy*, 31:279–299, 2000.

Muscheler, R., Joos, F., Beer, J., Müller, S. A., Vonmoos, M., and Snowball, I. Solar activity during the last 1000 yr inferred from radionuclide records. *Quaternary Science Reviews*, 26:82–97, 2007.

Nasmyth, J. On the structure of the luminous envelope of the Sun, *Memoirs of the Literary and Philosophical Society of Manchester*, 3:407–411, 1862.

Nasmyth, J. Manchester Literary and Philosophical Society – The Willow Leaves. *Astronomical register*, 3:81–82, 1865.

Pastorff, C. J. G. Beobachtung der am 24. May und 21, Juny 1828 zur bedeutendsten Größe angewachsenen sehr großen Sonnenflecke. *Astronomische Nachrichten*, 6:471–472, 1828.

Pettit, E. A Telescope of Galileo. *Publications Astronomical Society Pacific*, 51: 147–150, 1939.

Picolet, G. La correspondence de Jean Picard avec Johann Hevelius. *Revue d'histoire des Sciences*, 31:3–41, 1978.

Reeves, E. *Painting the Heavens: Art and Science in the Age of Galileo.* Princeton University Press, 1997.

Reeves, E. *Galileo's Glassworks.* University of Harvard Press, 2008.

Ribes, E., Ribes, J. C., and Barthalot, R. Evidence for a larger sun with a slower rotation during the seventeenth century. *Nature*, 326:52–55, 1987.

Ribes, J. C. and Nesme-Ribes, E. The solar sunspot cycle in the Maunder minimum AD1645 to AD1715. *Astronomy and Astrophysics*, 276:549–563, 1993.

Sawyer Hogg, H. Out of Old Books (Schwabe's Determination of the Sun-spot Cycle). *Journal Astronomical Society of Canada*, 42:42–46, 1948.

Schroter, J. H. *Beobachtungen uber die Sonnenfackeln und Sonnenflecken : samt beylaeufigen Bemerkungen ueber die scheinbare Flaeche, Rotation und das Licht der Sonne.* Bey Georg Adam Keyser, Erfurt 1789.

Schwabe, M. Sonnenbeobachtungen im Jahre 1843. Von Herrn Hofrath Schwabe in Dessau. *Astronomische Nachrichten*, 21:233–234, 1844.

Seymour, C. Dark Chamber and Lightfilled Room. Vermeer and the Camera Oscura. *The Art Bulletin*, 46:323–331, 1964.

Shea, W. R. and Artigas, M. *Galileo in Rome: The Rise and Fall of a Troublesome Genius.* Oxford University Press, 2004.

Sheehan, W. *The Planet Mars: A History of Observation and Discovery.* University of Arizona Press, 1996.

Shirley, J. W. *Thomas Harriot : A Biography.* Oxford, Clarendon Press, 1983.

Smith, A. M. Galileo's proof for the Earth's motion from the movement of sunspots. *Isis*, 70:543–551, 1985.

Sobel, D. *Galileo's Daughter.* Walker and Company, 1999.

Sobel, D. *Galileo's Daughter.* Penguin Books, 2000.

Soon, W. and Yaskell, S. H. *The Maunder Minimum and the Variable Sun-Earth Connection.* World Scientific, 2003.

Spörer, G. Über die periodizität der Sonnenflecken seit dem jahre 1618. *Nova Acta der Ksl. Leop. Carol. deutschen Akademie der Naturforscher*, 53:283–324, 1889.

Staiger, R. C. *Thomas Harriot: Science Pioneer.* New York, Clarion Books, 1998.

Steadman, P. *Vermeer's Camera: Uncovering the Truth behind the masterpieces.* Oxford University Press, 2001.

Stone, E. J. Mr. Nasmyth's Willow-Leaves. *Monthly Notices Royal Astronomical Society*, 24:124, 1864.

Thiele, T. N. De Macularum Solis antiquioribus quibusdam obstervationibus Hafniae institutis, scripsit Th. N. Thiele, astr. stud. *Astronomische Nachrichten*, 50: 257–261, 1859.

Tóth, L., Mezõ, G., and Gerlei, O. Haynald observatory photosphere observations 1880–1919. *Journal for the History of Astronomy*, 33:278, 2002.

Trouvelot, E. L. *The Trouvelot Astronomical drawings Manual.* C. Scribner Sons, 1882.

Udías, A. *Searching the Heavens and the Earth: The History of Jesuit Observatories.* Kluwer Academic Publishers, 2003.

Usoskin, I. G., Kovaltsov, G. A., and Mursula, K. An upper limit of sunspot activity during the Maunder minimum. *Solar and Stellar Activity Cycles, 26th meeting of the IAU, Joint Discussion 8, 17-18 August 2006, Prague, Czech Republic*, 8, 2006.

Usoskin, I. G., Solanki, S. K., and Kovaltsov, G. A. Grand minima and maxima of solar activity: new observational constraints. *Astronomy and Astrophysics*, 471:301–309, 2007.

Vaquero, J. M. On the solar activity during the year 1784. *Solar Physics*, 219: 379–384, 2004.

Vaquero, J. M. Historical sunspot observations: A review. *Advances in Space Research*, 40:929–941, 2007.

Vaquero, J. M., Gallego, M. C., and Sánchez-Bajo, F. Reconstruction of a Monthly Homogeneous Sunspot Area Series Since 1832. *Solar Physics*, 221:179–189, 2004.

Vaquero, J. M., Sánchez-Bajo, F., and Gallego, M. C. On the reliability of the de la rue sunspot area measurements. *Solar Physics*, 209:311–319, 2002.

Vaquero, J. M., Trigo, R. M., Gallego, M. C., and Moreno-Corral, M. A. Two Early Sunspots Observers: Teodoro de Almeida and José Antonio Alzate. *Solar Physics*, 240:165–175, 2007.

Von Braunmühl, A. *Christoph Scheiner als Mathematiker, Physiker und Astronom.* Bayerische Bibliothek, Band 24, Bamberg, 1891.

Wallace, W. *Galileo and his Sources: The Heritage of the Collegio Romano in Galileo's Science.* Princeton University Press, 1984.

Watson, F. *Stargazer: The Life and Times of the Telescope.* Da Capo Press, Perseus Books Group, 2005.

Wilson, A. Observations on the solar spots, *Philosophical Transactions* (1683–1775), 64:1–30, 1774.

Wilson, A. An Answer to the Objections Stated by M. De la Lande, in the Memoirs of the French Academy for the Year 1776, against the Solar Spots Being Excavations in the Luminous Matter of the Sun, Together with a Short Examination of the Views Entertained by Him upon That Subject, *Philosophical Transactions of the Royal Society of London*, 73:144–168, 1783.

Wöhl, H. On the solar rotation elements i and Omega as determined by Doppler velocity measurements of the solar plasma. *Astronomy and Astrophysics*, 62: 165–177, 1978.

Yallop, B. D., Hohenkerk, C., Murdin, L., and Clark, D. H. Solar Rotation from 17th-century Records. *Quarterly Journal Royal Astronomical Society*, 23: 213–219, 1982.

Young, C. *The Sun*, D. Appleton, New York, 1894.

4

Solar Eclipses

4.1 The Basics of Solar Eclipses

A solar eclipse occurs when the Moon blocks the Sun from view. This is strictly correct, however other interesting conditions must be analysed for a better understanding of the phenomenon. The distances and diameters involved in the problem are listed in Table 4.1. Solar eclipses can only happen at a new moon phase when the Sun and the Moon are apparently very close in the sky observed from a place on the Earth's surface. A typical popular graphical explanation of the solar eclipse phenomenon appears in Figure 4.1. However, eclipses do not occur every month (every new moon) because the lunar orbit is tilted. Thus, the apparent diameters of the Sun and the Moon and the position of the Moon in its orbit play fundamental roles in the production of solar eclipses.

The orbit of the Moon is inclined with respect to the plane of the ecliptic by $5° 8' 43''$. Therefore, there are two points called *nodes* where the orbit of the Moon crosses this plane. The Moon passes the *ascending node* when it crosses the ecliptic travelling from the southern to the northern hemisphere. And it passes the *descending node* when it crosses from the northern hemisphere. Figure 4.2 shows a three-dimensional view of the lunar orbit showing the Moon situated just at the ascending node.

The axis between the ascending and descending nodes is called the "line of the nodes". This line moves in a retrograde fashion during the lunar revolution and turns once in a period of 18 years and 11 days approximately. This period is known as the saros.

A solar eclipse is possible when the Moon's penumbral, umbral, or antumbral[1] shadows sweep across the Earth's surface. There are four types of solar eclipse:

[1] Antumbral shadow is defined by the projection of the cone of the umbra.

J.M. Vaquero, M. Vázquez, *The Sun Recorded Through History*,
Astrophysics and Space Science Library 361, DOI 10.1007/978-0-387-92790-9_4,

Table 4.1. Diameters and distances for the Sun, Moon, and Earth (the unit is kilometres)

	Sun	Moon	Earth
Diameter	1 392 000	3 476	12 756
Mean distance from Earth	149 598 000	384 400	–

Fig. 4.1. Graphical explanation of solar eclipses, from Guillemin (1883).

1. *Annular*: The Moon's antumbral shadow traverses the Earth. The Moon is too far from Earth to completely cover the Sun.
2. *Total*: The Moon's umbral shadow traverses the Earth. The Moon is close enough to Earth to completely cover the Sun.

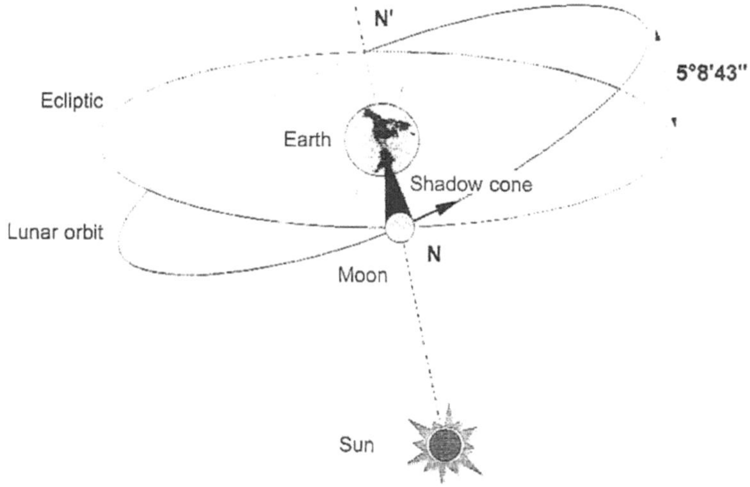

Fig. 4.2. A three-dimensional view of a solar eclipse (Guillermier and Koutchmy, 1999).

3. *Hybrid*: The Moon's umbral and antumbral shadows traverse the Earth. Thus, the eclipse appears annular or total along different sections of its path.
4. *Partial*: The Moon's penumbral shadow traverses the Earth, and the umbral and antumbral shadows completely miss Earth.

Observations of partial eclipses have relatively low interest with respect to annular and total eclipses. Figures 4.3 and 4.4 correspond to the configurations of Sun, Moon, and Earth that produce a total and an annular solar eclipse, respectively. We must emphasize that the relative sizes and distances of these bodies in these figures are not to scale. In Figure 4.3, the entire surface of the Sun is blocked from view within the umbra of the Moon (dark cone). Moreover, a fraction of the sunlight is blocked in the penumbra (lighter cone),

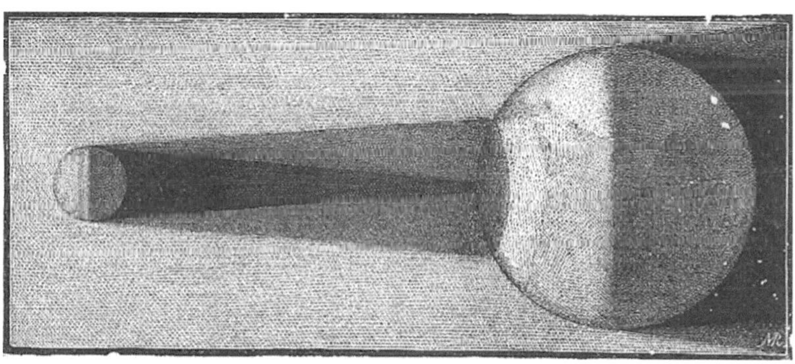

Fig. 4.3. Graphical explanation of total solar eclipses, from Guillemin (1883).

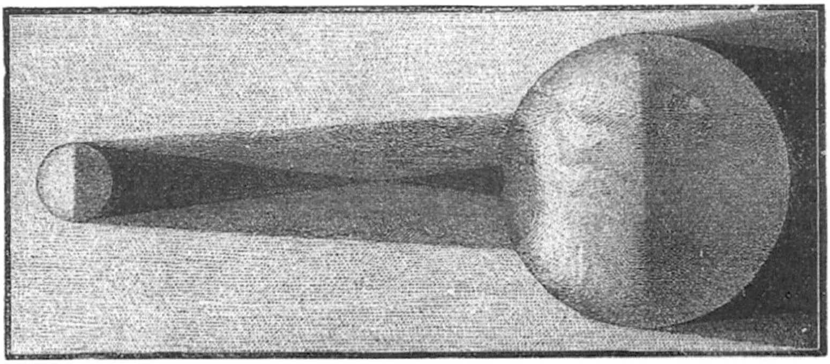

Fig. 4.4. Graphical explanation of annular solar eclipses, from Guillemin (1883).

resulting in a partial eclipse. In Figure 4.4, one can see what happens when the umbra of the Moon ends in space. An annular eclipse occurs and the region in which it is observed is defined by projecting the cone of the umbra onto the surface of Earth.

Table 4.2 lists the maximum, minimum, and mean values of the Moon's distance from Earth (centre to centre) and the length of the Moon's shadow cone (umbra). One can appreciate that the shadow of the Moon is too short to reach the Earth in average conditions. Therefore, total solar eclipses would occur less often than annular solar eclipses. Also, Table 4.2 lists the values of solar and lunar diameters observed from Earth. The angular diameter of the Moon can exceed the angular diameter of the Sun by as much as 6.6% (2.1 arcminutes) producing a total solar eclipse. The angular diameter of the Sun can exceed the angular diameter of the Moon by as much as 10.7% (3.1 arcminutes) producing an annular solar eclipse. Thus, the same conclusion can be reached: total solar eclipses would occur less often than annular solar eclipses.

The eclipse magnitude is defined as the ratio between the apparent angular diameters of the Moon and the Sun during the eclipse. Therefore, the magnitude of a total solar eclipse will always be greater than unity, and the magnitude of partial and annular eclipses will be less than unity.

The illumination of the Moon by the Sun from different angles due to its motion through the sky brings about the familiar phenomenon of phases.

Table 4.2. Maximum, minimum, and mean values of important parameters during solar eclipses (unit for distances is kilometres)

	Maximum	Minimum	Mean
Moon's distance from Earth	406 700	356 400	384 400
Length of Moon's shadow cone	379 870	367 230	373 540
Angular diameter of the Sun	$32' 31.9''$	$31' 27.7''$	$31' 59.3''$
Angular diameter of the Moon	$33' 31.8''$	$29' 23.0''$	$31' 05.3''$

The cycle of lunation, the time between two successive new moon phases, lasts approximately 29 days, 12 hours, 44 minutes, and 3 seconds (29.5305884 days). It is termed the *synodic period.* However, two consecutive passages of the Moon through lunar perigee last 27 days, 13 hours, 18 minutes, and 33 seconds (27.5545501 days). This interval is called the *anomalistic period.* The difference between the two periods is due to the fact that the Earth is orbiting the Sun. Moreover, the *draconic* or *nodical* period is the average interval between two successive transits of the Moon through its ascending node and lasts 27 days, 5 hours, 5 minutes, and 36 seconds (27.212220 days).

A common multiple of these three periods will indicate a time interval for the geometrical configuration of the Sun, Moon, and Earth to be nearly repeated: the Moon has the same phase, is located at the same position with respect to the nodes, and is at the same distance from the Earth. A quasi-common multiple is known as the saros cycle. It lasts 65851/3 days and is related to the aforementioned periods in the following form:

- 223 synodic months: $223 \times 29.5306 = 6585.32$ days.
- 239 anomalistic months: $239 \times 27.55455 = 6585.54$ days.
- 242 draconic months: $242 \times 27.2122 = 6585.35$ days.

Thus, if one observes an eclipse, another eclipse with similar characteristics will be observed 65851/3 days after. However, the saros cycle (18.031 years) is not equal to the precessional period of the lunar orbit (18.60 years), and therefore the Moon will be in a different position with respect to the fixed stars. Moreover, the saros period is not an integer number of days, but contains a fraction of approximately 1/3 of a day. Thus, two successive eclipses in a saros cycle will occur separated by an integer number of days plus 8 hours. In the case of an eclipse of the Sun, the region of visibility will shift westward by 120° and it will not be visible from the same place on Earth. Of course, if one waits three saros cycles them the local time of day of an eclipse will be nearly the same. This period of three saros cycles (almost 19 756 full days) is known as a triple saros or exeligmos.

Here, our intention has been to give an overview of the basic ideas about eclipses of the Sun. More detailed information about this subject can be found in numerous books, such as Zirker (1995), Littmann, Willcox and Espenak (1999), Guillermier and Koutchmy (1999), Brunier et al. (2000), and Steel (2001).

4.1.1 Total Solar Eclipse Step by Step

In this section we shall describe briefly the different phenomena or stages that can be observed during a total solar eclipse. This will help us to understand the descriptions of the solar eclipses that appear in historical documents.

The solar eclipse begins with the *first contact* when a tiny notch appears in the limb of the Sun. This notch increases gradually but there is no noticeable decrease in sunlight until half the solar disk is covered. The Moon progresses and the Sun is reduced to *crescent images.* These images can be seen on the ground if one is standing near a tree because the spaces between the tree-leaves act as pinhole cameras.

Baily's beads can be seen before totality. This phenomenon is produced because the lunar limb is uneven and the sunlight shines through the valleys and is blocked by mountains. The *diamond ring* is produced for a few seconds just before *second contact* when the last part of the Sun shines out against the lunar limb.

Another two phenomena can be observed just before totality. The *shadow bands* are wavy lines seen across the surface of the Earth that appear just before and just after the total phase of solar eclipses. These "bands" shift quickly and are not always seen. In fact, they are a purely atmospheric phenomenon, and are best displayed against bright surfaces such as whitewashed walls. Codona (1986) proposed a theory to explain the phenomenon suggesting that shadow bands are a scintillation effect. Recent observations (Gladysz et al., 2005) have shown excellent agreement with this theory. Also the shadow of the Moon can be seen sweeping across the landscape as the totality phase approaches. The shadow travels at over 1600 km h^{-1}.

When *second contact* occurs, the totality phase begins and the Sun is fully covered by the Moon. The degree of darkness varies from one eclipse to other because of atmospheric conditions. The *chromosphere* – the red "colour-sphere" – is striking as the totality phase begins. Also, it is possible to see *prominences* such as "red-flames" above the Sun's limb. However, the *corona* – the wonderful "pearly mist" – is the most outstanding of all phenomena of the totality phase. The shape of the corona varies depending on the state of solar activity at that moment.

Totality is over when *third contact* occurs. Then, one can again see the *diamond ring, Baily's beads*, and sunlight returns to the landscape. A collection of *decrescent images* can be observed at this stage of the eclipse. When the last fraction of the lunar disk leaves the Sun, *fourth contact* occurs. The eclipse has finished.

During the observation of an annular eclipse, some of these phenomena cannot be seen (i.e., *Baily's beads*, the *diamond ring*, the *chromosphere, prominences*, the *corona*, etc.). However, one can observe a great *solar ring* during the annularity phase. Figure 4.5 shows the different phases of an annular solar eclipse. There are several books devoted to eclipse observation. Some recent examples are Harrington (1997), Maunder and Moore (1998), and Mobberley (2007).

Fig. 4.5. This composition captures various phases of the 3 October 2005 annular eclipse observed at Madrid, Spain, by J. M. Vaquero.

4.1.2 Some Mathematics

The computation of solar eclipses is a difficult task reserved for specialists. However, in this section we shall present some simple mathematical treatments that provide us with a better comprehension of the eclipse phenomena. The Sun and the Moon are represented in Figure 4.6, with S and L indicating the centres of the Sun and the Moon, respectively. Comparing triangles ABA_1 and BCL, we can write

$$\frac{AA_1}{BL} = \frac{A_1B}{CL}$$

In the same way, comparing the triangles $SA''L$ and $C'D'L$, we can write

$$\frac{SA''}{C'D'} = \frac{A''L}{D'L}$$

where AA_1 and SA'' are the difference and the sum of the radii of the Sun and Moon, respectively. Expressed in terrestrial radii, we have:

$$AA_1 = 109.1 - 0.2726 = 108.8274$$
$$SA'' = 109.1 + 0.2726 = 109.3726$$

Solar eclipses only occur at new moon, therefore $A_1B(= SL)$ is the difference between the Sun–Earth and Moon–Earth distances. These distances, expressed in terrestrial radii (R_T), can be replaced by $1/\sin\pi_{Sun}$ and $1/\sin\pi_{Moon}$, with π_{Sun} and π_{Moon} being the parallaxes of the Sun and the Moon. Making x = LC and y = C′D, the proportions can be written as

$$\frac{108.83}{0.2726} = \frac{1/\sin\pi_{Sun} - 1/\sin\pi_{Moon}}{x}$$

and

$$\frac{109.37}{y - 0.2726} = \frac{1/\sin\pi_{Sun} - 1/\sin\pi_{Moon}}{LC}$$

Solving for x and y, we obtain

$$x = 0.002505(\frac{1}{\sin\pi_{Sun}} - \frac{1}{\sin\pi_{Moon}})$$

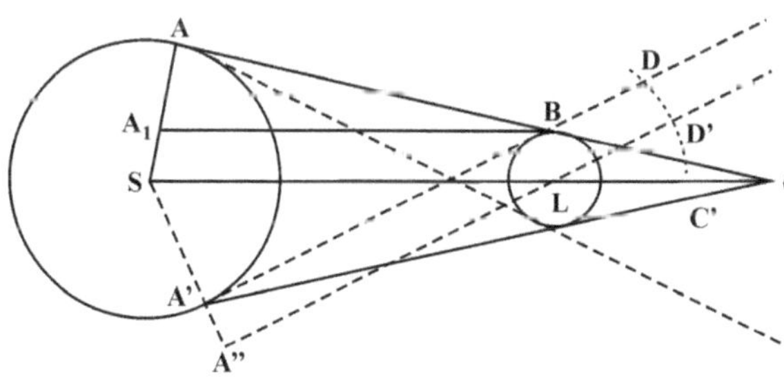

Fig. 4.6. Sketch of the basics of a solar eclipse showing the Sun and the Moon (from Aller, 1957).

and

$$y = 0.2726 + \frac{109.37 \cdot LC'}{\frac{1}{\sin \pi_{Sun}} - \frac{1}{\sin \pi_{Moon}}}$$

The length x of the cone of the Moon's shadow, calculated with the π_{Sun} and π_{Moon} values corresponding to the date of each eclipse, will indicate whether the eclipse is total or annular. If x is greater (smaller) than the distance from the centre of the Moon to the points of the surface of the Earth located in the vicinity of the axis of the cone the eclipse will be total (annular).

The value of y serves as a reference to obtain the radius of the penumbra. If we take the values of the Sun–Earth and Earth–Moon distances as $23440R_T$ and $60R_T$, respectively, we will have $x = 58.56R_T$ and $y = 0.55R_T$.

The radius z of the section produced in the shadow cone by a plane that goes through the point C′ can be expressed as

$$\frac{z}{0.2726} = \frac{CC'}{x}$$

If we write Δ for the Earth–Moon distance expressed in terrestrial radii and take

$$CC' = x - (\Delta - 1)$$

the value of z will be the section of the cone on the Earth's surface at a point of the equator, supposing the centres of the three bodies to be aligned. Therefore, z can be expressed as

$$z = \frac{0.2726(x - \Delta + 1)}{x}$$

Hence, when Δ is a minimum, z is equal to 132 kilometres. When Δ is a maximum, the numerator is negative, C′ is to the right of C in the Figure 4.6 and the eclipse is annular (Aller, 1957).

4.1.3 Canons and Statistics

T. von Oppolzer (1841–1886) made one of the greatest efforts in computational astronomy of the 19th century in publishing his *Canon der Finsternisse* (Canon of Eclipses) in 1887. This "catalogue" of eclipses includes the computation of 8000 solar eclipses (and 5200 lunar eclipses) for the period 1207 BC–AD 2161 (von Oppolzer, 1887, 1962). Moreover, maps showing the approximate positions of the central lines were also printed although a number of approximations were used in the calculations and maps. The Earth's variable rotation rate and the secular acceleration of the Moon were not taken into account. Other eclipse canons were published for shorter time intervals or for limited geographic regions after Oppolzer's remarkable work. Examples are Ginzel's *Spezieller Kanon* (1899) (period 900 BC–AD 600) and Schroeter (1923) (only Europe, period AD 600–1800).

The appearance of the electronic computer allowed more complex calculations to be made, and more quickly. Different canons were published successively. Meeus, Grosjean and Vanderleen (1966) published the Besselian elements of all solar eclipses occurring between AD 1898 and 2510 with central line tables and maps. Mucke and Meeus (1983) published Besselian elements and maps of all 10 774 solar eclipses during the period 2003 BC–AD 2526. The canon of Meeus, Grosjean and Vanderleen was intended mainly to provide data on future eclipses whereas that of Mucke and Meeus was for historical research. In the late 20th century, two important canons have been published. On the one hand, Stephenson and Houlden (1986) computed an atlas of annular and total eclipses visible from East Asia during the period 1499 BC–AD 1900. And on the other, Espenak (1987) published detailed maps and central path data for all solar eclipses from AD 1986 to 2035.

Recently, Espenak and Meeus (2006) have published the *Five Millennium Canon of Solar Eclipses: 1999 to* +3000 *(2000 BCE to 3000 CE)*. This work presents detailed and accurate maps for 5000 years of solar eclipses covering the historical period of eclipses and one millennium into the future. The computation is based on modern theories of the movements of the Sun and the Moon constructed at the Bureau des Longitudes (Paris) rather than the older Newcomb (1895) and Brown (1905) ephemerides. Moreover, the printed maps were drawn using the most up-to-date determination of the historical values of ΔT.

According to this last canon, 11 898 eclipses of the Sun occurred during the period 1999 BC–AD 3000. This large number of events allowed some interesting statistics on solar eclipse occurrence to be constructed. Here, we shall only show two examples. Readers interested in those statistical analyses over long time intervals may consult the work by Espenak and Meeus (2006).

The statistical distribution of the four basic solar eclipse types during this time interval is given in Table 4.3. Partial and annular eclipses have similar rates of occurrence (35.3 and 33.2, respectively). Total eclipses are rarer, and hybrid eclipses are very scarce. One could also ask how many solar eclipses occur during the same year. The number of solar eclipses per year in the Espenak and Meeus canon gives the results of Table 4.4. In general, two solar eclipses occur in one year. A lucky solar eclipse observer can enjoy five eclipses during the same year only once every 200 years (on average).

Table 4.3. Distribution of basic eclipse types, from Spenak and Meeus (2006)

Eclipse Type	Number	Percent
All eclipses	11898	100.0
Partial	4200	35.3
Annular	3956	33.2
Total	3173	26.7
Hybrid	569	4.8

Table 4.4. Number of solar eclipses per year, from Spenak and Meeus (2006)

Number of Eclipses	Number of years	Percent
2	3625	72.5
3	877	17.5
4	473	9.5
5	25	0.5

4.2 Historical Solar Eclipse Observations

One cannot be surprised by finding descriptions of eclipses of the Sun in historical sources. The observation of a total eclipse of the Sun is a spectacular experience that was seen as a supernatural manifestation or a prodigy. Therefore, the idea that eclipses influenced human activities was part of many cultures until relatively recent times.

Through the centuries, eclipses were incorporated in diverse ways into the myths, beliefs, and customs of cultures. However, the different explanations provided by the folklore of each culture can be grouped into four principal types (Littmann et al., 1999):

1. A celestial being that is generally a monster (i.e. a dragon) tries to destroy the Sun.
2. The Sun fights with the Moon.
3. The Sun and Moon make love. They discreetly hide themselves in darkness.
4. The Sun-god grows angry, sad, sick, or neglectful.

With the scientific revolution of the 16th century, eclipses were studied systematically. But one cannot forget the work carried out by astronomers during earlier centuries, as well as that of scribes, astrologers, court analysts, official historians, and monks who registered a number of observations that are now interesting to us. One has to be prudent, however, when working with these observations. In many cases, the historical references are vague or confused. Table 4.5 lists reliable dates of eclipse observations until AD 1000. R.R. Newton (1970) distinguishes three kinds of spurious eclipses:

1. *The assimilated eclipse.* A chronicler may shift the date of an eclipse by a year or more to relate it to some other event consciously or unconsciously.
2. *The literary eclipse.* In a work of pure fiction there appears an eclipse, and later it is taken as a real eclipse by some over-eager reader.
3. *The magical eclipse.* Solar eclipses and other celestial signs appear during important battles, deaths of great personages, or beginnings of wonderful enterprises.

Also eclipses have been reflected in the arts. Olson and Pasachoff (2002) discuss the influence of solar eclipses on the works of the Italian painter and

Table 4.5. Reliable dates of eclipse observations until AD 1000 (adapted form Littmann et al., 1999)

Date	Place	Date	Place
2165–1948 BC	China	136 April 15 BC	Babylon
1375 May 3	Ugarit	89 September 29	China
1330 June 14	An-Yang	80 September 20	China
1131 September 30	Gibeon	28 June 19	China
763 June 15	Nineveh	2 February 5	China
709 July 17	Chu-Fu	2 November AD 23	China
601 September 20	Ying	59 April 30	Armenia
549 June 12	Chu-Fu	65 December 6	Kuang Ling
442 March 11	China	120 January 18	Lo Yang
431 August 3	Athens	243 June 5	China
424 March 21	Athens	360 August 28	China
392 August 14	Chaldonea	429 December 12	China
382 July 3	China	484 January 14	Athens
364 July 13	Thebes	516 April 18	Nan-Ching
322 September 26	Babylon	522 Jan 10	Nan-Ching
310 August 15	Sicily	590 October 4	Mediterranean
300 July 26	China	840 May 5	Bergamo
198 August 7	Chang-An	912 June 17	Cordoba
188 July 17	China	968 December 12	Constantinople
181 March 4	Chang-An	975 August 10	Kyoto
147 November 10	Chang-An		

architect Taddeo Gaddi (c. 1300–1366), active during the early Renaissance and a pupil of Giotto (c. 1267–1337), who may have been partially blinded by a solar eclipse. Figure 4.7 shows the composition *Astronomers Studying an Eclipse* by Antoine Caron (1521–1599), a major figure in the history of French Renaissance painting (the age of Mannerism) and a painter at the Court of France during the period from 1559 to 1589. A solar eclipse seems to be represented in this oil painting on wood. However, there were no total or annular solar eclipse passing through Paris during Caron's lifetime (Kerr Reaves & Reaves, 1965). Another interesting example is the case of representations of eclipses by Cosmas Damian Asam (1686–1739), painter and architect of the *Gesamtkunstwerk* (southern German illusionistic religious decoration and architecture in the Grand Manner). Figure 4.8 shows a fragment of a painting that was completed in 1735 by Asam entitled *Vision of St. Benedict* (1735). The phenomenon represented in the figure is not only a total solar eclipse, but shows the solar corona and the "diamond ring" effect. Olson and Pasachoff (2007) have hypothesized that Asam himself may have seen at first hand one or all of the total solar eclipses of 12 May 1706, 22 May 1724, and 13 May 1733.

In the following subsections, we shall briefly review the most important historical sources of solar eclipse observations. The most important records

Fig. 4.7. *Astronomers Studying an Eclipse* by Antoine Caron around 1571, oil on panel (original in the J. Paul Getty Museum, Los Angeles).

of eclipse observations that survived were made in Babylon, China, and the Arab world.

4.2.1 Babylon and Greece

Babylonian astronomers observed celestial phenomena in a regular form. In particular, both solar and lunar eclipses of the Sun were frequently recorded (Steele, 2000). Ptolemy (ca. AD 83–161) preserved in his *Almagest* a few Babylonian eclipse reports of the period 721–381 BC. We now have a number of astronomical Babylonian texts preserved as clay tablets with syllabic

Fig. 4.8. Vision of St. Benedict (1735) by Cosmas Damian Asam, oil on canvas, original at Benedictine Abbey, Weltenberg, Germany. Imaged by Jay M. Pasachoff (Williams College, Williamstown, Massachusetts, USA) and Roberta J. M. Olson (New-York Historical Society, New York City, USA) and described in Olson & Pasachoff (2007). It is also reproduced in Golub & Pasachoff (2009).

cuneiform script for the period 700–50 BC. These documents (2000 clay tablets very much broken into fragments in general) are part of the records of the official astronomers who observed from Babylon. Sachs and Hunger (1988) provide most of the dated texts. The Late Babylonian astronomical texts that contain eclipse observations can be classified into three main types: astronomical diaries, "goal-year texts", and texts devoted wholly to eclipses (Sachs, 1955). The texts belonging to the first type contain observations of celestial phenomena recorded on a daily basis. These diaries were used to produce texts to assist in making predictions for a specific year and texts devoted to specific phenomena (such as lists of eclipses). The Babylonians had developed a well-known lunisolar calendar. Thus, it is relatively easy to convert the dates of their calendar. The duration of the phases of an eclipse was usually expressed in a unit of time denominated US ("time degree") equal to 4 minutes. Probably, these measurements were made with the aid of a clepsydra (water clock).

Steele et al. (1997) have made a comparison between the timing made by late Babylonian astronomers during eclipses and modern computed values. They concluded that the Babylonians were able to rate their clocks with a typical error of 9% and read off the time with an accuracy of 2 degrees (8 minutes approximately).

Aaboe (1972) indicated that Babylonian texts contain excellent accounts of the motions of the Sun and the Moon in longitude and latitude. These data are the basis on which to make predictions of lunar eclipses. He encountered an understanding and control of the daily parallax of the Moon to predict solar eclipse visibility for a given place. Therefore Babylonian astronomers had some understanding of the mechanism of eclipses. A study of the predictions of solar eclipse times recorded in late Babylonian astronomical texts was carried out by Steele (1997). The accuracy of the prediction of solar eclipses visible at the latitude of Babylon was typically under two hours, which is poorer than their predictions of lunar eclipses (Steele and Stephenson, 1997). We must consider that solar eclipse prediction is much more complicated that lunar eclipses because the local circumstances must be considered.

Recently, Baikouzis and Magnasco (2008) have studied in detail a probable eclipse described in poetic form in the *Odyssey*. They performed an exhaustive search of possible dates in the period 1250–1115 BCE and speculated that 16 April 1178 BCE may be the date of this specific eclipse.

The possible prediction by Thales of Miletus of a solar eclipse has been the object of great interest (Ginzel, 1899; Hartner, 1969; Panchenko, 1994, Stephenson and Fatoohi, 1997). Herodotus mentioned in his *History* (I, 74) that Thales had predicted a loss of daylight just as indeed happened during a battle between the Lydians and the Medes. The text of Herodotus says:

> After this, seeing that Alyattes would not give up the Scythians to Cyraxes at his demand, there was war between the Lydians and the Medes five years [...] They were still warring with equal success, when it chanced, at an encounter which happened in the sixth year, that during the battle the day was turned to night. Thales of Miletus had foretold this loss of daylight to the Ionians, fixing it within the year in which the change did indeed happen. So when the Lydians and the Medes saw the day turned to night, they ceased from fighting, and both were the more zealous to make peace. Those who reconciled them were Syennensis the Cicilian and Labnetus the Babylonian.[2]

[2] Translation by A.D. Godley, *Herodotus* (London, 1975), i, 351.

The vagueness of Herodotus's account was emphasized by Neugebauer (1957) who discredited this statement about Thales as solar predictor. Two additional vague reports describing the event are available. Pliny does not mention the battle in his *Naturalis Historia*. However, we can read that "*the original discovery* (of the cause of eclipses) *was made in Greece by Thales of Miletus, who in the fourth year of the 48th Olympiad* (585/4 BC) *foretold the eclipse of the Sun that occurred in the reign of Alyattes, in the 170th year after the foundation of Rome* (584/3 BC)."[3] Diogenes Laertius[4] also cite the prediction by Thales in his *Life of Thales*: "*He* (Thales) *seems by some accounts to have been the first to study astronomy, the first to predict eclipses of the Sun and to fix the solstices.*"[5]

Stephenson and Fatoohi (1997) analysed the dates of several possible candidates and the visibility from the approximate location of the battle taking account the long-term changes in the Earth's rotation rate. They established that the only plausible date for the eclipse of Thales is 28 May 585 BC.

The last words of this section are devoted to the two great figures of Greek philosophy – Plato and Aristotle – who constructed cosmological theories but had no interest in eclipses. For example, Gregory (2000) indicated that Aristotle's ignition theory,developed in his *De Caelo*, has disastrous implications for the causes of eclipses or the phases of the Moon. Similarly, Plato allowed only regular circular motion (or combinations) for the heavens in his *Timaeus*. However, if the Sun and the Moon were permanently in the same plane as the Earth, then there would be a total lunar eclipse every full moon and a solar eclipse every new moon. Undoubtedly, the personal interest of Plato and Aristotle lay far more in teleological cosmology than astronomy.

4.2.2 Mediaeval Arabic Records

Eclipse observations recorded in Mediaeval Arabic documents can be studied using two clearly distinguishable kinds of sources. First, Mediaeval Muslim historians occasionally recorded remarkable celestial phenomena such as eclipses, comets, and meteors. And second, we have astronomical texts written by scholars interested in astronomy. Moreover, Muslim astronomers were capable of making solar eclipse predictions. For example, the technique used by Yaḥya ibn Abi Mansur (9th century) to predict eclipses was explained by Kennedy and Faris (1970).

A large number of interesting observations of solar eclipses can be found in mediaeval Arabic historical chronicles, as also in mediaeval European annals (see Newton, 1972), although the descriptions in these chronicles lack precision. Said et al. (1989) made an extensive search of references to eclipses of the

[3] Translation by H. Rackham, *Pliny: Natural History* (London, 1938), i, 203.

[4] A biographer of ancient Greek philosophers who was probably living in the first half of the third century AD.

[5] Translation by R.D. Hicks, *Diogenes Laertius: Lives of Eminent Philosophers* (New York, 1972), i, 25.

Sun recorded in published texts of Arabic chronicles. They found observations of nearly 30 solar eclipses during the period AD 833–1513. Full translations of the eclipse reports are available and the authors provided a chronological and astronomical commentary. As an example, an unusually detailed description of an eclipse on 20 June AD 1061 was recorded by Ibn al-Jawzī:

> *[A.H. 453] On Wednesday, when two nights remained to the completion of Jumādā al-Aula, two hours after sunrise, the Sun was totally eclipsed. There was darkness and the birds fell while flying. The astrologers claimed that one-sixth of the Sun should have remained [uneclipsed] but nothing of it did so. The Sun reappeared after four hours and a fraction [of an hour]. The eclipse was not in the whole of the Sun [i.e. it was not total] in places other than Baghdad and its provinces.*

The task of comparing modern astronomical computations with an extensive literature survey was carried out by Said and Stephenson (1991). They converted all Muslim dates to the Julian calendar using a computer program based on the Freeman-Grenville (1977) tables. Their comparison between observed and computed magnitudes for solar eclipses recorded in mediaeval

Table 4.6. Observed and computed magnitudes for solar eclipses recorded in mediaeval Arabic chronicles (adapted from Said and Stephenson, 1991)

Julian Date	Place	Observed mag.	Computed mag
17 Sep 833	Cordoba	great eclipse; darkness	total
17 Aug 882	Baghdad	—	0.10
27 Jun 903	Cordoba	—	0.93
17 Jun 912	Cordoba	total; stars seen	total
19 Jul 939	Toledo	almost total	total
19 Jul 939	Cordoba	—	0.995
20 Aug 993	Cairo	total; stars seen	0.96
24 Jan 1004	Cordoba	—	0.30
20 Jun 1061	Baghdad	total; darkness	total
26 Nov 1174	Baghdad	—	0.91
11 Apr 1176	Baghdad	—	0.92
11 Apr 1176	Cizre	total; darkness; stars seen	total
11 Apr 1176	Orontes R	darkness; stars seen	total
13 Sep 1178	Baghdad	—	0.82
06 Oct 1241	Cairo	total; darkness; stars seen	total
17 Jul 1376	Cairo	—	0.95
10 Jan 1377	Cairo	more than half eclipsed	0.88
01 Jan 1386	Cairo	large	0.80
01 Jan 1386	Damascus	—	0.84
09 Nov 1398	Cairo	about half eclipsed	0.68
29 Oct 1399	Cairo	—	0.50

Table 4.6. (cont.,)

29 Oct 1399	Damascus	—	0.55
26 Mar 1419	Cairo	—	0.88
12 Feb 1431	Granada	7/8 eclipsed	0.94
17 Jun 1433	Aleppo	—	0.995
17 Jun 1433	Cairo	2/3 eclipsed; darkness; stars	0.88
19 Sep 1438	Cairo	nearly 2/3 eclipsed	0.52
18 May 1463	Cairo	eclipsed greatly; darkness	0.85
06 Mar 1467	Cairo	—	0.12
27 Apr 1473	Cairo	eclipsed generally; darkness	0.93
08 May 1491	Cairo	eclipsed totally	0.40
01 Oct 1502	Cairo	—	0.55
07 Mar 1513	Damascus	darkness	0.98
07 Mar 1513	Cairo	large	0.91

Arabic chronicles is presented in Table 4.6. The calculated dates of eclipses in the Julian calendar, the places of observation, and a brief description provided by the chronicles are listed in the first, second, and third columns, respectively. The computed eclipse magnitudes are given in the fourth column. Although recorded estimates of the magnitudes are of low precision, there is reasonable agreement with the calculations in general. A notable exception is the total eclipse reported at Cairo in AD 1491.

Careful observations of solar eclipses are recorded in Mediaeval Islamic astronomical handbooks (called *zijes*). Measurement of the local time of the beginning and end of an eclipse and an estimate of the eclipse magnitude are included in a typical record. The time of the mid-eclipse is also reported in some cases. Muslim astronomers had two main reasons to make eclipse observations: one was to test the reliability of contemporary eclipse calculations, and the other was to determine the difference between the geographical longitudes of two places (Stephenson and Said, 1991).

The *al-Zij al-Hakimi al-kabir* compiled by Ibn Yunus (950–1009) contains most of the accessible observations. The Muslim astronomers Al-Battani (850–929) and Al-Biruni (973–1048) also recorded a few eclipses in their works. Nearly 30 solar eclipses ranging in date from 833 to 1513 were compiled by Said and Stephenson (1991). Said and Stephenson (1996, 1997) made a careful investigation of the accuracy of the dates of eclipses, and a study of the precision with which the various observations were made. They found the dating accuracy to be high, with errors seldom exceeding a single day. Eclipse magnitudes were only crudely estimated. Muslim astronomers used the unaided eye, and therefore the precision is mediocre (about 1 digit). However, the measurements of altitude and time during eclipses are among the most accurate observations in the whole pre-telescopic period. These records are thus of considerable value in modern astronomical studies.

4.2.3 Chinese Observations

Many eclipse observations made in China during the last 4000 years have survived until today. The earliest observation is recorded in *Shu-Chin* (The Book). According to Chen (1955), this eclipse occurred on 22 October 2137 BC, although Liu (1945) dated it as 23 October 2110 BC. Chu (1933, 1934) lists 916 different eclipses that appear in the literature of China during the period 2137 BC–AD 1785. Many hundreds of solar eclipse reports were recorded in the *Wu-hsing Chih* (Five Phases Treatise) and the *T'ien-wen Chih* (Astrological Treatise) of the dynastic histories (Steele, 1998). The ancient record "Tian-da-yi" (the sky darkened greatly) in the *Bamboo Chronicle* was identified by Liu (2002) with the solar eclipse that occurred on 31 May 976 BC. This identification was suggested through palaeographic, astronomical, and chronological analyses.

Solar eclipses were essential to the regulation of the Chinese calendar. Thus, one can find different systems of eclipse prediction in China. If a predicted eclipse did not occur, the virtue of the Emperor was assumed to have prevented it. Therefore, the Chinese astronomers predicted too many rather than too few eclipses. They used numerical cycles deduced from their past records of eclipse observations (Steele, 1998). However, many changes were introduced into the predictive system during the Sung dynasty (960–1279). In the 17th century, Jesuit missionaries introduced significant improvements in the solar eclipse prediction techniques, obtaining more accurate predictions than those of the official Chinese astronomer. The classical works by Bernard (1973), D'Elia (1960) and Rodrigues (1935) explain the great influence of the Jesuits on Chinese astronomy.

Steele and Stephenson (1998) have shown that the typical accuracy of the clocks used to time eclipses in pre-Jesuit China (AD 400–1600) was about 0.4 hours. They compared the calculated times (using modern ephemerides) with the observed times recorded in Chinese reports. Until the arrival of the Jesuit astronomers, timing errors of around 0.25 hours were typical in China (Stephenson and Fatoohi, 1995; Fatoohi and Stephenson, 1996; Shi, 2000).

4.2.4 From the Scientific Revolution to Photography (1450–1840)

Very few eclipses were recorded in Europe from 1450 to 1600, and even until the late 17th century when the telescope became widely disseminated reliable observations of eclipses are rare. This circumstance adds to the importance of the few records available. Johannes Müller (1436–1476), known by his Latin pseudonym Regiomontanus, and his student and associate Bernard Walther (1430–1504) were two of the pioneers of astronomical observation in Europe. They made many careful astronomical observations, including comets, lunar and solar eclipses, and measurements of the position of the Sun, Moon, and planets. The astronomical records made by Regiomontanus and Walther were published for the first time in 1544 by Johann Schoener in his *Scripta clarissimi M. Ioannis Regiomontani.* The eclipse observations made by them contained in Schoener's work have been analysed by Steele and Stephenson (1998). They give translations of the relevant parts of Regiomontanus and Walther's reports on eclipse observations, and indicate that the two astronomers were able to measure the time of an eclipse with an accuracy of about 7 minutes (comparable to the precision of the Mediaeval Arab astronomers).

Two eclipses that occurred in 1560 and 1567 were recorded by the German Jesuit mathematician and astronomer Christoph Clavius (1538–1612). He was very fortunate to observe two great eclipses in the space of seven years. Stephenson et al. (1997) translated the account by Clavius published in his book *In sphaeram Ioannis de Sacro Bosco Commentarius*:

> I shall cite two remarkable eclipses of the Sun, which happened in my own time and thus not long ago. One of these I observed about midday at Coimbra in Lusitania [Portugal] in the year 1559 [sic], in which the Moon was placed between my sight and the Sun with the result that it covered the whole Sun for a considerable length of time. There was darkness in some manner greater than night; neither could one see where one stepped. Stars appeared in the sky and (marvellous to behold) the birds fell down from the sky to the ground in terror of such horrid darkness. The other I saw at Rome in the year 1567 also about midday in which although the Moon was placed between my sight and the Sun it did not obscure the whole Sun as previously but (a thing which perhaps never before occurred at any other time) a certain narrow circle was left on the Sun, surrounding the whole of the Moon on all sides (Clavius, 1593, p. 508)

These eclipses seem not to have attracted attention among contemporary astronomers. Clavius recorded only the year of each observation. However, very large solar obscurations are rare at any one place to permit the precise dates of both events to be unambiguously calculated by modern astronomical computations. Stephenson et al. (1997) note that Clavius has mistaken the year of the first observation because the only large eclipse for many years around 1559 which was observable in Coimbra occurred on 21 August 1560. The year of the second eclipse is given correctly by Clavius, and the computed date for this event is 9 April 1567.

The dissemination of telescopic instrumentation in the early 17th century was a revolution in astronomy. The number of eclipse reports and the quality of the observations were constantly improving during the 17th century. Moreover, the appearance of scientific journals and societies (such as *Philosophical Transactions* and the *Royal Society of London*) provided places to present and publish the reports of eclipse observations. The English astronomer Edmund Halley played an important role in the study of eclipses. He was very interested in the calculation and in the reports of early observations of eclipses (and their scientific consequences). Thanks to his calculations, he predicted the total solar eclipse visible from the South of England and Wales in 22 April 1715. Also, he organized an exhaustive observation of the phenomenon. After the eclipse, several dozens of reports were available from diverse places of England and Wales. The analysis of these observations provided valuable information to study the size of the Sun (see Chapter 5). Moreover, there were European astronomers dispersed around the world, such as the Jesuit scientists recording solar eclipses from far-off places like China (Shi, 2000) and other countries.

During the 18th century, there began to be carried out the first scientific expeditions to observe astronomical phenomena of interest such as transits of Venus (see Chapter 5) or total solar eclipses. However, these expeditions were difficult to organize and, on occasions, they failed. The scientific expeditions to observe the eclipse of the Sun of 26 October 1753 from Iberia (Portugal and Spain) constitute an interesting example. The astronomers Le Vallois and Pedegache (French and Portuguese, respectively) tried to observe the eclipse from the city of Aveiro (Le Vallois and Pedegache, 1753). A group of Spanish astronomers of the Spanish Army Astronomical Observatory (in Cádiz) was displaced to the city of Trujillo (Truxillo, in ancient Spanish) to observe the eclipse (Lafuente and Sellés, 2000). The theoretical totality path crossed both places. However, these places were to the south of the path, and both expeditions could only observe the phenomenon as a partial eclipse.

During the 19th century, the observation of total solar eclipses was transformed into a habitual task of astronomers and major expeditions were organized (Pang, 2002). We also have records of observations by amateur astronomers, such as the drawings by the naturalist John Linnell (1792–1882) of the solar eclipse of 1816 (Olson and Pasachoff, 1992). The first attempts at photographing an eclipse date to the early 1840s (De Vaucouleurs, 1961) when G.A. Majocchi obtained a daguerreotype of the eclipse of 8 July 1842 (Thomas, 1997). However, it has not survived. There were other attempts but the expeditions to Spain to observe the total eclipse of the Sun of 18 July 1860 were certainly the first concerted attempt of photographic observation involving a number of participants (Hingley, 2001). Pang (1994, 1995) has reviewed the printing technology and the representation of the solar corona during the 1860s and 1870s.

4.3 Science Using Early Reports of Solar Eclipses

All this quantity of solar eclipse observations provides astronomers, solar and Earth scientists, and historians with a wonderful tool to apply in doing science. One of the most important applications of the early solar eclipse observations is the determination of the solar radius. Chapter 5 will be devoted to this interesting issue. In this section, we shall briefly review other applications – chronology, the Earth's rotation rate, and long-term changes of the outer layer of the Sun.

4.3.1 Chronology

One of the most elementary uses of historical eclipse observations is the calibration of calendars. If there exists a report of a certain eclipse observed in a place and on a date expressed in a local calendar of which its exact conversion to our calendar is unknown, then one can calibrate the calendar that the observers were using. Steel (2001) explains, for example, how one can calibrate the calendar of the Roman Republic.

On other occasions, the report of the eclipse includes the date expressed in various calendars. A wonderful example is the case of the dates of various reports by the Mediaeval Muslim astronomer Ibn Yunus. The Muslim eclipse reports are usually expressed in terms of the Islamic lunar calendar, but some of the eclipse observations cited by Ibn Yunus are expressed in Persian, Syrian, and Coptic calendars. Said and Stephenson (1996) indicated that all four of these calendars are used together for certain eclipse reports. Schove (1984) contains a chronology of eclipses from AD 1 to 1000 that can be consulted.

4.3.2 The Earth's Rotation Clock

Historical observations of eclipses can serve to study the Earth's rotational energy dissipation rate. This section will be based on the works of Stephenson and Morrison (1984, 1995) and on the monograph of Stephenson (1997). Some later works have been published (see Morrison and Stephenson, 2004, 2005).

From remote antiquity until the 18th century, the basic unit of time was the apparent solar day, i.e. the interval of time that elapses between two consecutive transits of the Sun across the same meridian. However, the development of precise clocks allowed the introduction of the mean solar day (MSD) that remains a fundamental standard for all practical tasks. The mean time of Greenwich (GMT) was introduced as a world standard in 1884, and later was transformed into Universal Time (UT).

After decades of speculation, Harold Spencer Jones (1890–1960) demonstrated conclusively that MSD is not an ideal unit because astronomical observations reveal fluctuations in the length of the day (LOD) of millisecond order (Spencer Jones, 1939). These fluctuations can be reconstructed for the last 2500 years using astronomical historical observations.

The mechanisms responsible for the variations of the rate of terrestrial rotation can be classified as external and internal. The external most important causes are the lunar and solar tides that produce a secular increment approximately in LOD of 2.3 ms cy^{-1}. The internal mechanisms are more diverse. They include changes in wind patterns (yearly time-scale), electromagnetic coupling between the fluid core of the Earth and the lower mantle (decade to centennial time-scale), global sea level changes associated with climatic variations (centennial and longer time-scale) and others.

To trace back the fluctuations of LOD, it is necessary therefore to have astronomical observations that can give us some information. During the last

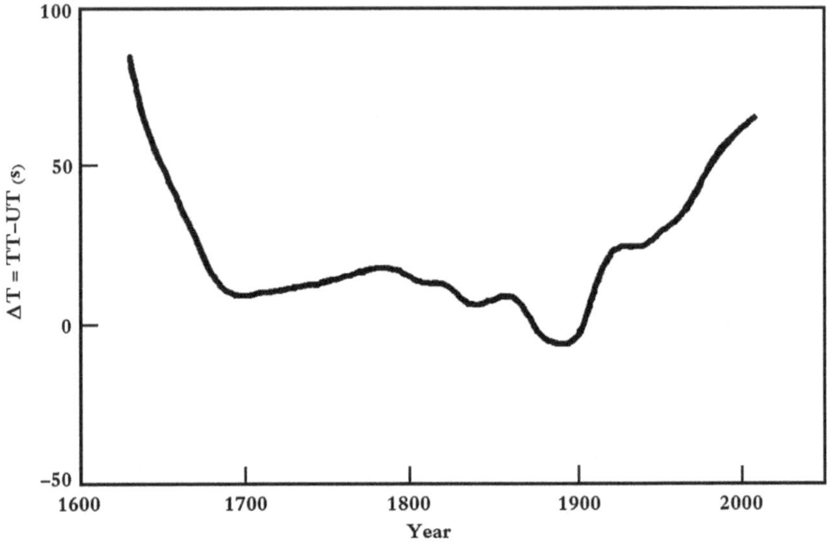

Fig. 4.9. Earth's rotational clock error during the last four centuries (adapted from Stephenson, 2003).

400 years, i.e. the telescopic era, the observations of occultations of stars by the Moon are the best source of information (Figure 4.9). However, to detect long-term trends it is unavoidable to use the astronomical rough naked-eye observations, especially observations of eclipses.

The first suggestions about the inconstancy of the rate of terrestrial rotation were made in the 18th century. However, the first estimates of the dissipation rate due to tides were not made until the 20th century.

Two quantities closely related to the rate of terrestrial rotation are the acceleration of the movement of the Moon and of the Sun. Cowell (1905) suggested that the solar acceleration was purely apparent and came from a gradual increment of the unit of time used (MSD). In 1939, Jones introduced Ephemeris Time (ET), a theoretically invariant time-system. The Sun has no acceleration in ET. Standard LOD in this time-scale was defined as the mean LOD in the system UT in the period AD 1750–1890 (mean epoch 1820). Thus, the difference between ET and UT (denominated ΔT) is a measure of the error of the rotational clock of the Earth.

The analysis of historical observations was crucial for the development of the studies on the variations of the rate of terrestrial rotation. In the late 19th century, S. Newcomb and F. Ginzel made calculations of the lunar acceleration using historical observations of eclipses. In the early 20th century, J.K. Fotheringham also used a wide variety of antique observations, especially European. In the 1970s, Newton (1970, 1972) calculated the orbital acceleration of the Moon and the rotational acceleration of the Earth using historical observations of eclipses from Europe and the Arab world. Later, Stephenson and Morrison (1984, 1995) and Stephenson (1997) have continued exhaustive investigations on the topic.

There are numerous observations that could be of utility to determine variations in the terrestrial rotation rate. However, only the eclipse observations have actually been useful for this task. The observations of star occultations by the Moon are also very important but are only sufficiently numerous in the telescopic period. However, the observations of eclipses are relatively numerous in certain cultures from antiquity: Babylonia, China, Europe, and the Arab world. In these documentary sources, one can find three types of useful reports of eclipses:

1. Timed measurements of limb contacts between the Moon and Sun. Timed observations are very interesting. All these observations were carried out by astronomers. The quantity of this type of record is relatively low, and few astronomers had the possibility of observing a total (or near-total) eclipse of the Sun.
2. Untimed descriptions of total or near-total solar eclipses. The reports of total (or near-total) solar eclipses were made by chroniclers, and were usually very qualitative. In any case, the total phase is so well-defined that even a qualitative description of the complete disappearance of the Sun is useful.
3. Untimed observations that narrate the setting of the Sun or Moon while eclipsed. Observations of this type are of relatively low precision. They can be affected by anomalous refraction and a horizon level is required. Therefore, one only has the first two types for practical analysis.

Timings of the various phases of solar or lunar eclipses enable specific values of the parameter ΔT to be determined. Untimed solar eclipse accounts in which the Sun was either said to disappear completely (or was reduced to a slender crescent) are also of interest because reports of totality provide sharp limits within which the value of ΔT must lie and, when an eclipse fell a little short of totality, a range of ΔT is excluded.

A spectacular case of an eclipse observed during the pre-telescopic era corresponds to the Babylonian records of a solar eclipse in 136 BC. The Babylonian astronomers recorded and timed the phases and noted that the eclipse was total. Stephenson (2003) provided an English translation using two damaged texts conserved in the British Museum collection:

> *(Year) 175, month XII_2. The 29th (day), at 24 time-deg after sunrise, solar eclipse; when it began on the south-west side, in 18 time-deg of daytime it was entirely total; Venus, Mercury, and the Normal Stars were visible; Jupiter and Mars, which were in their period of invisibility, were visible in that eclipse. It threw off (the shadow) from south-west to north-east; 35 time-deg (duration) of onset, maximal phase and clearing (trans. H. Hunger).*

The date is expressed relative to the Seleucid era and corresponds to 15 April 136 BC in the Julian calendar. Stephenson (2003) computed the local time of sunrise at Babylon (2.53 h in UT scale). The three contact instants can be deduced as 4.19, 5.39, and 6.52 h from the relation 1 time-deg equal to 4 minutes. One can compute the TT of these phases, obtaining 7.69, 8.76, and 9.93, respectively. We can obtain three values for ΔT from the differences between the corresponding TT and UT values. The results are 12 600 s (3.50 h), 12 100 s (3.37 h) and 12 250 s (3.41 h).

If we compute this eclipse on the TT time-scale, we only obtain a small partial obscuration of the Sun at Babylon. However, the documentary sources (two fragmentary clay tablets) show clearly a total eclipse. The computed track of totality for the eclipse of 15 April 136 BC is drawn in Figure 4.10 as a shadow band assuming $\Delta T = 0$. The track passes far to the west of Babylon. Values of ΔT in the interval 11 200-12 150 s (3.11–3.38 h) provide a total solar eclipse at Babylon. Therefore, the three individual values obtained from timed phases and the interval derived from the totality agree remarkably. Similarly, we can trace the variation of ΔT during 700–50 BC using numerous further timings of both lunar and solar eclipses from Babylonian documents.

Another interesting example comes from the few European chronicles that give descriptions of very large partial solar eclipses. The Annals of the monastery of Aula Regia, located in Zbraslav near Prague, describe an eclipse on 16 July 1330: *1330. In this same year on the Ides of July at the 8th hour of the day, the Sun was so greatly obscured that of its great body only a small extremity like a three-night-old Moon was seen.*

The estimated time is not of value because it is too imprecise. However, the fact that the eclipse was observed to be only partial establishes limits to the values of ΔT at this date: $\Delta T < 890$ s or $\Delta T > 1210$ s. Intermediate values would produce a total eclipse at this site and they must be excluded according to the report.

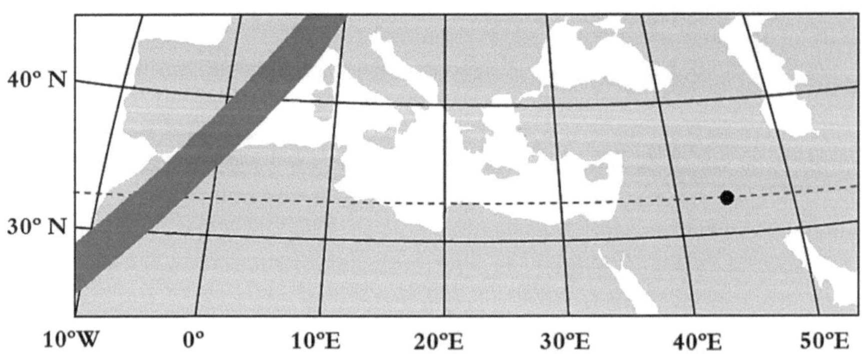

Fig. 4.10. Computed track of totality for the eclipse of 15 April 136 BC (shadow band) assuming $\Delta T = 0$. Totality phase was observed from Babylon. The latitude of this place is marked with a dashed line (adapted from Stephenson, 2003).

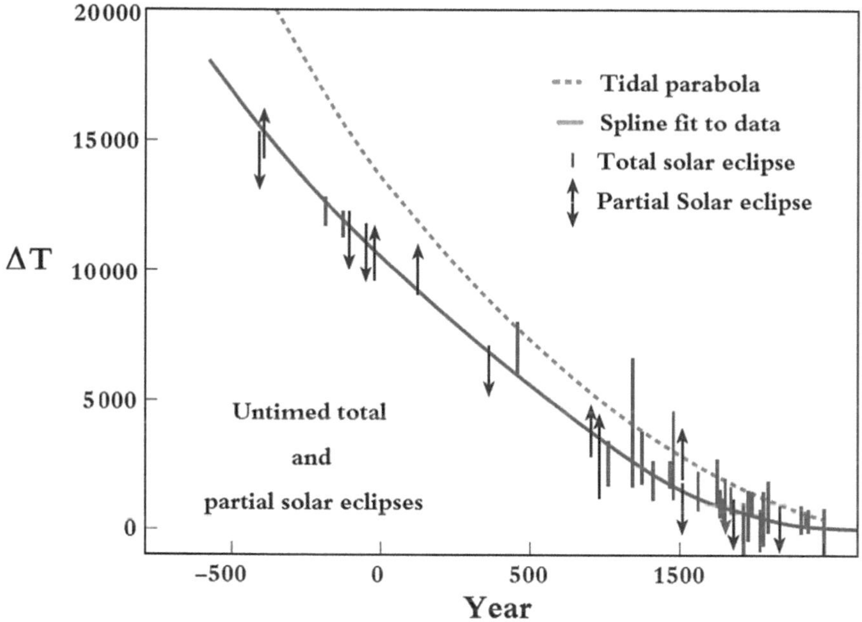

Fig. 4.11. Limits of ΔT during the pre-telescopic period using untimed observations of total (arrows) and near-total solar eclipses (bars). The tidal parabola (dashed line) and the cubic spline fit (continuous line) are also represented (adapted from Stephenson, 2003).

Short-term variation in ΔT can only be traced using the telescopic occultations of stars by the Moon (Figure 4.9). However, the precision of observation is quite insufficient during the pre-telescopic period to trace any short-term variations. Instead, the eclipse data provide important evidence of long-term trends over more than two millennia as one can see in Figure 4.11 where only results from the most reliable observations are displayed. A single vertical bar in the figure denotes the ΔT range defined by a total solar eclipse observation. When in a report the totality is expressly denied, then there are two possible solutions separated by an exclusion zone. In this case, a pair of vertical lines with arrowheads is used. The dashed line in the figure represents the mean tidal parabola of the equation $\Delta T = 42\tau$ where τ is measured in Julian centuries from AD 1820 (an epoch determined by the definition of ET). The solid line represents a cubic spline fit to the data that satisfies all of the constraints imposed by the untimed observations. Table 4.7 presents the cubic spline fits to measurements for ΔT from 500 BC to AD 1950 including the standard error for each value (Morrison and Stephenson, 2004, 2005).

The study of the Earth's long-term rotation has made considerable progress during the last decades, but more historical observations are needed to trace variations in LOD with greater accuracy. The early observations indicate long-term variations in the Earth's spin rate which are of non-tidal origin, but the

Table 4.7. Values of ΔT derived from historical records (from Morrison and Stephenson, 2004)

Year	ΔT (s)	Standard error (s)	Year	ΔT (s)	Standard error (s)
−500	17190	430	900	2200	70
−400	15530	390	1000	1570	55
−300	14080	360	1100	1090	40
−200	12790	330	1200	740	30
−100	11640	290	1300	490	20
0	10580	260	1400	320	20
100	9600	240	1500	200	20
200	8640	210	1600	120	20
300	7680	180	1700	9	5
400	6700	160	1750	13	2
500	5710	140	1800	14	1
600	4740	120	1850	7	< 1
700	3810	100	1900	-3	< 1
800	2960	80	1950	29	< 0.1

contributions of different mechanisms are not determined. The cooperation of historians and further studies would be more than welcome.

4.3.3 The Outer Layers of the Sun

Descriptions of eclipses in historical documents are very simple. Thus, the solar corona or the prominences are generally not described. This fact led Eddy (1976) to suspect that during historical eclipses of the Sun these structures had not been observed because the solar activity was significantly attenuated. Since this work, astrophysicists and historians began to search for those descriptions.

The shape of the solar corona observable during total eclipses varies with the phase of the 11-year cycle of solar activity. During the maximum, the corona extends uniformly around the solar disk. However, the brilliance of the corona is less intense, and only extends in a fringe on both sides of the solar equator during the minimum of solar activity. Some drawings of the solar corona made during different stages of solar cycle are collected in Figure 4.12 where the temporal evolution of the sunspot number is also represented.

The shape of the solar corona observed in a total eclipse during a maximum of solar activity clearly resembles a halo. It is interesting to note that the Christian images usually show Jesus Christ or the saints with halos. The use of this symbol comes from the Greek and Roman pagan cults. With the halo, they represented the God of the Sun (Helios or Apollo). Later, Christian artists continued using it as a symbol of divinity.

The winged Sun is another symbol related to divinity (in the Ancient Near East and other ancient cultures). The elongated coronal streamers observed

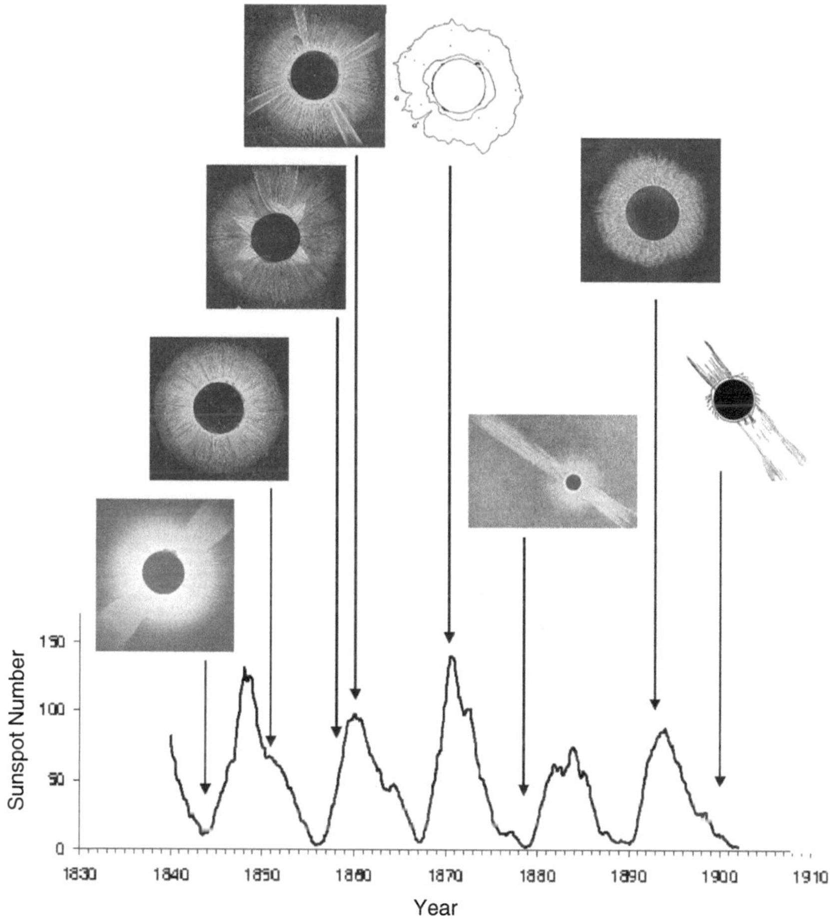

Fig. 4.12. Different solar corona drawings during solar cycles 8–13. The observers and the year of the eclipse are the following (chronological order): Anonymous, 1842; Dawes, 1851; Liais, 1858; Felitzch, 1860; Lindsay, 1870; Langley, 1878; Deslandres, 1893; Johnson, 1900. Sources: Guillemin (1880), Lindsay (1871), Deslandres (1893), and Johnson (1900).

during a total solar eclipse at the minimum of an 11-year cycle resemble this symbol. Figure 4.13 shows an Assyrian representation of the winged Sun. In early Egyptian religion, this symbol represented *Horus*, later identified with *Ra*.

Several authors have noticed the great similarity between this symbol and the solar corona during the minimum of solar activity. Also, one's curiosity increases when one finds that historical Egyptian records of total solar eclipses do not exist, in spite of the fact that they prepared words to name the eclipses and about 28 total solar eclipses crossed the Nile valley from 2837 BC until AD 493. Some authors explain this apparent anomaly in indicating that many

Fig. 4.13. Detail of the stele to Assurnasiripal II at Nimrud (9th century BC) showing the winged Sun.

chronicles have disappeared due to the fragility of the papyri. Other authors indicate that the sudden disappearance of the god Ra during an eclipse was not socially correct, and the memory of eclipses was associated with the symbol of the god Ra. Changes in Egyptian governments may have occurred after the observation of eclipses (Sellers, 1992).

A reference to this symbol can be found in the Bible, in the book of Malachi, where the prophet referred to a winged "Sun of righteousness": *But unto you that fear my name shall the Sun of righteousness arise with healing in his wings* (Malachi 4:2).[6]

Wang and Siscoe (1980) presented three dated and 13 undated cases of coronal observations during eclipses recorded in Chinese literature. As an example of possible solar corona observations, we can cite the text that appears in *Han Shu* (Official History of the Han Dynasty): *In the 3rd year of Duke Huan (713 BC) there was a total solar eclipse. Jin Fan (77–37 BC) said during the eclipse in the 3rd year of Duke Huan the Sun was completely dark in the centre from top to bottom, but was yellowish outside. This meant that the subjects tried to murder the King but would not be successful.* They conclude that the descriptions compiled show coronal structures and prominences in spite of the metaphorical and astrological language used in Chinese literature.

Some possible descriptions of solar corona can be found in some Greek classical texts. Stothers (1979) provides some examples such as "the visible

[6] Or Malachi 3:20, because the book of Malachi is divided into three chapters in the Hebrew Bible and the Greek Septuagint and four chapters in the Latin Vulgate. The fourth chapter in the Vulgate consists of the remainder of the third chapter starting at verse 3:19.

light about the rim of the eclipsed Sun" (Cleomedes, *De motu circulari corporum caelestium* 2.105; Plutarch, *De facie quae in orbe lunae apparet* 932B) and "the comet that once was seen near the Sun when the latter was eclipsed" (Posidonius in Seneca, *Naturales quaestiones* 7.20.4; Ptolemy, *Tetrabiblos* 2.9; Arrianus Meteorologicus in Stobaeus, *Eclogae* 1.28.2). Unfortunately, we have no further details of these observations.

The corona was also described briefly by some early observers: Julius Firmicus Maternus probably described a prominence during the eclipse that occurred on 17 July 334 observed from Sicily. In Corfu, during the eclipse of 22 December 968, the corona was also observed (Maunder and Moore, 1998, p. 55). However, this type of description is very ambiguous.

Foukal and Eddy (2007) draw attention to the earliest historical reports of a red flash during the 1706 and 1715 eclipses. The presence of the red flash during total solar eclipses requires the existence of an extended chromosphere and therefore of a chromospheric magnetic network that gives rise to spicules. The observations of the 1706 and 1715 eclipses imply a substantial, widespread photospheric magnetic field during at least the last decade of the Maunder minimum and are consistent with reports of a persistent photospheric field throughout the Maunder minimum from analyses of the ^{10}Be radioisotope record. Moreover, Judge and Saar (2007) estimated the solar chromospheric, transition region, and coronal emissions that might have existed during the Maunder minimum. They suggested that the radiative output of the Maunder minimum chromosphere, transition region, and corona were similar to (or at least not much less than) those observed under conditions close to current solar minima.

The first scientific description of solar prominences was made by B. Wassenius[7] (1687–1771). He began studying mathematics, physics, and astronomy in Uppsala where he obtained a philosophy degree in 1722. Wassenius made his famous description of solar prominences (and earthshine) while he observed the total solar eclipse of 13 May 1733 from Göteborg (Vassenio, 1733). This is one of the unambiguous historical descriptions of solar prominences, and Wassenius suggested that prominences were clouds in the Moon's atmosphere. Charbonneau (2007) provides a short biography of Wassenius.

The eclipse of 24 June 1778 was observed from different parts of the world (Vaquero, 2003). The path of totality began in the Pacific Ocean, crossed Mexico, the south-eastern USA, the Atlantic Ocean, Morocco, and Algeria, to end at the current frontier between Libya and Chad. It was observed as total from Mexico by the Mexican astronomer Antonio de León y Gama (1735–1802), who published his calculations and observations (León y Gama, 1778). Unfortunately, he had to observe between clouds. Although the eclipse was also observed from Europe, the reports of European scientists have nothing to say about the characteristics of the corona as it was not total. Alexandre Guy Pingré observed from Sainte-Geneviève, Edme-Sébastien

[7] Also called Vassenius.

Jeaurat from the Paris Royal Observatory, and Charles Messier from the Paris Naval Observatory (Pingré, 1781; Jeaurat, 1781; Messier, 1781). Pierre-Charles le Monnier (1715–1799) also observed from Paris (Monnier, 1781), adding in his report information provided by Spanish scientist Antonio de Ulloa (1716–1795) and by Desoteux, a French cavalry officer who observed the total eclipse from Salé (Morocco) in Africa. The eclipse was also observed by W. Ludlam (1779) from Leicester and W. Wales (1779) from London.

One of the most important observations was that made by Ulloa from the Atlantic Ocean on board the ship '*El España*', en route from the Americas to Spain. Ulloa's ship had the good fortune to be situated within the path of totality, so that the crew observed the eclipse (and the solar corona) from the sea. Antonio de Ulloa was a well-regarded scientist who had taken part in the Paris Academy of Sciences' expedition to the equator to measure a terrestrial meridional arc (López Piñero et al., 1983). Ulloa published a pamphlet in Spanish in which he described the circumstances of the observation and the observed characteristics of the eclipse (Ulloa, 1779a). Ulloa also wrote a brief summary of his pamphlet for the *Royal Society of London* which was published in English and French (Ulloa, 1779b). With respect to the Sun's corona, one can read:

> "Five or six seconds after Immersion occurred there began to show itself around the Moon a very brilliant circle of light that could be looked at without hurting one's sight [...]. This light increased [...] until the two centres coincided, or were very close, being then perceived in all its strength and beauty: its breadth then was two digits or one sixth the diameter of the Moon. All around its circumference this luminous ring shot out rays of light, perceptible to the distance of one diameter of the Moon, some longer than others"[8] (Ulloa, 1779a, pp. 4–5).

We have information on the Sun's corona during the eclipse of 24 June 1778 thanks to Ulloa's description and to Ulloa and Desoteux's drawings (Figures 4.14 and 4.15) which were published in the *Philosophical Transactions* together with Monnier's article (Monnier, 1781). This information is of a corona formed by streamers at all solar latitudes, both at the equator and at the poles. The angular length of the streamers was comparable to a lunar diameter, i.e. approximately 30 arcminutes. These descriptions of the corona would be indicative of a high level of solar activity.

[8] The original text, in old Spanish, is: "Cosa de 5. ú 6. segundos despues que la Inmersion sucedió, empezó á descubrirse al rededor de la Luna un círculo de luz mui brillante, que sin ofender la vista, se dexaba ver [...]. Esta luz aumentó [...] hasta que los dos centros coincidieron, ó estubieron en la mayor inmediacion, que se percibió en toda su fuerza y hermosura: entonces era su grosor de dos dígitos, ó la sexta parte del diámetro de la Luna. Este anulo luminoso despedia rayos de luz por toda su circunferencia, perceptibles hasta la distancia de un diámetro de la Luna, los unos algo mas largos que los otros." (Ulloa, 1779a, p. 45).

Fig. 4.14. Solar corona observed by Antonio de Ulloa during the 1778 total solar eclipse.

The most complete current information that we have on solar activity at the time of this eclipse is given by the number of sunspots. The 24 June 1778

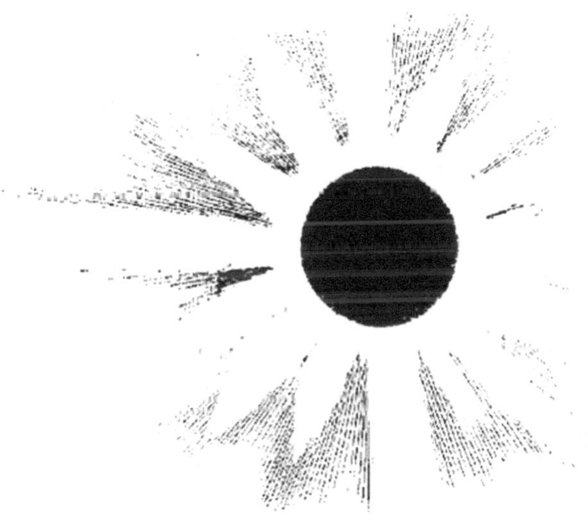

Fig. 4.15. Solar corona observed by Desoteux during the 1778 total solar eclipse.

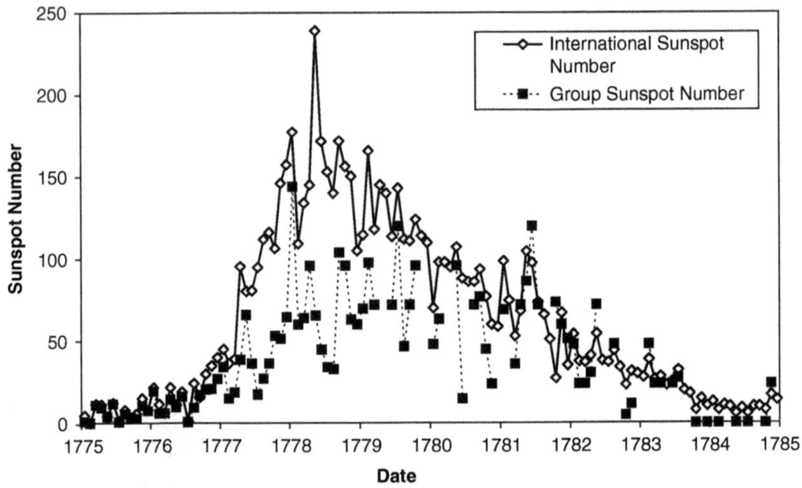

Fig. 4.16. Sunspot numbers for the period 1775–1785.

eclipse occurred during a maximum of the Sun's 11-year cycle, in particular during that known as cycle number 3. There do exist, however, certain discrepancies between the Wolf Sunspot Number, R_Z, and the Group Sunspot Number, R_G (see Figure 4.16). Particularly noticeable is the difference between them for the month of June, with the value of R_Z being very high, almost reaching a major peak, whereas the value of R_G is lower and is close to a shallow minimum of the Sun's activity within the solar maximum. Despite these differences, the overall behaviour of the indices clearly shows the presence of a solar maximum in 1778. We also have information available on the Sun's activity from the series of occurrences of aurorae, especially of aurorae observed at low latitudes. By way of example, three aurorae were observed from the Iberian Peninsula in 1778: 25 February, 28 June, and 11 September (Vaquero, 2003).

The different observations of the Sun's corona during the eclipse of 24 June 1778 showed characteristics that are typical of a period of high solar activity, with streamers at all solar latitudes. This is in agreement with the classical data on sunspots which indicate that the eclipse occurred during the maximum of cycle number 3. The scarcity of drawings of this event, and the lack of detail in those that we do have, make it impossible to detect the presence of other details of interest such as those contributed by Eddy (1973, 1974) for other total solar eclipses.

The corona was established to be the extended and tenuous atmosphere of the Sun from observations made during the eclipse of 16 June 1806. The Spanish astronomer José Joaquín de Ferrer y Cafranga (1763–1830) observed the eclipse from Kinderhook (near Albany, New York). He published a paper describing his timings of the eclipse phases and the details of the observation (Ferrer, 1809). Figure 4.17 shows the plate included in Ferrer's work.

Fig. 4.17. Solar corona observed by José Joaquín Ferrer in 1806.

He described the solar corona saying: "*The disk had round it a ring or illuminated atmosphere, which was of a pearl colour, and projected 6' from the limb, the diameter of the ring was estimated at 45'. The darkness was not so great than that of the Full Moon. From the extremity of the ring, many luminous rays were projected to more than 3 degrees distance. The lunar disk was ill defined, very dark, forming a contrast with the luminous corona*" (Ferrer, 1809, pp. 266–267). This is probably the first scientific work where the Spanish word "corona" was used to describe the solar corona. Moreover, he made estimates of the extent of the corona, reaching the conclusion that it is the solar atmosphere. He noted that if it were a lunar atmosphere then it must be about 50 times larger than that of the Earth.

The final proof of the origin of the corona did not come until the first eclipse photographs obtained in the 1840s and 1850s. The first photograph of the corona was made by Berkowski during the eclipse of 28 July 1851 from Poland.

4.3.4 CMEs and Comets

A great emission of coronal mass was observed on 21 August 1973 with the coronograph installed in the Skylab space laboratory. Thus began the systematic investigation of Coronal Mass Ejections (CMEs) (MacQueen et al., 1974). J. Eddy immediately realized that this type of solar phenomena must have been observed previously during the totality phase of solar eclipses. Eddy's search was successful, because many drawings of the solar corona of the total eclipse of the Sun of 18 July 1860 showed a spectacular characteristic in the southwest quadrant (Eddy, 1974).

Eddy was able to reconstruct the evolution of the CME based on the drawings and the different observers' comments located at different places of the curve of totality. The shadow crossed Canada, the Atlantic Ocean, Spain, Northern Africa, and Nubia.

Eddy found two descriptions of the solar corona by Canadian observers of the eclipse. R.N. Ashe probably described an early phase of the CME although he observed under poor meteorological conditions. Spain was the place of observation chosen by several astronomers. The majority of the descriptions of the solar corona showing the CME were made from Spain. Eddy found the last description of the event from observers in Algeria. Figure 4.18 shows a collection of drawings of the solar corona during this eclipse.

We can extend even more the history of the CME with the observations of the total solar eclipse carried out from Nubia, the last station where the eclipse was reported. Figure 4.19 shows a drawing of the solar corona during the phase of totality from Dongolah (Nubia) made by an Egyptian astronomer called Mahmoud-Bey. We can observe in Figure 4.19 the last phase of the CME that was completely developed in the records made in Spain (see Figure 4.18). A brief report with the observations from Nubia was published in Paris (Mahmoud-Bey, 1861).

This example of a CME observed during the totality phase of the 1860 eclipse shows us that the observations of the solar corona during eclipses have great interest. Eddy (1973) also studied a linear coronal ray, centred on a large helmet, observed during the 20 September 1922 eclipse from Australia. This coronal ray extends from near the limb out to four solar radii and was related to a coronal neutral sheet which separates areas of opposite magnetic polarity in the two surface regions.

There is no doubt that total solar eclipses are an opportunity to observe astronomical objects that are very close to the Sun. Comets very close to

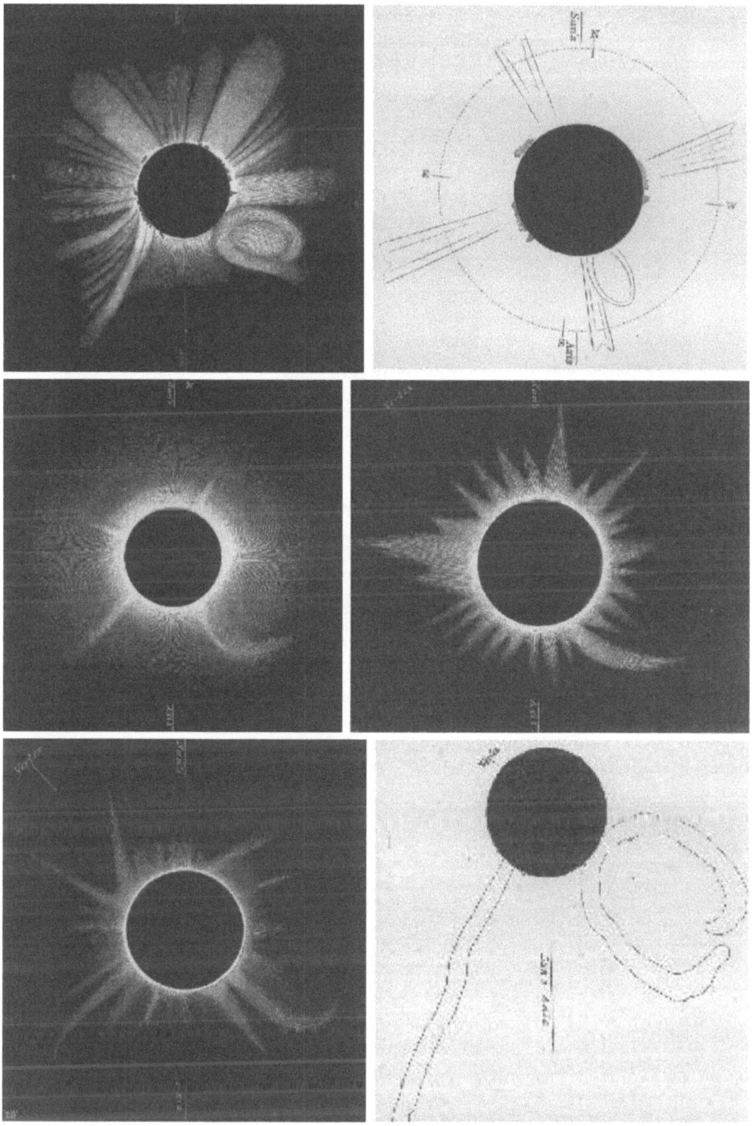

Fig. 4.18. Collection of solar corona drawings of the 18 July 1860 total solar eclipse. The observers and places of observation are (from left to right and from top to bottom): G. Tempel (Torreblanca, Alicante), von Feilitzsh (Castellón de la Plana), F.A. Oom (Alto de Urbaneja), E.W. Murray (Llodio, Logroño), F. Galton (Colina de la Guardia), and von Wallenberg (Valencia). All sites are in Spain.

Fig. 4.19. Drawing of the solar corona during the 18 July 1860 eclipse from the Nubia station.

Fig. 4.20. Total solar eclipse on 17 May 1882 painted by M. Ranyard from Sohag (Egypt). The 1882 eclipse comet is also depicted. Source: *Journal L'Astronomie* (volume 12, 1893, p. 123).

the Sun are interesting objects for observation during eclipses. Indeed, there have been a number of eclipse comets found or observed during the last two centuries (Solc, 1999). For example, the "Eclipse comet Tewfik" X/1882 K1 was observed during the eclipse on 17 May 1882 (Figure 4.20). With some exceptions, eclipse comets belong to the Kreuz family of sungrazing comets.

On some occasions, it is difficult to differentiate between CMEs and comets very near to the sun during the totality phase of solar eclipses. Cliver (1989) made a re-evaluation of the observations made on 16 April 1893 of a "comet" located very close to the Sun. J.M. Schaeberle of the Lick Observatory observed the eclipse from Mina Bronces, Chile. He obtained seven large-scale plates of the corona during the totality phase. A coronal feature recorded on one plate resembled a comet (Schaeberle, 1894). Cliver (1989) suggested that this feature is a disconnected mass ejection.

The study of sungrazing comets has advanced rapidly with the aid of the SOHO spacecraft. In the last 10 years more than 1000 sungrazing comets have been discovered and SOHO become the most successful comet discoverer in history. Bemporad et al. (2007) reviewed the SOHO/UVCS observations of sungrazing comets.

References

Aaboe, A. Remarks on the theoretical treatment of eclipses in antiquity. *Journal for the History of Astronomy*, 3:105–118, 1972.

Aller, R. M. (ed.). *Introducción a la Astronomía*. Consejo Superior de Investigaciones Científicas, Madrid, 2nd edition, 1957.

Aristotle. *On the Heavens*. London, 1939. Edited and English translation by W. K. C. Guthrie.

Baikouzis, C. and Magnasco, M. Is an eclipse described in the Odyssey? *PNAS*, 105:8823–8828, 2008.

Bemporad, A., Poletto, G., Raymond, J., and Giordano, S. A review of SOHO/UVCS observations of sungrazing comets. *Planetary and Space Science*, 55:1021–1030, 2007.

Bernard, H. *Matteo Ricci's Scientific Contribution to China*. Westport, Conn., 1973.

Brown, E. W. Theory of the motion of the Moon. *Memoirs, Royal Astronomical Society*, LVII(II):136–141, 1905.

Brunier, S., Luminet, J.-P., and Dunlop, S. *Glorious Eclipses : Their Past, Present and Future*. Cambridge, Cambridge University Press, 2000.

Charbonneau, P. Wassenius [vassenius], birger. In T. Hockey (ed.), *The Biographical Encyclopedia of Astronomers*, pp. 1196–1197. Springer, 2007.

Chen, T.-W. *A Brief History on the Ancient Astronomy of China*. Shanghai, China, Shanghai Publishers, 1955, in Chinese.

Chu, W.-S. *Li Tai Jih Shih K'ao (Dating Solar Eclipses of All Dynasties)*. Shanghai, China, Shang-Wu press, 1934, in Chinese.

Chu, W.-S. *Tien Wen Kao Ku Lu (An Archeology of Astronomy)*. Shanghai, China, Shang-Wu press, 1933, in Chinese.

Clavius, C. *In sphaeram loannis de Sacro Bosco commentarius*. Lyon, 1593.

Cliver, E. W. Was the eclipse comet of 1893 a disconnected coronal mass ejection? *Solar Physics*, 122:319–333, 1989.

Codona, J. L. The scintillation theory of eclipse shadow bands. *Astronomy and Astrophysics*, 164:415–427, 1986.

Cowell, P. H. On the secular acceleration of the Earth's orbital motion. *The Observatory*, 28:454–456, 1905.

De Vaucouleurs, G. *Astronomical Photography; From the Daguerreotype to the Electron Camera.* London, Faber and Faber, 1961.

D'Elia, P. M. *Galileo in China.* Cambridge, MA, 1960.

Deslandres, H. Observation de l'éclipse totale de soleil du 16 avril. *L'astronomie*, 12:246–248, 1893.

Diogenes Laercius. *Lives of eminent philosophers.* Harvard University Press, New York, 1972. The Loeb Classical Library. English Translation by R. D. Hicks.

Eddy, J. A. Observation of a possible neutral sheet in the Corona. *Solar Physics*, 30:385–394, 1973.

Eddy, J. A. A nineteenth-century coronal transient. *Astronomy and Astrophysics*, 34:235–240, 1974.

Eddy, J. A. The Maunder Minimum. *Science*, 192:1189–1202, 1976.

Espenak, F. *Fifty Year Canon of Solar Eclipses: 19862035.* Sky Publishing Corp., Cambridge, 1987. NASA RP-1178.

Espenak, F. and Meeus, J. *Five Millennium Canon of Solar Eclipses: 1999 to* +3000 *(2000 BCE to 3000 CE).* NASA/TP2006214141, 2006.

Fatoohi, L. J. and Stephenson, F. R. Accuracy of lunar eclipse observations made by Jesuit astronomers in China. *Journal for the History of Astronomy*, 27:61–67, 1996.

Ferrer y Cafranga, J. J. Observations of the eclipse of the Sun, June 16th, 1806, made at Kinderhook, in the State of New-York. *Transactions of the American Philosophical Society*, 6:264–275, 1809.

Foukal. P., and Eddy, J. Did the Sun's Prairie Ever Stop Burning? *Solar Physics*, 245:247–249, 2007.

Freeman-Grenville, G. S. P. *The Muslin and Christian Calendars.* Oxford, 1977.

Ginzel, F. K. *Spezieller Kanon der Sonnen- und Mondfinsternisse.* Berlin, 1899.

Gladysz, S., Redfern, M., and Jones, B. W. Shadow bands observed during the total solar eclipse of 4 December 2002, by high-resolution imaging. *Journal of Atmospheric and Solar-Terrestrial Physics*, 67:899–906, 2005.

Golub, L. and Pasachoff, J. M. *The Solar Corona*, 2nd edition, Cambridge University Press, 2009.

Gregory, A. Plato and Aristole eclipses. *Journal for the History of Astronomy*, 31:245–259, 2000.

Guillemin, A. *Le Soleil.* Hachette, Paris, 1880.

Guillemin, A. *El mundo físico.* Montaner y Simón, editores, Barcelona, 1883. Tome II.

Guillermier, P. and Koutchmy, S. *Total Eclipses. Science, Observations, Myth and Legends.* Chichester, Springer-Praxis Series in Astronomy, 1999.

Harrington, P. S. *Eclipse! The What, Where, When, Why, and How. Guide to Watching Solar and Lunar Eclipses.* John Wiley & Sons, 1997.

Hartner, W. Eclipse Periods and Thales' Prediction of a Solar Eclipse: Historic Truth and Modern Myth. *Centaurus*, 14:60–71, 1969.

Herodotus. *Herodotus.* William Heinemann, London, 1975. The Loeb Classical Library. English Translation by A. D. Godley. In four volumes.

Hingley, P. D. The first photographic eclipse? *Astronomy and Geophysics*, 42:1.18–1.22, 2001.

Jeaurat, E. S. Observation de l'Éclipse de Soleil, du 24 juin 1778. *Histoire de l'Academie Royale des Sciences*, MCCLXXVIII:39–43, 1781.

Johnson, S. J. 1900 May 28, observed at Navalmoral, Spain. *Monthly Notices Royal Astronomical Society*, 60:590–593, 1900.

Judge, P. G. and Saar, S. H. The Outer Solar Atmosphere during the Maunder Minimum: A Stellar Perspective. *Astrophysical Journal*, 663:643–656, 2007.

Kennedy, E. S. and Faris, N. The Solar Eclipse Technique of Yahya b. Abi Mansur. *Journal for the History of Astronomy*, 1:20–38, 1970.

Kerr Reaves, M. and Reaves, G. Antoine Caron's Painting "Astronomers Studying an Eclipse". *Publications of the Astronomical Society of the Pacific*, 77:153–157, 1965.

Lafuente, A. and Sellés, M. *El Observatorio de Cádiz (1753-1831).* Ministerio de Defensa, Instituto de Historia y Cultura Naval, 2000.

Le Vallois and Pedegache. *Observation de l'eclipse de Soleil du 26 Octobre 1753 faite au College Royal de S. Antoine l'Hermite par Messieurs Le Vallois & Pedegache.* [s.l.], [s.i.], 1753. 3 fol.

León y Gama, A. L. *Descripcion Orthographica Universal del Eclipse de Sol del Dia 24 de Junio de 1778.* Imprenta nueva matritense de D. Felipe de Zúñiga, Mexico, 1778.

Lindsay, L. The Solar eclipse of Dec. 21–22, 1870: Cadiz. *Monthly Notices Royal Astronomical Society*, 31:49–60, 1871.

Littmann, M., Willcox, K., and Espenak, F. *Totality. Eclipses of the Sun.* Oxford, New York, Oxford University Press, second edition, 1999.

Liu, C.-Y. [n.a.]. *Bull. Chinese Studies*, 5:1, 1945.

Liu, C.-Y. An Ancient Solar Eclipse Record "Tian-da-yi" in the 10th Century BC. *Chinese Journal of Astronomy and Astrophysics*, 2:391–395, 2002.

López Piñero, J. M., Glick, T. F., Navarro Brotóns, V., and Portela Marco, E. *Diccionario histórico de la Ciencia Moderna en España.* Península, Barcelona, 2 vols, 1983.

Ludlam, W. An Eclipse of the Sun June 24, 1778, observed at Leicester. *Philosophical Transactions*, 68:1019–1021, 1779.

MacQueen, R. M., Eddy, J. A., Gosling, J. T., Hildnor, E., Munro, R. H., Newkirk, G. A., Jr., Poland, A. I., and Ross, C. L. The Outer Solar Corona as Observed from Skylab: Preliminary Results. *Astrophysical Journal*, 187:L85–L88, 1974.

Mahmoud-Bey. *Rapport a Son Altesse Mohammed Saïd, Vice-Roi D'Égypte, Sur L'Éclipse Totale de Soleil observée a Dongolah (Nubie), le 18 Juillet 1860.* Paris, Mallet-Bachelier, 1861.

Maunder, M. and Moore, P. *The Sun in Eclipse.* Springer, 1998.

Meeus, J., Grosjean, C. C., and Vanderleen, W. *Canon of Solar Eclipses.* Oxford, Pergamon Press, 1966.

Messier, C. Éclipse de Soleil du 24 juin 1778. *Histoire de l'Academie Royale des Sciences*, MCCLXXVIII:36–38, 1781.

Mobberley, M. *Total Solar Eclipses and How to Observe Them.* Springer, 2007.

Monnier, P. C. L. Observation de l'Éclipse de Soleil du 24 juin 1778. *Histoire de l'Academie Royale des Sciences*, MCCLXXVIII:62–65, 1781.

Morrison, L. V. and Stephenson, F. R. Historical values of the Earth's clock error ΔT and the calculation of eclipses. *Journal for the History of Astronomy*, 35:327–336, 2004.

Morrison, L. V. and Stephenson, F. R. Addendum: Historical values of the Earth's clock error. *Journal for the History of Astronomy*, 36:339, 2005.

Mucke, H. and Meeus, J. *Canon of Solar Eclipses: 2003 to +2526*. Astronomisches B'uro, Vienna, 1983.

Neugebauer, O. *The exact sciences in Antiquity*. Providence, R.I., second edition, 1957.

Newcomb, S. *Tables of the Motion of the Earth on its Axis and Around the Sun*. Bureau of Equipment, Navy Dept., Washington, 1895.

Newton, R. R. *Ancient Astronomical Observations and the Accelerations of the Earth and Moon*. Baltimore, Johns Hopkins University Press, 1970.

Newton, R. R. *Medieval Chronicles and the Rotation of the Earth*. Baltimore, Johns Hopkins University Press, 1972.

Olson, R. J. M. and Pasachoff, J. M. The 1816 Solar Eclipse and the Comet 1811I in Linnell's Astronomical Album. *Journal for the History of Astronomy*, 23:121–133, 1992.

Olson, R. J. M. and Pasachoff, J. M. Comets, meteors, and eclipses: Art and science in early Renaissance Italy. *Meteoritics and Planetary Science*, 37:1563–1578, 2002.

Olson, R. J. M. and Pasachoff, J. M. St. Benedict Sees the Light: Asam's Solar Eclipses as Metaphor. *Religion and the Arts*, 11:299–329, 2007.

von Oppolzer, T. R. *Canon der Finsternisse*. Wien, 1887.

von Oppolzer, T. R. *Canon of Eclipses*. New York, Dover Publications, 1962. Translated by O. Gingerich.

Panchenko, D. Thales's prediction of a solar eclipse. *Journal for the History of Astronomy*, 25:275–288, 1994.

Pang, A. S.-K. Victorian observing practices, printing technology, and representations of the Solar Corona, (1): The 1860s and 1870s. *Journal for the History of Astronomy*, 25:249–274, 1994.

Pang, A. S.-K. Victorian observing practices, printing technology, and representations of the Solar Corona, (2): The age of photomechanical reproduction. *Journal for the History of Astronomy*, 26:63–75, 1995.

Pang, A. S.-K. *Empire and the Sun: Victorian Solar Eclipse Expeditions*. Stanford, CA, Stanford University Press, 2002.

Pingré, A. G. Observation de l'Éclipse de Soleil du 24 juin 1778. *Histoire de l'Academie Royale des Sciences*, MCCLXXVIII:61, 1781.

Plato. *Timaeus*. Penguin Books, 1965. English translation by H. D. P. Lee.

Pliny. *Natural History*. London, Harvard University Press, 1938. The Loeb Classical Library. English Translation by H. Rackham.

Rodrigues, F. *Jesuitas portugueses astronomos na China*. Tipografia Porto Medico, Porto, 1935.

Sachs, A. H. and Hunger, H. *Astronomical diaries and related texts from Babylonia*. Vol. I. Wien: Osterreichischen Akademie der Wissenschaften, 377 pp. 1988.

Sachs, A. J. *Late Babylonian Astronomical and Related Texts*. Providence, 1955.

Said, S. S. and Stephenson, F. R. Accuracy of eclipse observations recorded in medieval arabic chronicles. *Journal for the History of Astronomy*, 22:297–310, 1991.

Said, S. S. and Stephenson, F. R. Solar and Lunar eclipse measurements by medieval muslim astronomers I: Background. *Journal for the History of Astronomy*, 27:259–273, 1996.

Said, S. S. and Stephenson, F. R. Solar and Lunar eclipse measurements by medieval muslim astronomers, II: Observations. *Journal for the History of Astronomy*, 28:29–48, 1997.

Said, S. S., Stephenson, F. R. and Rada, W. S. Records of solar eclipses in Arabic chronicles. *Bulletin of the School of Oriental and African Studies*, University of London LII, 38–64, 1989.

Schaeberle, J. M. The Coronal Comet of April 16, 1893. *Publications of the Astronomical Society of the Pacific*, 6:237–238, 1894.

Schove, D. J. *Chronology of Eclipses and Comets AD 1 to 1000.* The Boydell Press, 1984.

Schroeter, J. Fr. *Sonnenfinsternisse von 600 bis 1800 n. Chr.* Kristiana, 1923.

Sellers, J. B. *The Death of Gods in Ancient Egypt.* Penguin Books, 1992.

Shi, Y. Eclipse observations made by Jesuit astronomers in China: A reconsideration. *Journal for the History of Astronomy*, 31:135–147, 2000.

Solc, M. A note on eclipse comets. *Contributions of the Astronomical Observatory Skalnate Pleso*, 28:296–299, 1999.

Spencer Jones, H. The rotation of the earth, and the secular accelerations of the sun, moon and planets. *Monthly Notices Royal Astronomical Society*, 99:541, 1939.

Steel, D. *Eclipse. The celestial phenomenon that changed the course of history.* Washington, Joseph Henry Press, 2001.

Steele, J. M. Solar eclipse times predicted by the Babylonians. *Journal for the History of Astronomy*, 28:133–139, 1997.

Steele, J. M. Predictions of eclipse times recorded in Chinese history. *Journal for the History of Astronomy*, 29:275–285, 1998.

Steele, J. M. *Observations and Predictions of Eclipse Times by Early Astronomers.* Dordrecht, Kluwer Academic Publishers, 2000.

Steele, J. M. and Stephenson, F. R. Lunar eclipse times predicted by the Babylonians. *Journal for the History of Astronomy*, 28:119–131, 1997.

Steele, J. M. and Stephenson, F. R. Astronomical evidence for the accuracy of Clocks in Pre-Jesuit China. *Journal for the History of Astronomy*, 29:35–48, 1998.

Steele, J. M. and Stephenson, F. R. Eclipse observations made by Regiomontanus and Walther. *Journal for the History of Astronomy*, 29:331–344, 1998.

Steele, J. M. Stephenson, F. R. and Morrison, L. V. The accuracy of eclipse times measured by the Babylonians. *Journal for the History of Astronomy*, 28:337–345, 1997.

Stephenson, F. R. *Historical Eclipses and Earths Rotation.* Cambridge University Press, 1997.

Stephenson, F. R. Historical eclipses and Earth's rotation. *Astronomy and Geophysics*, 44(2):22–27, 2003.

Stephenson, F. R. and Fatoohi, L. J. Accuracy of solar eclipse observations made by Jesuit astronomers in China. *Journal for the History of Astronomy*, 26:227–236, 1995.

Stephenson, F. R. and Fatoohi, L. J. Thale's prediction of a solar eclipse. *Journal for the History of Astronomy*, 28:279–282, 1997.

Stephenson, F. R. and Houlden, M. A. *Atlas of Historical Eclipse Maps, East Asia 1500BCAD 1900.* Cambridge, New York, Cambridge University Press, 1986.

Stephenson, F. R. Jones, J. E. and Morrison, L. V. The solar eclipse observed by Clavius in A.D. 1567. *Astronomy and Astrophysics*, 322:347–351, 1997.

Stephenson, F. R. and Morrison, L. V. Long-term changes in the rotation of the Earth: 700 B.C. to A.D. 1980. *Royal Society of London Philosophical Transactions Series A*, 313:47–70, 1984.

Stephenson, F. R. and Morrison, L. V. Long-term fluctuations in the Earth's rotation: 700 BC to AD 1990. *Royal Society of London Philosophical Transactions Series A*, 351:165–202, 1995.

Stephenson, F. R. and Said, S. S. Precision of medieval Islamic eclipse measurements. *Journal for the History of Astronomy*, 22:195–207, 1991.

Stothers, R. Solar activity cycle during classical antiquity. *Astronomy and Astrophysics*, 77:121–127, 1979.

Thomas, A. (ed.). *Beauty of Another Order; Photography in Science.* New Haven, Yale University Press, 1997.

Ulloa, A. *El eclipse de sol con el anillo refractario de sus rayos.* Imprenta de Don Antonio de Sancha, Madrid, 1779a. There is a facsimile edition published by the Universidad Nacional Autónoma de México in 1992. There is a French translation by Auguste Darquier de Pellepoix printed in Toulouse in 1780.

Ulloa, A. Observation on the total (with duration) and annular Eclipse of the Sun, taken on the 24th of June, 1778. *Philosophical Transactions*, 69:105–117, 1779b.

Vaquero, J. M. The Solar Corona in the eclipse of 24 June 1778. *Solar Physics*, 216:41–45, 2003.

Vassenio, B. Observatio Eclipsis Solis totalis cum mora facta Gothoburgi Sveciae. *Philosophical Transactions*, 38:134–135, 1733.

Wales, W. Observations on the solar eclipse which happened June 24, 1778. *Philosophical Transactions*, 68:1013–1018, 1779.

Wang, P. K. and Siscoe, G. L. Ancient Chinese observations of physical phenomena attending solar eclipses. *Solar Physics*, 66:187–193, 1980.

Zirker, J. B. *Total Eclipses of the Sun.* Princeton, Princeton University Press, 1995.

5

The Solar Diameter and the Astronomical Unit

Since the beginnings of Astronomy, the measurement of the sizes of the different bodies in the Solar System and the distances between them has been a primary challenge. The determination of the Earth – Sun distance was the first step for many of the subsequent works: not in vain is this parameter called the "astronomical unit". Here we will describe two topics related to this essential parameter: First, the determination of the solar radius, and second, the description of planetary transits across the solar disk. As in the rest of the book, we will concentrate our study on historical documents and visual observations.

We will be discussing measurements of well-defined parameters (angles, sizes), and avoiding issues involving imagination and vague descriptions, consequently this chaptor will be more technical than the previous ones. The reader can find an adequate background on spherical astronomy in the texts of Smart (1977), Green (1985) and Taft (1991).

5.1 The Earth's Orbit

Seen in the sky, the Sun traces a circle. Its position is usually determined by the coordinates (α, δ) in the equatorial system of reference. The hour circle is a great circle passing through the poles of the celestial sphere and intersecting the celestial equator at right angles. The declination, δ, is the angular distance to the celestial equator and the right ascension to the vernal point (see Figure 5.1).

The orbital parameters describe the path of the Earth with respect to a reference plane: the ecliptic. They are (Figure 5.2):

- Semimajor axis.
- Eccentricity ($0 < e < 1$), determining the shape of the orbit.

J.M. Vaquero, M. Vázquez, *The Sun Recorded Through History*, Astrophysics and Space Science Library 361, DOI 10.1007/978-0-387-92790-9_5,

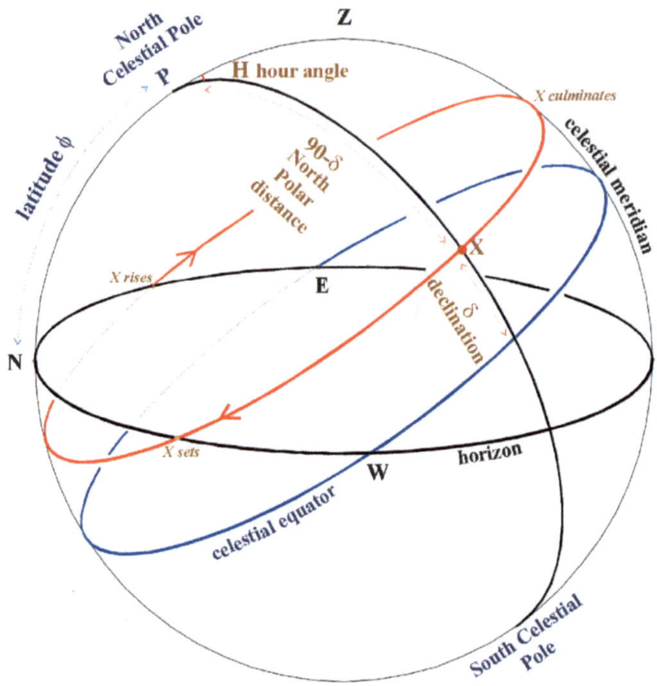

Fig. 5.1. The equatorial system of reference.

- The inclination, i, orients the orbital plane with respect to the plane of the ecliptic. Imagine the angle being formed by pivoting the orbital plane through an axis of rotation coinciding with the line of nodes.

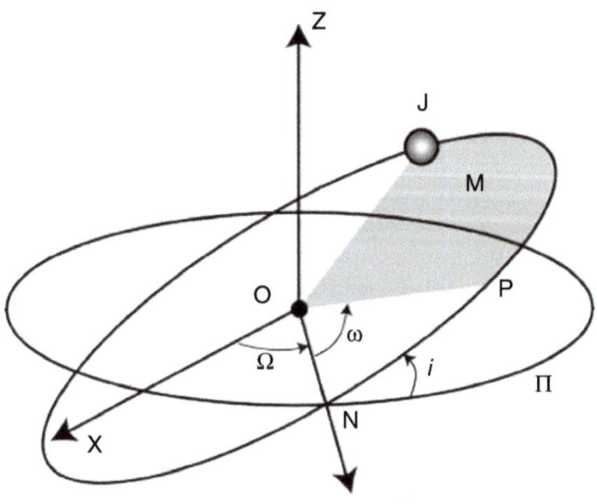

Fig. 5.2. The orbital elements.

- The longitude of the ascending node, $0 < \Omega < 360°$, orients the ascending node with respect to the vernal point. Imagine the angle being formed by pivoting the orbital plane through an axis of rotation perpendicular to the plane of the ecliptic and passing through the centre of mass.
- The argument of periapsis (perihelion) (ω) orients the semimajor axis with respect to the ascending node. Imagine the angle being formed by pivoting the orbital plane through an axis of rotation perpendicular to itself and passing through the centre of mass.
- The true anomaly, v, also written T, orients the celestial body in space. It is the angle between the direction of periapsis and the current position of the object on its orbit, measured at the focus of the ellipse.

The rotation axis is fixed in direction with respect to the stars, not with respect to the Sun. This means that, as the Earth orbits the Sun, the axis continues to point in the same direction in 3D space. Thus, the tilt of the rotation axis to the Earth's orbit implies that the Sun, as viewed from the Earth, will appear to move north and south of the celestial equator through the year by a maximum of ± 23.5 degrees. The annual variation of this parameter can be easily determined by observing, at a particular latitude ϕ, the minimum zenith distance z (Figure 5.3):

$$\delta = \phi - \mathrm{z}$$

This variation can be calculated by the formula

$$\delta = -23.45 \cos \left[\frac{360}{365} . (\mathrm{N} + 10) \right]$$

where N is the number of days spent since 1 January. The solar declination determines the hours of daylight and the angle at which sunlight strikes the Earth's surface at a given location.

From Kepler's first law, we know that the path of the Earth around the Sun is an ellipse of eccentricity e, the position of the Sun being at the focus of the ellipse. The perihelion distance is $a(1-e)$ and the aphelion $a(1+e)$, where

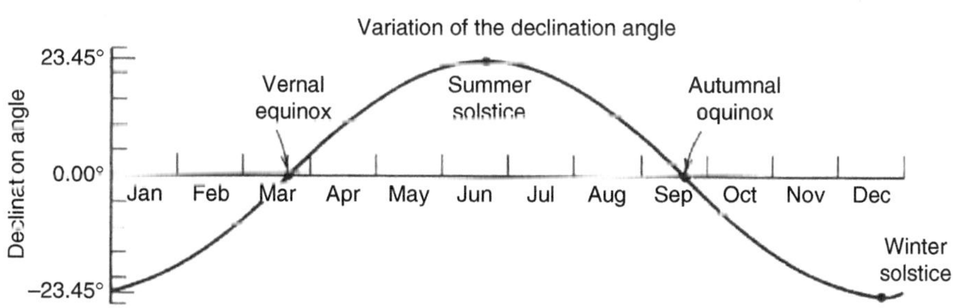

Fig. 5.3. Annual variation of solar declination: the Sun's distance from the celestial equator.

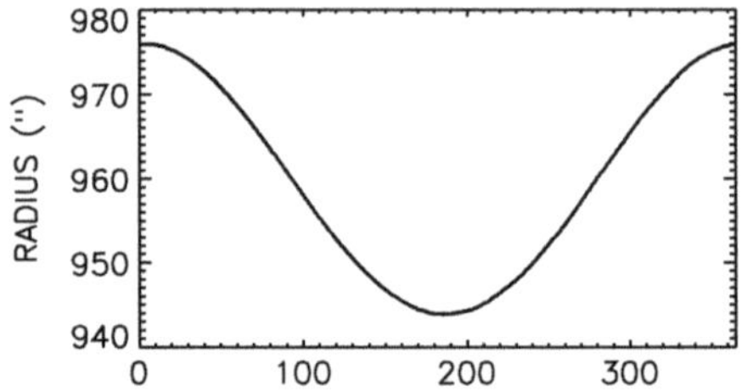

Fig. 5.4. Annual variation of the apparent solar radius.

a is the semimajor axis of the ellipse. As a consequence, the apparent solar radius as measured from the Earth will undergo annual variation (Figure 5.4). At aphelion:

$$R_{min} = \frac{R}{1+e} \simeq R - R_e = 944'' \, (1'' = 738\text{km})$$

At perihelion:

$$R_{max} = \frac{R}{1-e} \simeq R + R_e = 976'' \, (1'' = 713\text{km})$$

The measurement of this variation was essential for discriminating between the Copernican and Ptolomean models of the solar system. Nowadays, this can be easily reproduced by comparing full disk images of the Sun taken at different times of the year.

5.2 Measuring the Known World

5.2.1 Trigonometry

Geometry was the basis of one of the first tasks of science, that of being able to locate things in space. Trigonometry is the part of geometry that studies the relationship between the length of the sides of a triangle and the size of its angles. The development of this science goes back to ancient Egypt and Greece, and was mainly created to aid in the study of astronomy. Angles were defined by taking the Babylonian measurement of 360 degrees. The main trigonometrical functions (sine, cosine) were loosely defined by Hipparcus (190–120 BC) circa 150 BC. In Antiquity, the most important book was the *Almagest* of Claudius Ptolemy, which included the first trigonometrical tables (see Van Brummelen, 2009).

5.2.2 The First Measurements

Parallax is the change of angular position of two stationary points relative to each other as seen by an observer, due to the motion of the observer. By observing parallax, measuring angles, and using geometry, the distance to various objects can be determined. This is also the case for the distance from the Earth to the Sun.

The distance from the Earth to the Sun, d_S, is usually expressed in terms of the solar parallax $\sin p_S = R_E/d_S$ or $p_S = R_E/d_S$ because the angle p is very small. Given this distance, the solar radius can be calculated. However, this solar radius cannot be measured directly, since, due to the brightness of the Sun, it is impossible to observe it directly against the background field of stars. Hence, indirect methods must be used.

Aristarchus (310–c. 230 BC) was the first to propose a heliocentric model of the solar system. In the only work which has survived, *On the sizes and Distances of the Sun and the Moon*, he calculates the sizes of Moon and the Sun as well their distances from the Earth (Heath, 1981).[1]

He assumed that, when the Moon was exactly halflit (waxing or waning quarter moon), it forms a right triangle with the Sun and the Moon. By observing the other angles of the triangle he could derive the ratio (Figure 5.5)

$$\frac{\mathrm{SE}}{\mathrm{ME}} = \frac{1}{\cos\alpha} = \sec\alpha$$

Aristarchus used a value of $\alpha = 87°$ (which is actually 89.86°) obtaining $18 < \mathrm{SE}/\mathrm{ME} < 20$, therefore finding that the Sun was an object extraordinarily distant from the Earth. Then, by measuring the angular size of the Sun at the sky, the solar diameter could be determined. He was also the first to abandon mere speculation in favour of rational and empirical arguments (Neugebauer, 1975) and provided bounds for the measured values. His observations were carried out in Alexandria, where at that time there was an important school of astronomy.[2]

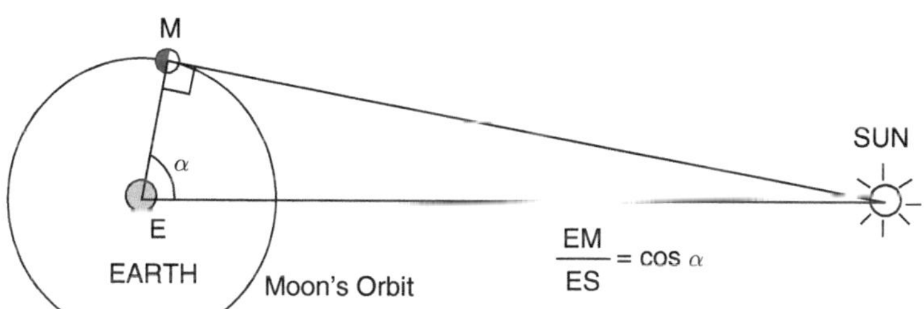

Fig. 5.5. The method used by Aristarchus, lunar dichotomy, to determine the distance of the Sun. The diagramme is greatly exaggerated, because S = 390M.

[1] Reprint of the 1913 original edition.
[2] Timocharis (320–260) and Aristyllus also worked in Alexandria at this time and compiled the first catalogue of the fixed stars, unfortunately lost. Observations of 18 stars survived in Ptolemy's Almagest.

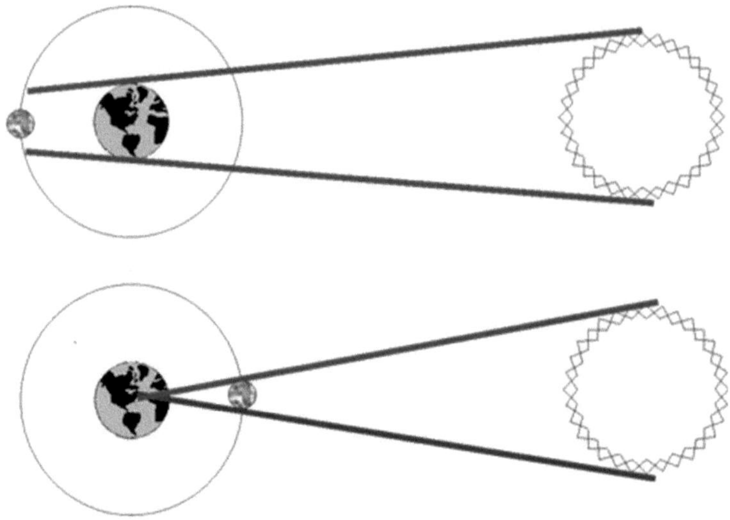

Fig. 5.6. The method used by Hipparcus and Ptolemy.

Hipparcus (190–120 BC) correctly reasoned that if the Sun were just as large as the Earth, the shadow of a lunar eclipse would be exactly equal to the Earth's diameter. The first calculation was based on the solar eclipse of 14 March 189 BC, during which time the Sun was totally eclipsed near the Hellespont. The same method was later applied by Ptolemy (Figure 5.6). For a study on the accuracy of Greek astronomical observations, see Maeyama (1984).

Present-day textbooks give a value of 696 400 km for the solar radius (Cox, 2000). Obtaining the measurement of this parameter was really a battle to obtain smaller angular resolutions on the sky, and in order to do so new instruments and methods of correction of the disturbing effects produced by the terrestrial atmosphere were developed (Van Helden, 1985; Chapman, 1996).

5.3 Observing Methods of the Solar Diameter

Personal bias continues to be one of the main problems of solar observing methods and is very difficult to estimate. Moreover, there is no reason why this bias should remain constant over time. Overviews of observing methods are given by Wittmann and Debarbat (1990) and Ribes et al. (1991).

5.3.1 Direct Estimation

One simple method is to hold the hand above the horizon with your arm stretched out. The width of a finger is an angle just over 1.5 degrees. We can

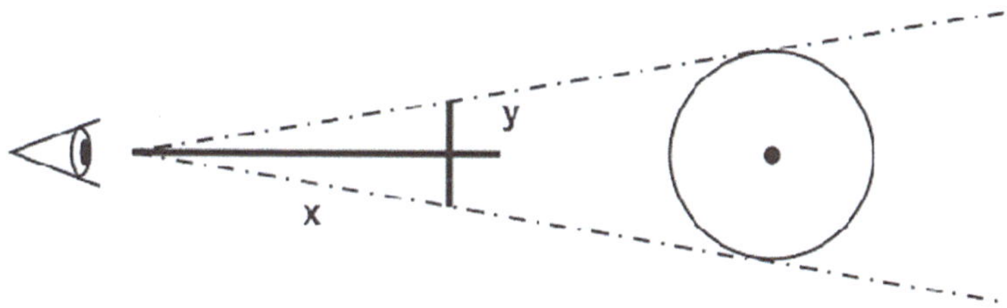

Fig. 5.7. Scheme of a "Jacob's staff", an early instrument used to measure angular distances.

do the same with an instrument: Jacob's staff, also called a cross-staff, was one of the first instruments used to measure angular distances of an object above the horizon, mainly used for navigation purposes. The first description of this instrument comes from Levi Ben Gerson (1288–1344)[3] (Goldstein, 1985). At first, the instrument was used facing the Sun and therefore required the observer to look straight into the bright sunlight. The problem was solved with the invention of the back staff developed around 1594 by John Davis (Figure 5.8). The observer placed the staff on his shoulder and stood with his

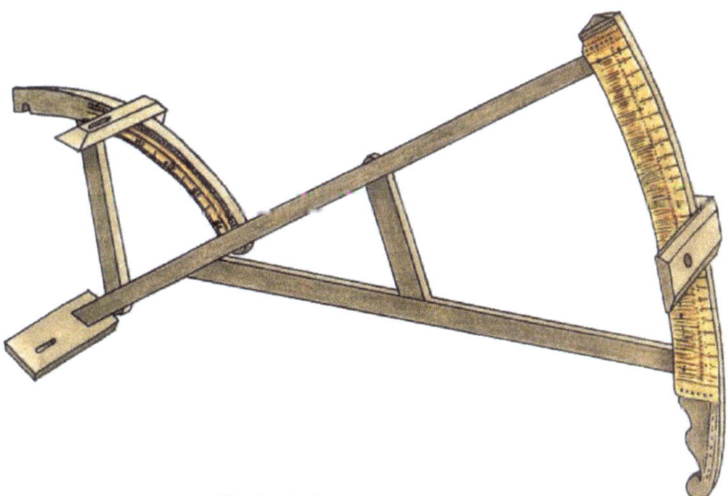

Fig. 5.8. The backstaff of John Davis. It consisted of two-half-crosses, which divided an accurate scale into two parts. It looked like a large triangle equipped with a 30° arc at one end and a small 60° arc at the other. One scale was engraved on its upper side towards the front of the staff, the other was on its underside and at the back.

[3] The instrument is described in his book *Book of the Wars of the Lord* as consisting of a staff of $4\frac{1}{2}$ feet long and about one inch wide, with six or seven perforated tablets which could slide along the staff, each tablet being an integral fraction of the staff length to facilitate calculation, used to measure the distance between stars or planets, and the altitudes and diameters of the Sun, Moon and stars.

back to the Sun. With the horizon vane lined up with the horizon, he slid the half-cross back and forth until the shadow of its vane fell across the slit in the bottom vane while the horizon was visible through the slit. By doing this, the observer was able to sight both the Sun and the horizon while his back was towards the Sun.

The solar radius is easily determined (Figure 5.7).

$$\mathrm{R(arcseconds)} = 206264.8\,\mathrm{arctg(y/x)}$$

Al Battani (858–929), also known as Albatenius, proved the annual variation of the apparent angular diameter of the Sun and the possibility of annular eclipses. He gave the first accurate estimate of solar radius (974″).

Based on this principle, more precise instruments were developed. Astronomical quadrants are essentially a graduated quarter of a circle and came in two forms: mural quadrants are fixed to a meridian wall and used to measure meridian altitudes and altazimuth quadrants measure altitude and azimuth, simultaneously.

In 1699, Isaac Newton discovered the principle of the doubly reflecting instrument, the sextant, but never published it. John Hadley (1682–1744) and Thomas Godfrey (1704–1749) rediscovered it around 1730. The scale of a sextant has a length of 1/6 of a full circle, hence its name. A variant is the octant (1/8 of a circle or 45°).

Tycho Brahe developed a method for calculating the solar diameter with a pinhole camera. The method consisted of correcting the measured diameter of the image by subtracting the diameter of the aperture. This idea had already been suggested by Guillaume de Saint Cloud in 1292 and Levi Ben Gerson in fragments of his *Milhamot Adonai*, written in Hebrew in 1328 (Goldstein, 1985; Mancha, 1993). Recommendations that the camera obscura could also be used to measure the solar diameter appeared in the works by Egidius of Baisiu (Mancha, 1989), Henry of Hesse (1325–1397) and Erasmus Reinhold (1511–1553) as well.

In 1591, Tycho Brahe performed eleven measurements of the solar diameter using a closed tube 2-m long mounted on a side of a wooden quadrants.[4] At the upper end of the tube was a square pinhole opening of 2 cm. He measured the diameter of the circular image produced on the lower screen by calculating the ratio (Sigismondi and Fraschetti 2001):

$$\theta_\mathrm{S} = \frac{\mathrm{d_m} - \mathrm{d_p}}{\mathrm{f}}$$

where d_m is the diameter of the image on the screen, d_p the diameter of the entrance pinhole, and f the focal length of the system.

[4] Description included in his *Opera Omnia*.

Johannes Kepler analysed the data of Tycho but also developed his own instrument. His main conclusion was: *Sed res certa est et cuilibet obvia exploratu, diametrum Solis in apogeo 30'in perigeo 31' esse.*[5]

C. Scheiner noticed that as the Sun approached the horizon, it assumed an elliptical shape. He explained this in terms of atmospheric refraction, and made the first announcement of his discovery and theory in *Sol Elliptic* published in 1615. A second edition, entitled *Refractiones coelestes*, was published in Ingolstadt in 1617, but differs both in text and in illustration. Pages 26–30 contain further observations of sunspots and a defence of Scheiner's earlier works against the attacks Galileo made on them. In the *Rosa Ursina* (Parss. II Cap. V, p. 581) he describes under the title "Observations Visualium Solis Diametrorum" his own determinations of the solar diameter made in Rome over the years 1625, 1626 and 1627.

M. Maestlin (1550–1631) recommended the use of great distances between the aperture and the screen and apertures as small as possible. These suggestions led to the building of large meridian circles in cathedrals to determine the passage of the Sun in different seasons. The objective was to improve the calendar and check the date of Easter. The monograph of Heilbron (1999) constitutes an excellent reference on this topic.

The spot of light produced on the floor of a large church by the solar rays, which came into the interior through a small hole, made it possible to better define the positions of the Sun. The Dominican Egnazio Danti (1536–1586) built a large-scale gnomon[6] in the church of Santa Maria Novella in Florence around 1575. This camera obscura formed an image of the Sun in different parts of the temple during the year (Bartolini, 2006). The details are described in the document *Usus et Tractatio Gnomonis Magnis*, preserved at the private library of T.B. Settle.

A complete meridian is composed of the hole and a horizontal lane in the floor of the church running to the north from a point under the hole (see Figure 5.9). From this, it will be possible to find the diameter of the Sun, by measuring the diameter of the ellipse and the distance from the centre of the ellipse to the gnomon hole in the facade of the temple.[7]

Danti also built a meridian at the church of San Petronio in Bologna. However, in the year 1653, a wall was removed to extend the church, making the system inoperative. At that time, J.D. Cassini (1625–1712) had the chair of Astronomy in Bologna University and proposed the construction of a new and improved meridian. The project was approved by the Senate of the city and soon Cassini started to work. The summer solstice of 1655 was used for the first calibration and finally inaugurated on the occasion of the winter

[5] It is a certain thing that the solar diameter at aphelion is $30'$ and at perihelion $31'$.

[6] A second hole was perforated above the rose window in the marble with the unusual shape of a pear.

[7] To measure the exact duration of the year, Danti also built a quadrant on the facade of the temple with eight solar clocks and an equinoctial ring.

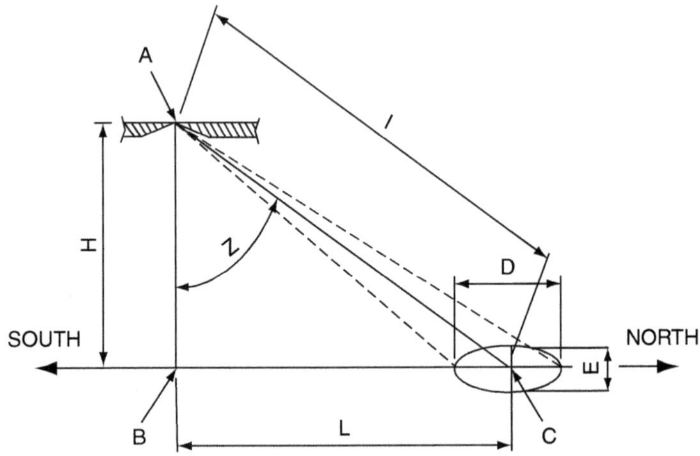

Fig. 5.9. Scheme of a meridian. All the parameters change during the year, excepting the height H.

solstice.[8] Soon he was able to demonstrate that the Sun's apparent diameter decreased as its distance from Earth increases, verifying Kepler laws (Bònoli et al., 2006). Cassini moved to Paris in 1669 and the meridian was first restored by Domenico Guglielmini (1655–1710) in 1695 and later by E. Zanotti (1709–1782) in 1776, leading to the instrument that can be seen today.

Finally, we can also mention the meridian line built by Danti at the Vatican Tower of the Winds (Mancinelli and Casanovas, 1980). The system consists of two separate pieces. One lies on the floor; it is a perfectly horizontal rod running due north for about 67 meters from a spot under one of the side chapels to the front door of a chapel. The hole is permanently open so as to give free access to the Sun's rays around noon throughout the year. A book containing 4500 observations of the solar diameter, spanning 80 years, was published in 1736 by Eustachio Manfredi (1674–1739).[9] Figure 5.10 shows a summary of these uncorrected measurements. The motion of the image during the measurements and the low contrast of the solar image to define the solar edges are the main causes of errors. However, diffraction plays a major role, enlarging the thickness of the solar limb.[10]

[8] The original document, dated 1695, is included in a booklet published in Italian by Arnoldo Forni Editore (2005), with a transcription and introduction by G. Paltrinieri.

[9] Together with Cassini, many other astronomers such as Geminiano Montanari (1633–1687) and G.B. Riccioli (1598–1671) worked on the project.

[10] The diffraction of light was discovered by F.M. Grimaldi (1613–1663), decades after Kepler's measurements.

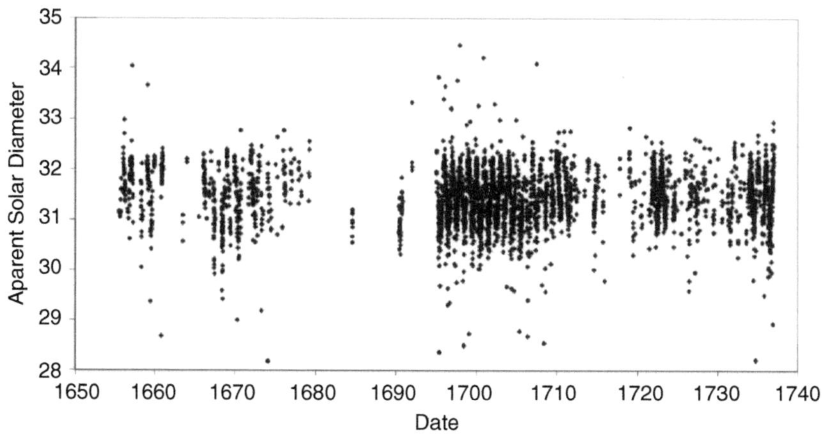

Fig. 5.10. Measurements of solar radius, in arcminutes, made in San Petronio from 1655 to 1736.

5.3.2 Transits of the Sun in the Sky

The method consists of measuring the difference between the times at which two opposite points on the solar limb cross a fixed reference line. This line may be the horizon, the meridian or another plane in the sky (see Wittmann and Neckel 1996).

Given a lens with focal length f, the Sun with an angular size θ will have a linear projection on the focal plane of size $a = f.\theta$. If the Earth rotates with an angular velocity $\omega = 2\pi/T$ ($15''.041009$ per solar second) for a time t, the image will be displaced by a distance $b = f\omega t \cos\delta$, where δ is the geocentric declination of the Sun. Combining these expressions we get

$$\theta_{\mathrm{diameter}} = \omega\,\mathrm{t}\,(\cos\delta)\left(\frac{\mathrm{a}}{\mathrm{b}}\right) = \left(\frac{2\pi}{\mathrm{T}}\right)\mathrm{T_d}\,(\cos\delta)\left(\frac{\mathrm{a}}{\mathrm{b}}\right)$$

where T is the apparent rotation period of the Sun on the sky, 23.934 hours.[11] If now $\mathrm{T_d}$ represents the time that it takes for the solar image to be displaced by one solar diameter (in hours), then a = b and we have

$$\mathrm{R_S}(\mathrm{degrees}) = \frac{1}{2}\left(\frac{T_d}{23.934(hours)}\right)\cos\delta(1 - d\alpha/dt) \times 360$$

where $d\alpha/\mathrm{dt}$ is the proper motion of the Sun in right ascension, which is of the order of $1°$/day = 1/15 hr per 24 hours = 0.0028 and varies between 0.0025 and 0.0031 during the course of the year.

The observed solar radius R(r) must be reduced to unit distance (r = 1 astronomical unit AU) according to

[11] The time between two successive returns of the Sun to a given hour angle.

$$\mathrm{R(1AU) = asin\{rR(r)\}}$$

where r is the geocentric distance of the Sun at the time, t, of the observation.

The observational definition of the solar diameter is the angular distance between the points, on opposite limbs, where the brightness gradient is maximum. We should also take into account the fact that these values are topocentric, and must be converted into geocentric ones.

The main advantages of this method are that it is almost self-calibrated by the Earth's rotation and it does not require correction by refraction. Among the problems, we can mention: (a) distortion of the limb, mostly related with atmospheric turbulence and (b) systematic and random errors in the observer's timing. The constancy of the observer and the equipment are two factors positively influencing the reliability of this method.

The horizon was the first reference used by the classic Greeks such as Aristarchus and Cleomedes[12] (Figure 5.11).

$$\mathrm{R('')} = 7.5\mathrm{T_d}\cos\delta(1-\mathrm{d}\alpha/\mathrm{dt})\sqrt{1-\sin^2\phi/\cos^2\delta}$$

where ϕ is the latitude of the observing site and $\sin\psi = \sqrt{1-\sin^2\phi/\cos^2\delta}$.

When using a vertical reference (Figure 5.12), we have

$$\mathrm{R('')} = \frac{7.5\mathrm{T_d}\cos\delta(1-\mathrm{d}\alpha/\mathrm{dt})(\cos\delta\sin\phi-\sin\delta\cos\phi\cos\mathrm{H})}{\cos\mathrm{a}}$$

where a is the altitude of the crossing point above the horizon and H the hour angle.

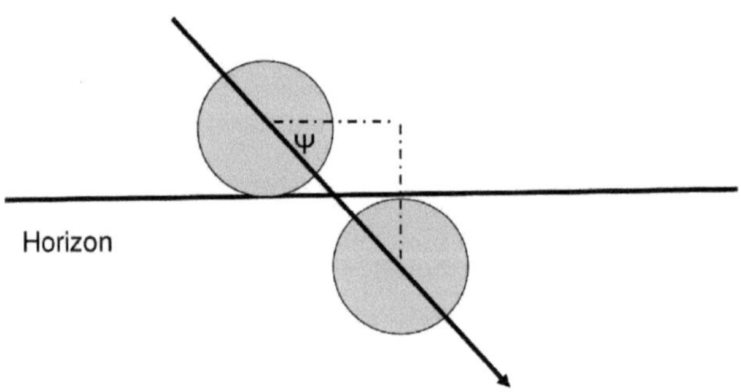

Fig. 5.11. Measurements of solar radius using the horizon as reference.

Almucantar Transits: The Astrolabe

The astrolabe is based on a stereographic projection, allowing the 3-D sphere to be drawn on a flat disk marked with a grid of curved lines. They are mainly

[12] In A.C. Bowen and R.B. Todd, 2004, *Cleomedes' Lectures on Astronomy*, University of California Press, we have an English translation of his work *Caelestia* (The Heavens).

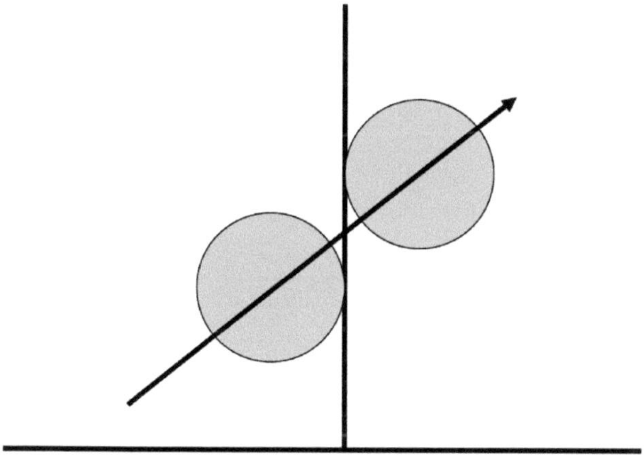

Fig. 5.12. Measurements of solar radius using a vertical reference.

used to show how the sky looks at a specific place at a given time. An early rudimentary astrolabe was invented in the Hellenistic world in either the first or second centuries BC and is often attributed to Hipparchus. The astrolabe was introduced to the Islamic world in the eighth and ninth centuries through translations of Greek texts. Later it passed through North Africa into Spain (Andalusia) where it was introduced to European culture through Christian monasteries in northern Spain.

The reduction of astrolabe observations is based on the method of equal altitudes (Débarbat and Guinot, 1970). Observing a transit requires the recording of the time when both images of the Sun's edge crosses the parallel of altitude (almucantar), of which the horizon is a special case. The measurement of the diameter will then be given as the difference between the zenith distances computed at the times of two successive crossings. The instrument used for this purpose is usually an astrolabe, a horizontal telescope looking through a device that produces two images of the Sun, one merging into the other when the star crosses a horizontal circle of a given zenith distance. M. Maestlin (1550–1631)[13] first used this method in 1577.

A solar astrolabe measures the times at which the limbs of the Sun are tangent to the almucantar (Figure 5.13). Kovalevsky (2002) describes this method in detail. Let C_1 and C_2 be the centres of the Sun at each of these times. We assume that the topocentric coordinates of these points (α_i, δ_i) and their variations $(\Delta\delta_{12} = \delta_2 - \delta_1)$ are known with great precision

At the two crossing times, the centre of the Sun is at the following zenith distances

$$z_1 = z + R_S;\ z_2 = z - R_S$$

In the triangle PCZi = 1, 2 the zenith angle z_i is given by

[13] He was a professor at the University of Tübingen and among his students was Johannes Kepler.

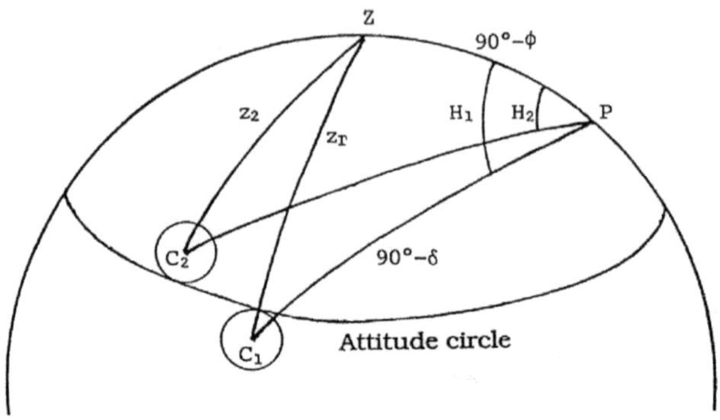

Fig. 5.13. The measurement principle of the astrolabe. H is the hour angle of the Sun ($H = ST - \alpha$), where ST is the sidereal time.

$$\cos z_i = \sin\phi \sin\delta_i + \cos\phi \cos\delta_i \cos H_i$$

The signs are for an eastern transit and should be reversed for a western transit. Assuming that the longitude and latitude of the observer are known, the difference between the theoretical zenith distance of observation and the one computed from the ephemerides for the same time, Dz, is

$$Dz = \cos S\Delta\delta + \sin A \cos\phi\Delta\alpha + \Delta z + R_S$$

where S is the parallactic angle[14] and A the azimuth reckoned from the south point.

Making successive observations of both transits we get successive equations for different azimuths. The instrumental errors can be estimated this way. We should also consider the variation of refraction between the two zenith distances, the effect of the curvature of the almucantar and a possible error in the determinations of $\Delta\alpha$ and $\Delta\delta$.

The first measurements were made observing at 30° zenith distance. Later, a series of additional prisms allowed the observation at larger zenith distances (Laclare et al., 1996). A. Danjon (1890–1967) improved the prismatic astrolabe for positional astronomy (Danjon, 1958). An equilateral glass prism is fixed in front of a horizontal telescope in such a way that its edges are horizontal and one of its faces is vertical. The main problem lies in the systematic effects due to visual observers, especially the estimate of the tangency of the Sun's images, mainly limited by atmospheric effects. The mean error of a single observation is 0″25, but astrolabes are able to perform a large number of observations every clear day.

Solar astrolabes currently distributed around the world are measuring radius variations with the solar cycle of $\simeq 0.1''$. A recent analysis (Badache-Damiani and Rozelot 2006) suggests that astrolabes are really measuring

[14] Angle between the vertical and the celestial meridian.

fluctuations of the Earth stratosphere. As these variations are driven by the solar cycle, a variable signal is recorded in the measurements.

Meridian Transits

The meridian is an imaginary line joining the poles and the local zenith. The measurement of time was the most critical point of classic astronomy. The duration of the day was determined as the time elapsed between two consecutive crossings of the Sun through the local meridian.

The first measurements of the Sun on the meridian were made using a gnomon, allowing the altitude of the Sun to be estimated. The style is the part of the gnomon that casts the shadow and can change as the Sun moves.

The meridian circle is an instrument for observing the time of stars passing the meridian and at the same time measuring their angular distance from the zenith. A serious problem with early measurements is that these could not be performed rapidly at the time of the Sun's passing through the zenith. The idea of having an instrument (quadrant) fixed in the plane of the meridian occurred even to the ancient astronomers and is mentioned by Ptolemy, but it was not put into practice until Tycho Brahe constructed a large meridian quadrant. In 1682, a telescope was fixed in the meridian plane.

Let T_d, expressed in sidereal seconds, be the measured transit time and δ the topocentric declination of the Sun. The horizontal semidiameter will be given by the expression we have given for the solar transit

$$R_H('') = \frac{15.041069}{2} T_d \cos\delta(1 - d\alpha/dt)$$

where $d\alpha/dt$ is the topocentric proper motion of the Sun in right ascension during the transit.

We have already mentioned Jean Picard in reference to sunspot observations. He applied telescopes and micrometers to graduated astronomical and measuring instruments as early as 1667, carrying out regular observations of the solar radius with a 38-inch quadrant and 6-foot sextant, both fitted with telescopic sights. The focal length of the refractor was 6 feet and the solar image was about 2 cm in diameter.

In the 17th century Paris archives, there are essentially two types of measurements. First, observations were carried out at the meridian, when the Sun was moving roughly parallel to the horizon and span the period from 1666 to 1672. The distance between the filars tangential to the edges of the solar image determined the vertical diameter. To measure horizontal diameters, the filars were placed perpendicular to the horizon. J. Picard could measure the apparent horizontal diameter with an accuracy of between 5 and 10 seconds of arc, at least six times better than the previous standard.[15]

[15] J. Cassini estimated that no telescope existing in 1655 could be relied upon to determine the diameter of the solar disk to within a minute of arc.

The second method consisted in timing the edge passages of the solar disk. This was done by Philipe La Hire, using the same equipment, from 1683 to 1718. Certain observations were performed independently in Lyon by Gabriel Mouton (1618–1694)[16]. See Ribes and Nesme-Ribes (1993) for details.

The first meridian circle equipped with a telescope was built by O. Roemer (1644–1710) in 1704 at his observatory near Copenhagen (Figure 5.14). He accumulated a great number of observations, including solar ones, but most of them perished in the Copenhagen Fire of 21 October 1728.

The Royal Greenwich Observatory (RGO) was one of the first institutions to run a regular programme of solar observations.[17] Since 1750 the observations had been made with a meridian transit telescope with the primary aim of determining the exact time the Sun crosses the prime meridian at noon. The telescope could swing up and down vertically but not

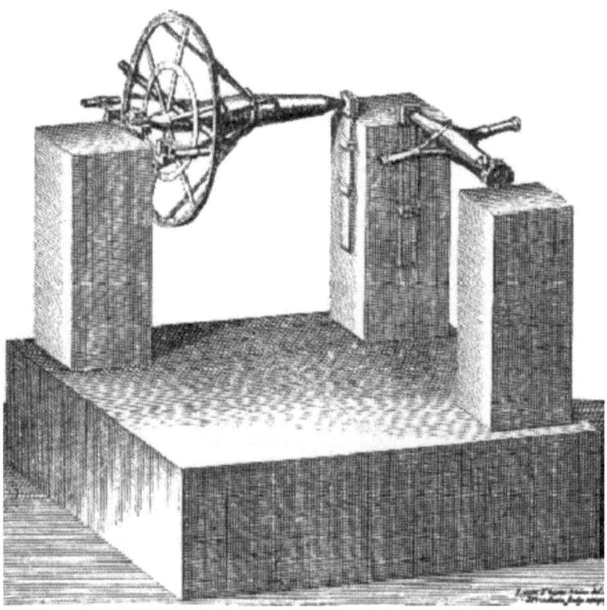

Fig. 5.14. The first meridian circle constructed by O. Roemer in Copenhagen (Observatorium Tusculanum). Source: P. Horrebow (1735) *Basis Astronomiae.*

[16] His 1670 book, *Observationes diametrorum solis et lunae apparentium*, came to form the basis of what was to become the metric system a hundred years later. Based on the measurements of the size of the Earth conducted by G. Riccioli of Bologna (at 321 815 Bologna feet to the degree), Mouton proposed a decimal system of measurement based on the circumference of the Earth, explaining the advantages of a system based on Nature.

[17] Founded in 1675 under the design of Christopher Wren (1632–1723) and Robert Hooke (1635–1703), it was closed in 1998.

horizontally, viewing the sky from a slit in the south side of the building. To find when the centre of the Sun crossed the meridian, the observer recorded first the moment of passage of the leading edge across that line and then the trailing edge. The centre's passage would be the midpoint and would give the horizontal diameter directly. Early observers would lie on a couch beneath the telescope's eyepiece looking at the pendulum clock as the Sun edge approached and then listen for the number of tic-tocs until the end. Improvements were made in 1850 and finally in 1953 the observatory was moved to a southern location. The vertical diameter was measured by swinging the telescope vertically to the Sun's upper and lower limits as its disk crossed the telescope. This was done during the two minutes the observer had between right- and left-limb passage. The discussion about the long-term variability of the solar radius opened up the possibility of studying in depth the errors made by the observers during the measurements. Surprisingly, it was verified that the vertical measurements are more accurate because the observer has more time to decide on the exact position of the solar edges in the telescope crosshairs.

In 1751 Tobias Mayer (1723–1762) became director of the Göttingen University Observatory, where solar and lunar tables were calculated to determine the longitude at sea with an accuracy of a degree. His observations of the solar radius were in the period 1756–1761 with a 6 feet mural

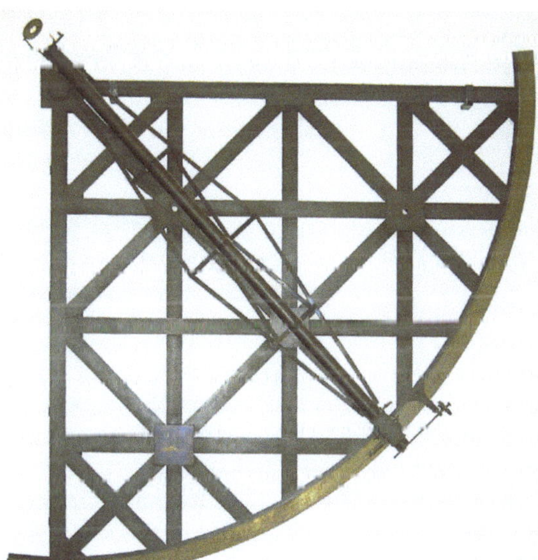

Fig. 5.15. The 6-ft mural quadrant of the former Göttingen Observatory. Courtesy: A. Wittmann.

Table 5.1. Observations by Tobias Mayer in Göttingen. Source: Wittmann (1998)

	Number of obs.	Solar Radius ($''$)
1756–1758	120	960.46 ± 0.11
1760–1761	13	959.85 ± 0.72

quadrant ($D = 5.0\,\mathrm{cm}$, $f = 1.86\,\mathrm{m}$) manufactured in London (Figure 5.15).[18] A precision pendulum clock designed by Franz Kampe was used. Forbes (1971, 1972) and Wittmann (1980) provide the most detailed account of his solar work. Wittmann (1998) re-analysed the original results and found that they could be classified into two different groups, the first proving to be much more reliable (Table 5.1).

5.3.3 Solar Eclipses

In the previous chapter, we discussed the basics of solar eclipses. They were used to determine solar diameter using a pinhole camera (Wlodarczyk, 2007). We can easily understand that the value of the solar diameter influences the duration of a total eclipse. A smaller radius produces a wider cone of totality and therefore a longer eclipse. In other words, measuring the eclipse umbra on the ground and given the diameter of the Moon, the solar diameter can be calculated. Thus,

$$R('') = \frac{T_3 - T_1}{2V_{rel}}$$

where V_{rel} is the relative Sun-Moon velocity estimated in $0.5''$/second, T_1 the instant when the Moon's disk starts to cover the Sun and T_3 the instant when the Moon's disk starts to come out of the Sun. Carefully designed observations could provide the solar diameter, relative to the lunar, with an accuracy of $0''.01$ (e.g. Kubo, 1993).

The solar diameter may be determined much better by fitting a theoretical eclipse to the observed one, with this parameter remaining the only unknown one. The accuracy of the determinations depends basically on two factors:

(a) Accurate geographic location of the observers near the edges (north and south) of the path of totality. Unfortunately, in old eclipses, astronomers tended to place the telescopes in the centre of the path to ensure being on target. An error of 10 metres in the geographical position of the observers gives an error of $0.006''$ in the solar diameter.
(b) Location and depth of the Moon valleys (second contact). It was not until 1963 that the terrain of the lunar limb was known well enough to make an accurate measurement (Watts, 1963). See Herald (1983) for a method of correcting the effect of these irregularities.

[18] A duplicate of the Göttingen mural quadrant was made in approx. 1750 by Bird (London) for the Spanish Observatory of Cádiz (now, San Fernando).

Observers are stationed near the umbral limits (shadow extremes) during a solar eclipse. By stationing the observers at the shadow limits, we only need to determine whether or not the observer actually has totality at that point. Being at the shadow limits and considering that the Moon's surface is full of craters, mountains and valleys, these observers will experience a long period of Baily's beads, sunlight shining through the lunar mountains and valleys. If no complete totality is seen (that is, Baily's beads are continuously visible) then the observer is outside the Moon's shadow. If totality is seen, then the observer is within the Moon's shadow. A line of observers can easily establish the geographical location of the edge of the Moon's shadow to within 100 metres. This uncertainty corresponds to an error in the solar diameter of 0.05 arcseconds. Early results using this technique showed a precision of better than 0.1 arcsecond, and although this is quite an impressive figure, it was not enough to detect any changes in the solar diameter. But in comparison with a historical eclipse observed and timed by Edmund Halley, the resultant change in the Sun's size was computed to be–0.34±0.2 arcseconds.

This technique was also used during the 17 April 1912 solar eclipse visible in the northwest corner of Spain. It was predicted to be a hybrid eclipse, also called annular-total with totality lasting no longer than a few seconds. Tomás de Azcárate (1849–1921) led an expedition of the Naval Observatory of San Fernando. In order to measure the duration of the totality, a group of observers, equipped with smoked glasses, were placed one every kilometre, forming three parallel rows of 10 km in extension in the area of centrality. The average value of visibility of the phase of totality was 0.5 seconds (Azcárate, 1914).

The difference in solar radius between the Maunder minimum and our times is still a matter of debate. This claim is supported by the observation of the 9 April 1567 eclipse by Cristoph Clavius (1538–1612) as annular in Rome[19], as described in his book *In sphaeram Ioannis de Sacro Bosco Commentarius: a certain narrow circle was left on the Sun, surrounding the whole of the Moon on all sides.* (Stephenson et al., 1997), when it would have been total with the modern value of the solar radius.

Fiala et al. (1994) have re-analysed determinations of the solar radius at the times of solar eclipses since 1715. The 3 May 1715 eclipse is especially important for the study of long-term trends, placed just at the end of the Maunder minimum. The totality path crossed Britain from Cornwall in the southwest to Norfolk in the east and E. Halley presented, for the first time, eclipse maps to the public in their common current form: looking down on the Earth's surface from above. All four of the Halley maps are in the Houghton Library, Harvard University.

[19] At that time, he was professor in the Roman College.

5.3.4 The Micrometer

An essential point was the development of instruments allowing the small angular distances to be read with increasing precision. In this way, the telescope was able to carry out quantitative measurements of astronomical parameters.

Micrometers were an essential component of the measuring system and classically consisted of cross-wires. William Gascoigne (1612–1644) was using a Keplerian telescope when part of a spider's web found its way inside the telescope. One small web line fell right at the focus, so both the thin line and the object Gascoigne was viewing were magnified together. This serendipitous discovery was a great help to astronomers for tracking objects and, for example, to determine more precisely the transit times of the solar limbs. He was probably the first to use a micrometer attached to a telescope. This micrometer was based on the the "nonius" of Pedro Nunes (1501–1577), later developed by Pierre Vernier (1580–1637). It let one read more precisely from an evenly divided straight or circular measurement scale and is fitted with a sliding secondary scale (see Figure 5.16).

He described the micrometer in a letter to the mathematician William Oughtred (1575–1660). However, Gascoigne died at the Marston-Moor battle[20] on 2 July, 1644, and his instrument fell into the possession of Richard Towneley (1629–1707), who probably exhibited an improved version at the meeting of the Royal Society held on 25 July 1667. Robert Hooke reported on this topic the same year, including an illustration of Gascoigne's micrometer (Hooke, 1667). However, this remained unknown to the rest of astronomers.

William Crabtree, while making a journey to Yorkshire in 1639 to see Gascoigne, writes thus to his friend Jeremiah Horrocks: *The first thing*

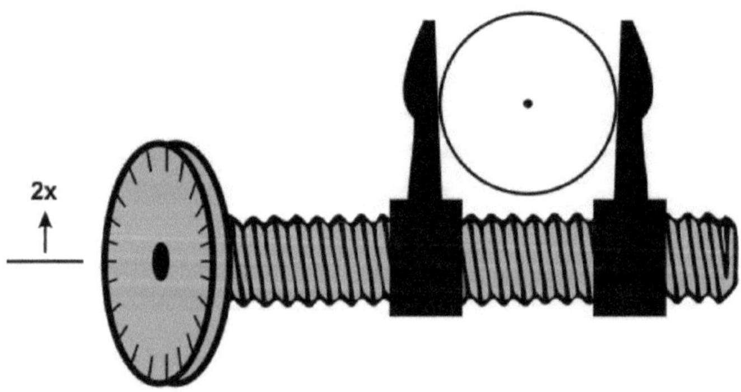

Fig. 5.16. A scheme of the ocular micrometer developed by W. Gascoigne.

[20] Part of the First English Civil War. Royalist troops (Charles I) under the Marquess of Newcastle and Prince Rupert vs. an allied army of Parliamentary and Scottish troops led by Sir Thomas Fairfax and Lord Manchester. This battle made the reputation of Oliver Cromwell.

Mr Gascoigne showed me was a large telescope amplified and adorned with inventions of his own, whereby he can take the diameters of the sun and moon, or any small angle in the heavens or upon the earth, most exactly through the glass, to a second. W. Derham published correspondence between Gascoigne and Crabtree to prove that the former was the inventor of the telescopic micrometer (Derham, 1717). J. Flamsteed, in the first volume of his *Historia coelestis*, inserted a series of measurements made by Gascoigne extending from 1638 to 1643. These included the mutual distances of some of the stars in the Pleiades, a few observations of the apparent diameter of the Sun, others of the distance of the Moon from neighbouring stars, and a great number of measurements of the diameter of the Moon. From his observations, Gascoigne deduced the greatest variation of the apparent diameter of the Sun to be 35″ (see Figure 5.4).

In his *Systema Saturnium*[21] (1659) C. Huygens described a micrometer, composed of a slip of metal of variable breadth inserted at the focus of the telescope, and observed at what part it exactly covered the object under examination; knowing the focal length of the telescope and the width of the slip at the point observed, he thence deduced the apparent angular breadth of the object. Another micrometer developed by A. Auzout (1622–1691)[22] was provided with silk fibres or silver wires instead of the edges of Gascoigne, but one of the silk fibres remained fixed while the other was moved by a screw. Auzout found values of the solar diameter between 31′35″ in aphelion and 32′45″ in perihelion with this instrument.[23] Later Jesse Ramsden (1735–1800) applied a microscope to a micrometer allowing a better accuracy in the reading.

The standard eyepiece reticule, when combined with a precision stage micrometer, provides a rapid, convenient, and accurate means of conducting measurements.

5.3.5 The Heliometer

A heliometer is a telescope which produces two images of the Sun. It was originally designed for measuring the variation of the Sun's diameter at different seasons of the year. In 1743, Servington Savery (1670–1744) sent a paper to the Royal Society describing a twin object-glass telescope for measuring the solar diameter (Savery, 1754). The idea remained dormant until James Short (1710–1768) extracted his account from the minutes of the Royal Society and published this account in 1753, at the same time drawing John Dollond's (1706–1761) attention to the device (Dollond, 1753). In fact, John and Peter

[21] Available in digital form at the Smithsonian Institution Libraries.

[22] Auzout, A., 1667, *Traite du Micrometer*, Paris. Reprinted in Memoires de l'Academie Royale des Sciences vol. 7, part I (1729).

[23] To be compared with the modern values of 31′28″ and 32′32″, respectively.

Fig. 5.17. Brass heliometer, made by John and Peter Dollond (1731–1820) of London, ca. 1758, and shipped to Harvard by Joseph Mico in 1765. Courtesy: Collection of Historical Scientific Instruments, Harvard University.

Dollond built one in 1758, improving on the Savery concept by separately mounting the two lens halves, allowing motion relative to each by a micrometer screw (Figure 5.17 and Dollond, 1754).

In the meantime, Pierre Bouguer (1698–1758) was able to develop, in 1748, a telescope whose objective lens had been split into two halves, the heliometer (Figure 5.18). Each half objective forms a solar image, and the two halves are moved until the two images are in contact (Bouguer, 1748; Geyer, 1985). The apparent solar radius is then proportional to the number of screw divisions, x. The solar radius is then given by

$$\mathrm{R}('') = \mathrm{C}.(\mathrm{x})$$

where C is a constant, whose calibration is the main problem to solve, including the small size of the Sun in the focal plane (full disk on the observing field).

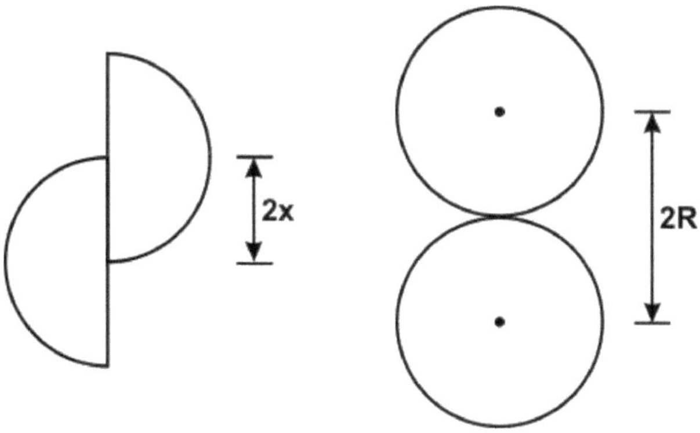

Fig. 5.18. The principle of the heliometer. Courtesy: A. Wittmann.

5.3.6 The Measurement of Time

The measurement of time was a constant challenge for astronomers, and it was a necessary tool for precise angular determinations. A sundial is an ancient clock that measures time by the position of the Sun (cf. Waugh, 1973). The most commonly seen designs cast a shadow on a flat surface marked with the hours of the day. As the position of the Sun changes during the day, the time indicated by the shadow changes. The "shadow-maker" of the sundial is called a gnomon and was imported from Babylonia by the Greeks. The Egyptians were apparently the next to formally divide their day into parts something like our hours. Obelisks (slender, tapering, four-sided monuments) were built as early as 3500 BC. Their moving shadows formed a kind of sundial, enabling people to partition the day into morning and afternoon. Obelisks also showed the year's longest and shortest days when the shadow at noon was the shortest or longest of the year. Later, additional markers around the base of the monument would indicate further subdivisions of time.

In Europe during most of the Middle Ages (roughly 500 CE to 1500 CE), technological advancement virtually ceased. Sundial styles evolved, but didn't move far from ancient Egyptian principles. G. Chaucer (1342–1400) wrote about the application of this method in the prologue of *The Canterbury Tales*:

It was four o'clock according to my guess
Since eleven feet, a little more or less
My shadow at the time did fall
Considering that I myself am six feet tall

As a modern example, Figure 5.19 shows an inclined sundial installed at the Teide Observatory (Tenerife).

In 1657, Christian Huygens revolutionized the measurement of time by creating the first working pendulum clock, although the first ideas about such an instrument were developed by Galileo. In fact, mechanical clocks had already appeared in Europe near the end of the 13th century.

Fig. 5.19. A sundial placed on an inclined wall at the Teide Observatory (Tenerife), based on a design by Mario Salomone.

J. Hevelius adapted Huygens' micrometer to measure planetary diameters, and at the end of his career, used Huygen's clocks to determine time (Lisicki, 1992). In 1679 Edmund Halley (1656–1742), commissioned by the Royal Society, visited Hevelius in Gdansk. He verified that Hevelius's determination of positions was as accurate as could be achieved at that time with the best British micrometers.

In 1721, George Graham (1673–1751) improved the pendulum clock's accuracy to 1 second per day by compensating for changes in the pendulum's length due to temperature variations. The development of clocks which were becoming more and more accurate was accelerated by the need to determine the geographical longitude at sea. See Burton (1992) for a history of clocks and watches.

5.4 Theoretical Background

Before we examine the possible temporal changes in the solar radius, it is reasonable to discuss the feasibility of such changes in the time-scale of centuries.

The solar luminosity L, one of the fundamental parameters, is dependent on the emitted radiation flux, F_{em}, and the radius via the expression

$$L = 4\pi R^2.F_{em}$$

The importance of changes in the solar radius is expressed by the index

$$W = \frac{\Delta \ln R}{\Delta \ln L}$$

This parameter provides some indication of the location of the process inside the Sun which is responsible for changing the solar luminosity (Gough, 1981, 2001).

If the radius and luminosity variations have the same origin, at some radius r_0, then W is a positively decreasing function of r_0 and virtually vanishes as $r_0 \longrightarrow R_\odot$. In particular, $W \approx 0.5$ when $r_0 = 0$, and $W = 0.2$ when the perturbing process is situated near the base of the convective zone; a modification to the upper superadiabatic convective boundary layer yields the very much smaller value $W \leq 10^{-3}$. Sofia et al. (2005) show that changes of solar diameter in response to large-scale magnetic fields are not homologous, i.e. the variation at the photospheric level is over 1000 times larger than the variation at a depth of 5 Mm.

Radius variations can be generated by (see Spruit, 1991; Endal et al., 1985):

- β effect: Changes in the non-gas component of pressure due to the variable magnetic component. The Sun should be larger (less gas pressure) at sunspot maximum (Thomas, 1979). Spiegel and Weiss (1980) proposed that only the layer where the magnetic field is stored is perturbed.
- α effect: The magnetic field is strong enough to start reducing the degrees of freedom of the convective flow. We have a direct blocking of the heat flux by the action of magnetic fields, causing changes in the convective efficiency. The effects at the surface depend on the depth of the perturbation.

Table 5.2 summarizes the theoretical determination of this parameter. Reliable measurements still need to be done.

Table 5.2. Estimates of the parameter W

W	Authors
$+7.5 \times 10^{-2}$	Sofia and Endal (1979)
$+5.0\ 10^{-3}$	Dearborn and Blake (1980)
$+8.1 \times 10^{-4}$	Gilliland (1980)
$+2.0\ 10^{-3}$	Spruit (1991)
$+2.0\ 10^{-3}$	Balmforth et al. (1996)
$7.5\ 10^{-2}$	Lefebvre and Rozelot (2001)

5.5 Long-Term Variations

It is a matter of current debate whether the variations of the solar radius are correlated with the 11-year solar cycle. A. Secchi and P. Rosa in 1872 (see Rosa, 1873) and J. Hilfiker (1851–1913) in 1878 were the first to find an anticorrelation between the solar radius and the number of sunspots. The visual records will help us to have a temporal perspective and to look for changes with longer periods. Parkinson et al. (1980) and Gilliland (1981) reported on a 76-year modulation in solar radius data during the period 1700–1980 when analysing results obtained by different techniques. A secular decrease of $0.1''$ per century was not excluded. Table 5.3 summarizes the main determinations of the solar radius.

A summary plot of all the observations since 1650 is shown in Figure 5.20, compiled from Toulmonde (1997) and Pap et al. (2001). We see clearly that since the 18th century, the values rapidly converged towards the current value. See, however, the comments in Rozelot (2001).

Analysing data obtained during the Maunder minimum, Ribes et al. (1987) suggested the existence of a larger Sun during this period and since then the radius has probably expanded at a rate of $0.1''$ per century . This claim was criticized by O'Dell and Van Helden (1987), who argued that the error in the measurements made by the Paris observers was larger than Ribes et al. had assumed. Ribes et al. (1989) showed clearly a modulation of the apparent horizontal radius, over a period of four activity cycles during the Maunder minimum, which corresponds to the period spanned by the Paris observers. Table 5.4 summarizes the main trends found in the literature.

In general we have a trend confirming the anticorrelation of sunspot number with the solar radius. The radius seems to be larger during times of

Table 5.3. First determinations of solar radius obtained with visual methods. Main Source: Wittmann and Debarbat (1990)

Method	$R_S['']$	Author (Year)
Sunset timing	900	Aristarchus (270)
Movil disk	898	Archimides ($\sim$ 230)
Parallactic linear	974	Al-Battani (880)
Meridian transittime	967	Maestlin (1577)
Heliotelescopium projection	1500	Scheiner (1620)
Eyepiece micrometer	967	Gascoigne (1640)
Gnomon shadow (upper/lower edge)	947	Cassini (1656)
Meridian transit time	961	Mouton (1670)
R(Sun)/R(Moon) during eclipse	967	Eimmart (1694)
Edge height with wall quadrant	961.5	Bradley (1753)
Mercury transit	961.7	Delisle (1756)
Heliometer (objective micrometer)	961.1	Lalande (1760)
Solar eclipse timing	960.3	du Séjour (1764)
Venus transit	958.02	Lalande (1771)
Venus transit	948.0	Lomosonov (1761)
Venus transit	958.4	Encke (1825)
Meridian transit	961.4	Fuhg (1875)

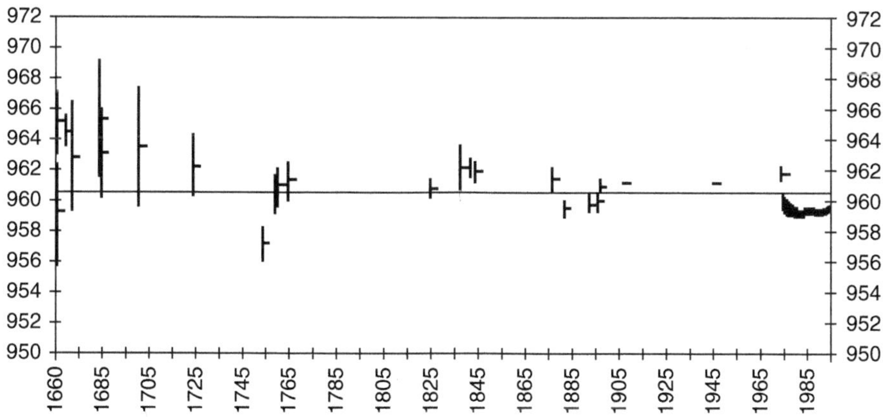

Fig. 5.20. Solar radius measurements observed from 1660. Adapted from Toulmonde (1997) and Pap et al. (2001)

Table 5.4. Long-term trends in solar radius determinations

Author	Method	Period	Trend $\dot{R}$ (″/century)
Dunham et al. (1980)	Eclipse timing	1715–1980	$-(0.13 \pm 0.08)$
Parkinson et al. (1980)	Mercury transits Eclipse timing	1715–1966	$-(0.14 \pm 0.08)$
Gilliland (1981)	Meridian Transits Mercury transits Eclipse timing	1715–1960	$-(0.1)$
Sveshnikov (2002)	Mercury transits	1631–1973	None

activity minimum. This was probably the case during the Maunder minimum (see, however, Toulmonde, 1997).

5.6 Planetary Transits

The recent transits of Venus and Mercury, together with the exoplanets discovered using this method, have strongly revived this field, including the historical research on former events. Figure 5.21 shows one of the pictures obtained during the 2005 transit of Venus.

5.6.1 Orbital Motion of the Inner Planets

The orbit of Venus is inclined by 3.4° to the Earth's. Transits occur when the two planets are in conjunction at (or very near) the points where their orbital planets cross, the so-called nodes (Figure 5.22).

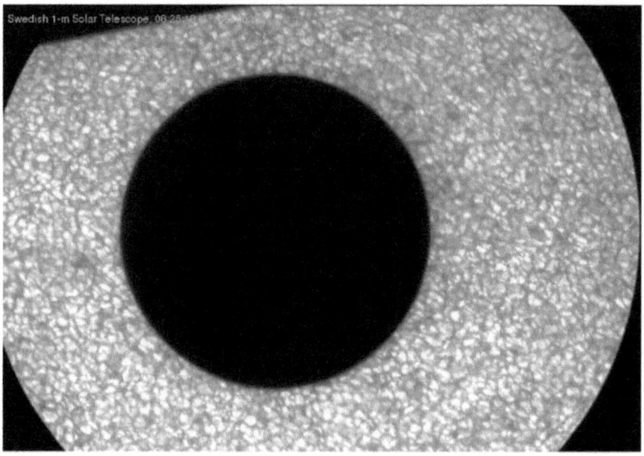

Fig. 5.21. Picture of the transit of Venus of 8 June 2004 crossing the solar disk obtained at the Swedish Solar Telescope at the Roque de los Muchachos Observatory (La Palma, Canary Islands).

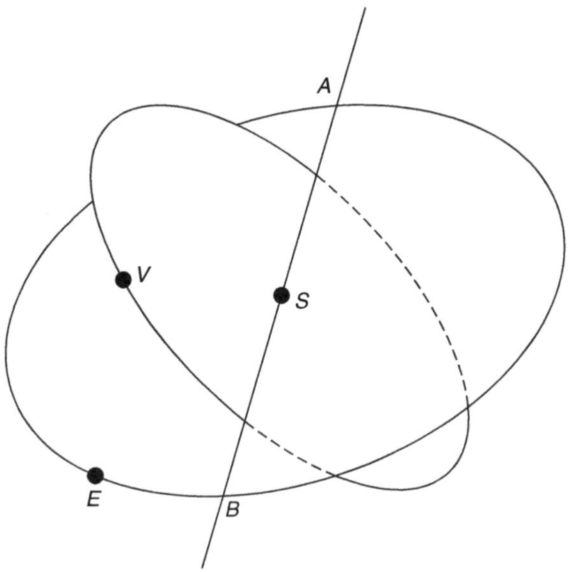

Fig. 5.22. Orbits of Venus and the Earth.

When a transit of Venus occurs, a second one often follows eight years later. This is because the orbital periods of Venus (224.701 days) and Earth (365.256 days) are in an 8-yr (2922 days) resonance with each other. In other words, in the time it takes Earth to orbit the Sun eight times, Venus completes almost exactly thirteen revolutions about the Sun. As a result, Venus and Earth line up in the same positions with respect to the Sun. The two orbital periods are actually not quite commensurate with each other since Venus arrives at the

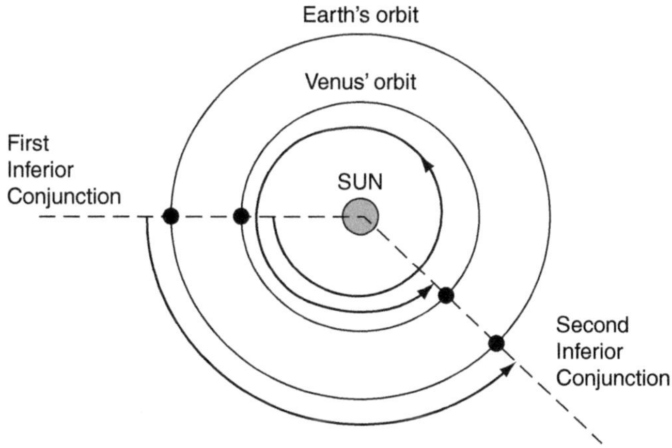

Fig. 5.23. Transit of Venus.

eight-year rendezvous about 22 hours earlier than Earth. By the third eight-year cycle, Venus arrives too early for a transit to occur. A pattern of transits repeats every 243 years, with pairs of transits eight years apart separated by long gaps of 121.5 and 105.5 years[24](Figure 5.23).

Figure 5.24 and Table 5.5 show the path of Venus across the solar disk and the relevant times for different transits, respectively.

The four contact times of the transit are defined as follows: (I) the instant when the planetary disk is externally tangent to the Sun. The transit begins, (II) the instant when the entire disk of the planet is first internally tangent to the Sun. The period from contact I to contact II is called Ingress, (III) the instant when the planet reaches the opposite limb of the Sun and it is again internally tangent to the Sun and (IV) the instant when the planet's disk is

Table 5.5. Transits of Venus between the years 1600 and 2012. Times are in UT. Data: Fred Espenak (NASA)

Date	First contact (I)	Mid–transit	Last contact (IV)	Duration
1631 Dec 07	03:51	05:19	06:47	2 h 56 m
1639 Dec 04	14:57	18:25	21:54	6 h 57 m
1761 Jun 06	02:02	05:19	08:37	6 h 35 m
1769 Jun 03/4	19:15	22:25	01:35	6 h 20 m
1874 Dec 09	01:49	04:07	06:26	4 h 36 m
1882 Dec 06	13:57	17:06	20:15	6 h 18 m
2004 June 8	05:13	08:20	11:26	6 h 13 m
2012 June 5/6	22:09	01:29	04:49	6 h 20 m

[24] 243 sidereal orbital periods of the Earth (365.25636 days) are 88753.3 days and 395 sidereal orbital periods of Venus (224701 days) are 88756.9 days.

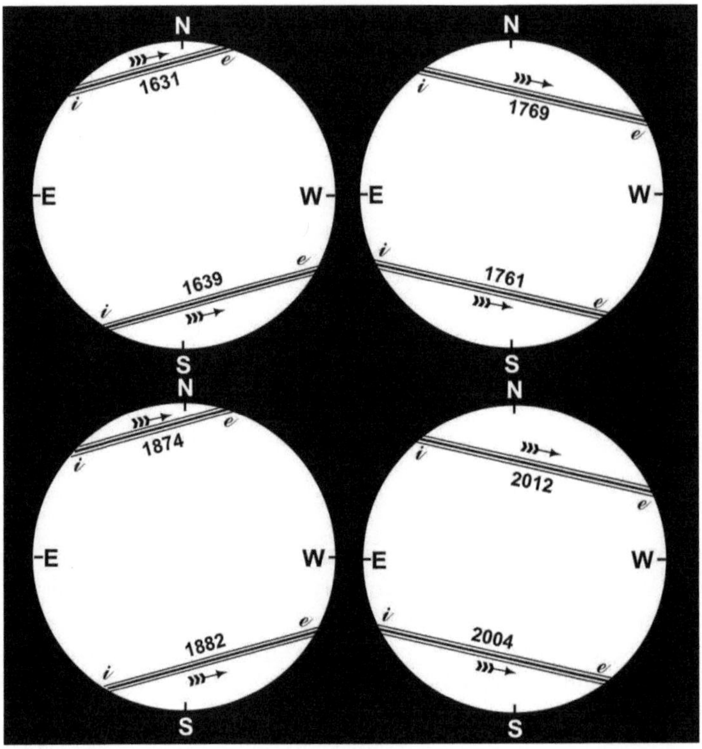

Fig. 5.24. Paths of Venus across the solar disk during the transits in the period 1631–2012. Adapted from Proctor (1882).

externally tangent to the Sun (transit ends). The period from contact III to IV is called Egress. In the case of Mercury, the internal contacts, at which Mercury appears completely inside the solar limb, are more definite than the external ones (Parkinson et al., 1980).

The orbital plane of Mercury is inclined 7° with respect to Earth's. Orbital crossings occur only in May and November. The relative velocity of Mercury with respect to the Sun is $0.07''$/s during the May transits and $0.10''$/s in November.

On average there are 13 transits of Mercury each century. The main transit repetition period is 46 years. The duration of 46 revolutions periods of the Earth relative to Mercury's nodes is 191 draconic periods of Mercury. The ratios of Mercury draconic and synodic periods produce additional periods of 7, 13 and 33 years.

Ephemerides of the transits are given in Meeus (1989, 1995). Table 5.6 gives the most relevant data of the Mercury transits in the period 1600–1850. To determine whether a transit is visible from a specific geographic location, it is a matter of calculating the altitude and azimuth of the Sun during each phase of the transit.

Table 5.6. Transits of Mercury between the years 1620 and 1820. Data: Fred Espenak (NASA/GSFC). The Gregorian Calendar is used. For discrepancies between reported transit dates, take into account that Great Britain did not adopt the Gregorian calendar until 1752

Date	First contact (I)	Mid-transit	Last contact (IV)	Duration
1615 May 03	06:41	10:09	13:36	6h 55m
1618 Nov 04	11:08	13:42	16:15	5h 07m
1628 May 05	14:19	17:32	20:44	6h 25m
1631 Nov 07	04:38	07:20	10:03	5h 25m
1644 Nov 09	22:53	00:57	03:00	4h 07m
1651 Nov 03	23:07	00:52	02:38	3h 31m
1661 May 03	13:05	16:54	20:43	7h 38m
1664 Nov 04	15:53	18:32	21:11	5h 18m
1674 May 07	21:56	00:16	02:37	4h 41m
1677 Nov 07	09:32	12:11	14:50	5h 18m
1690 Nov 10	03:57	05:43	07:29	3h 32m
1697 Nov 03	03:38	05:42	07:45	4h 07m
1707 May 05	19:34	23:32	03:30	7h 56m
1710 Nov 06	20:39	23:22	02:05	5h 26m
1723 Nov 09	14:25	16:59	19:32	5h 07m
1736 Nov 11	09:07	10:30	11:52	2h 45m
1740 May 02	21:34	23:02	00:29	2h 55m
1743 Nov 05	08:12	10:30	12:47	4h 35m
1753 May 06	02:16	06:13	10:09	7h 53m
1756 Nov 07	01:26	04:10	06:55	5h 29m
1769 Nov 09	19:21	21:46	00:12	4h 51m
1776 Nov 02	20:55	21:36	22:17	1h 22m
1782 Nov 12	14:35	15:16	15:57	1h 24m
1786 May 04	02:56	05:41	08:26	5h 30m
1789 Nov 05	12:51	15:19	17:46	4h 55m
1799 May 07	09:07	12:50	16:34	7h 27m
1802 Nov 09	06:14	08:58	11:43	5h 29m
1815 Nov 12	00:18	02:33	04:48	4h 30m

5.6.2 The Determination of the Solar Radius

For the case of a Mercury transit we have

$$R_S('') = \frac{T_3 - T_2}{2V_{rel}} + R_{Mercury}$$

One of the problems is the need to correct to an ideal case (diameter instead of a chord).

Sveshnikov (2002) collected more than 4500 contact-timing observations of Mercury transits from 1631 to 1973. From this material he obtained a secular decrease in the solar radius of $0''.06 \pm 0''.03$ attributed to systematic observational errors.

Shapiro (1980) analysed 23 transits of Mercury between 1736 and 1973, finding a decrease in this period of 0.3″ per century. Parkinson et al. (1980) estimated this decrease as 0.14 ± 0.08 arcseconds per century.

5.6.3 The Determination of the Sun's Distance

After the Copernican revolution, it was possible to build up a model of the Solar System, but without the scale. To ascertain the scale, it is necessary only to measure one distance within the solar system, e.g. the mean distance from the Earth to the Sun or the astronomical unit (AU). When done by triangulation, this is referred to as the solar parallax, the difference in position of the Sun as seen from the Earth's centre and a point one Earth radius away, i.e. π_S, the angle subtended at the Sun by the Earth's mean radius (Figure 5.25).

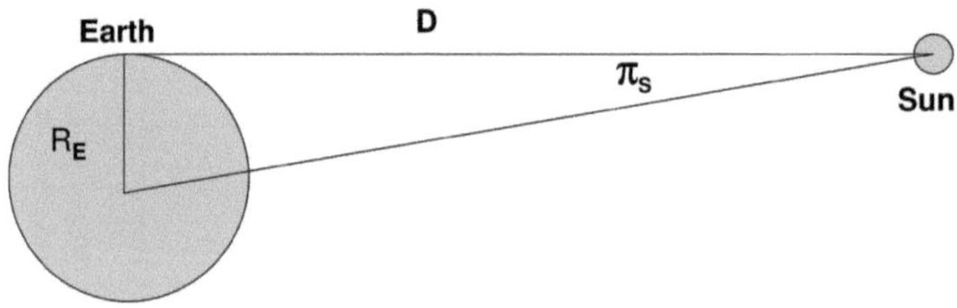

Fig. 5.25. The solar parallax. The angle subtended by Earth's equatorial radius at the centre of the Sun at the mean distance between Earth and the Sun (1 AU), equal to 8.794″.

The following equation holds

$$\sin \pi_S = \pi_S \,(\text{in radians}) = \frac{R_E}{D} \longrightarrow D = \frac{R_E}{\pi_S}$$

As the Earth's radius is known, the Sun–Earth distance, D, can be determined. However, this measurement cannot be done directly. We need an intermediate body.

James Gregory (1638–1675) suggested in an appendix to his *Optica Promota*, published in 1663, how the parallax of a planet can be determined by measuring the duration of its conjunction with another heavenly body. Some years later, in 1716, Edmond Halley described how planetary transits could be used to measure the mean distance of the Earth to the Sun, the astronomical unit.[25]

The main point of the method was to measure the duration of the transit from two stations (see Figure 5.26) sufficiently far apart, preferably at opposite

[25] In 1676, in his *Catalogue Stellarum Australium*, we find his first mention of the idea, first presented to the Royal Society on 3 October, 1691.

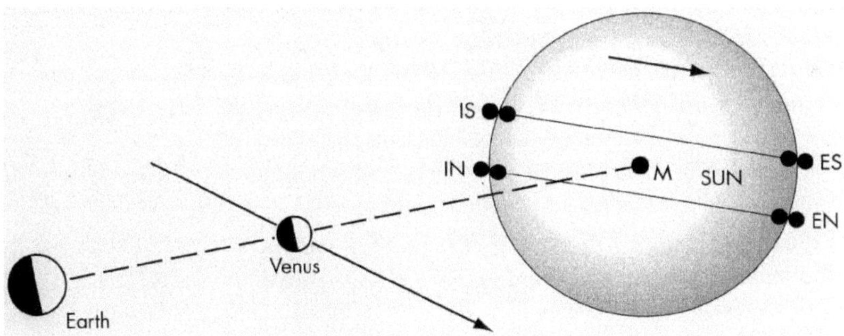

Fig. 5.26. Transit of Venus observed from different locations at the Earth. M is for a point on the Earth's equator and a bisecting transit across the solar equator. IS: Ingress at the Earth's South Pole. IN Ingress at the Earth's North Pole. ES: Egress at the Earth's South Pole. EN Egress at the Earth's North Pole. Adapted from Maunder and Moore (2000).

points with respect to the equator (Halley, 1717).[26] For a critical review of his paper and method see Browne (2005).

Because of the latitude difference of the observers, Venus would appear to move along chords of different length over the solar disk. The length of each chord would be proportional to the duration of the transit. Owing to its proximity to the Sun, Mercury is not adequate for this type of determination. The only prerequisite in longitude for the observers was to be located in the zone of visibility during the entire transit.

The detailed explanation of the formulas is beyond the scope of this book. Here we will provide a basic background based on Sellers (2001) and Mignard (2004). There is no evidence that this method was used by later astronomers, probably because the diffusion of the paper was not adequate.

For the interested reader, here follows a summary of the geometrical relations of the method. The geocentric timings of the contacts are T_i, i = 1,2,3,4 and the equatorial coordinates of Venus are (α_V, δ_V) and the Sun are (α_S, δ_S).

The difference τ of time between the contacts at the centre of the Earth (geocentric) and at a point of longitude λ and latitude ϕ, is

$$\tau = \mathrm{A}\alpha + \mathrm{B}\beta + \mathrm{C}\gamma$$

where A, B and C are numerical constants, and α, β and γ are the direction cosines of the observer with reference to the equator and the Greenwich meridian, given by

[26] Halley, E. *Methodus singularis qua Solis Parallaxis sive distantia a Terra, ope Veneris intra Sol conspiciende, tuto determinari poterit*, Philosophical Transactions of the Royal Society of London, xxix, (1714–1716), No. 348, 454–464. For an English translation see http://www.dsellers.demon.co.uk/venus/ven_ch8.htm

$$\alpha = \cos\phi \cos\lambda$$
$$\beta = \cos\phi \sin\lambda$$
$$\gamma = \cos\phi$$

Let two distant observers be located at (ϕ_1, λ_1) and (ϕ_2, λ_2). Assuming that their clocks are perfectly set and that they can express their measurements in UTC, the expected contact times for the first observer are

$$\mathrm{T}_1^{(\mathrm{i})} = \mathrm{t}_{\mathrm{geo}}^{(\mathrm{i})} + \overline{\mathrm{A}}\,\alpha_1 + \overline{\mathrm{B}}\,\beta_1 + \overline{\mathrm{C}}\gamma_1$$

for the second observer we have a similar equation

$$\mathrm{T}_2^{(\mathrm{i})} = \mathrm{t}_{\mathrm{geo}}^{(\mathrm{i})} + \overline{\mathrm{A}}\,\alpha_2 + \overline{\mathrm{B}}\,\beta_2 + \overline{\mathrm{C}}\,\gamma_2$$

Now, by subtracting these two equations, one gets the expected timing difference for the two observers

$$\mathrm{dt_C} = \overline{\mathrm{A}}(\alpha_2 - \alpha_1) + \overline{\mathrm{B}}(\beta_2 - \beta_1) + \overline{\mathrm{C}}\,(\gamma_2 - \gamma_1)$$

If the observers measure a difference dt_O and the only source of discrepancy is the solar parallax, then we obtain

$$\pi_{\mathrm{S}} = \pi_{\mathrm{ref}} \frac{\mathrm{dt_O}}{\mathrm{dt_c}}$$
$$\mathrm{d(AU)} = 149.60 \times 10^6 \frac{\delta \mathrm{t_{ref}}}{\delta \mathrm{t_{obs}}}$$

Joseph-Nicolas Delisle (1688–1768) suggested a variant to Halley's method, consisting in the measurement of timing of the same phase of the transit from several stations (Delisle, 1763). He showed how, at suitable selected stations, whose longitude had been accurately determined, the single observation of a contact, whether ingress or egress, would supply the means of determining the solar parallax. The advantage of this alternative was that observations from places where the transit would only be partially visible could also be used to establish the Sun's distance, thus increasing the number of potential observing sites. The main problem was the need to know the longitude of every station with high accuracy and furthermore the exact local time at which the transit begins and/or ends. In this case, they can determine the difference dt_c simply with their watches. It is important to stress that the reference value of the parallax is totally irrelevant.

Table 5.7 summarizes the different determinations of the solar parallax after the 7th century.

Several monographs offer an excellent overview of the historical transits of Venus and Mercury: Proctor (1882), Woolf (1959), Meeus (1989), Maunder (2000), Sheenan (2004), Maor (2004), Sterken (2005) and Kurtz (2005).

As a curious aside, we can also mention some of the transits observed of the hypothetical intramercurial planet Vulcan (Baum and Sheehan, 2003).

Table 5.7. Determinations of the solar parallax, π_S, after the 7th century

Method	Author	Value ($''$)
Venus transits 1761 and 1769		$8{,}43''$ - $8{,}80''$
Venus transits 1761 and 1769	Encke (1824)	$8{,}5776''$
Mars Parallax	Hall (1862)	$8{,}841''$
Parallax of asteroid Flora	Galle (1875)	$8{,}873''$
Mars Parallax	Gill (1881)	$8{,}78''$
Venus transits from 1761	Harkness (1889a)	$8,795'' \pm 0.016''$
	Harkness (1889b)	$8.759''$
Venus transits of 1874 and 1882	Newcomb (1890)	$8{,}79''$
Parallax of asteroid Eros	Hinks (1904)	$8{,}806''$
Parallax of asteroid Eros	Spencer Jones (1940)	$8{,}790''$
Radar echo from Venus	Butrica (1997)	$8{,}79415''$

Edmond Modeste Lescarbault (1814–1894) claimed to have seen the supposed planet crossing the solar disk on 26 March 1859. Based on these observations, Urbain Le Verrier calculated the parameters of its orbit. The measured perturbations in the orbit of Mercury were later explained in the framework of Einstein's theory of relativity.

5.6.4 Individual Transits

Abu Ishaq al Bitr (? –1204), known in the West as Alpetragius, was born in Morocco although he later emigrated to Al-Andalus (Seville, Spain). He commented that since he had never seen Mercury transiting the solar disk, the planet must be transparent. He did not realize that Mercury is too small to be seen with the naked eye.

The observation of pre-telescopic Venus and Mercury transits has been discussed by Johnson (1882), and Stephenson (1990). A small part of the Venus transit of 1275 may have been seen by Mayan observers from the city of Mayapan just before sunset on May 25 (Galindo Trejo and Allen, 2005). The excessive duration of the transit or the lack of coincidence with the ephemerides has led other attributed transits to be rejected (Goldstein 1969). Most of them were probably just sunspots visible to the naked eye (see Chapter 2).

Following the introduction of the telescope, three transits of Mercury took place, each of which were visible in Europe (3 May, 1615; 4 Nov., 1618 and 5 May, 1628). Although instruments with the necessary resolution were available (see Chapter 3), no reports of these events have been found.

The First Observed Transit: Mercury 6 December 1631

The transit of Mercury was first successfully observed by the French astronomer Pierre Gassendi (1592–1655) in Paris. He proudly describes the event in a letter to Wilhelm Schickard (1592–1635):[27] *Le rusé Mercure voulait passer sans être aperu, il était entré plus tôt qu'on ne s'y attendait, mais il n'a pu s'chapper sans être découvert, je l'ai trouvé et je l'ai vu; ce qui n'était arrivé á personne avant moi, le 7 novembre 1631, le matin.*[28] This first observation ever of the passage of a planet over the Sun was also made known by Gassendi in the booklet *Mercurius in Sole Visus et Venus Invisa* published in Paris in 1631.

For the transit, Gassendi used the same technique as the one used for observing sunspots. In a darkened room, the image of the Sun was admitted through a Galilean telescope and projected onto a piece of paper. He adjusted the distance so that the projected image was two-thirds of a Parisian foot in diameter and drew a circle with that diameter on the paper. He divided the diameter of this circle into 60 equal parts, so each division would correspond to 30 seconds of arc. An assistant was stationed in the room below, where he was to measure the Sun's altitude with a quadrant each time Gassendi stamped his foot (Van Helden, 1976).[29]

Other successful observers were Johann-Baptist Cysat (1588–1657) in Innsbruck (Austria) Johannes Remus Quietanus in Rouffach (Alsace) and an anonymous Jesuit in Ingolstadt (Germany), although they did not publish the observations (Humbert, 1936).

In his *Admonitio ad Astronomos*, published in 1630, Kepler had predicted that Venus, if observed on the Sun, would show a size almost a quarter of the solar diameter. So Gassendi expected that Mercury should then occupy a tenth part. In his own words, *I was far from suspecting that Mercury would project such a small shadow. For such was its smallness that its diameter hardly appeared to exceed half of one of the divisions marked. I thought rather that it was a sunspot which I had not noticed on the Sun the previous day, but which could have grown since that time, just as I had seen others do.* Their observations provided the first clear proof that the apparent diameter of Mercury was much smaller than had hitherto been believed, leading to a reassessment of the distances within the Solar System. In fact, Joseph Gaultier de la Valette (1564–1647) had noticed nothing on the solar disk during his observations at Aix en Provence. This was probably due to the fact that he was using a simple camera obscura with an aperture too large to show Mercury.

---

[27] Professor at the University of Tübingen, he built the first calculating machine in 1623.

[28] Astute Mercury wanted to cross without being seen, it came sooner than anyone had expected but it could not escape being discovered, I found it and I saw it; it happened to nobody before me on 7 November 1631 in the morning.

[29] Other authors stated that Gassendi admitted the light into the room through a small opening in the window. Van Helden (1976) correctly remarked that apart from the fact that Gassendi mentions the use of a telescope, it is inconceivable that he could have obtained an eight-inch image of the Sun in the space of a room, using a simple camera obscura.

The revised apparent diameters of the planets were reported by G.B. Riccioli in his *Almagestum Novum*, published in 1651.

Along with Galileo's observation of the phases of Venus, the transit of Mercury was among the first pieces of direct evidence in favour of the Copernican over the Ptolemaic (geocentric) conception of the Solar System.

Venus: December 1631

In 1629, as Kepler was preparing a set of astronomical ephemerides for the years 1629 to 1636, based on his new laws of planetary motion published in his *Rudolphine Tables* (Ulm, September 1627), he noted that a transit of Mercury would take place on 7 November 1631 and a transit of Venus on 6 December of the same year. Although his calculations indicated that the latter transit would best be visible from the American continent, he cautioned European astronomers in his pamphlet *De raris mirisque Anni 1631* (Leipzig, 1629) to be watchful as well.

Despite careful vigilance, Gassendi and other astronomers failed to see the transit of Venus in the following month. According to modern calculations, the transit actually ended about 50 minutes before sunrise (7:35 UT) at Paris although observers in most of Italy and along the Eastern Mediterranean should have been able to view the last stage of the transit.

Venus: 4 December 1639

The Rudolphine tables of J. Keppler predicted no further Venus transits until 1701.[30] However, Jeremiah Horrocks (1619–1641) born in Toxteth, near Liverpool, doubted the accuracy of Kepler's predictions and proceeded to predict, in early November 1639, that a further transit would occur on 4 December 1639. His close friend William Crabtree (1610–1644) was one of the few to get information on the coming transit. In a letter from Horrocks to him dated 26 October, Horrocks says: *My reason for now writing is to advise you of a remarkable conjunction of the Sun and Venus on the 24th of November, when there will be a transit. As such thing has not happened for many years past, and will not occur again in this century, I earnestly entreat you to watch attentively with your telescope in order to observe it as well as you can.*[31] (Aughton, 2004).

He had a small telescope, probably a Galilean with D = 4 cm. On that day in his home in Much Hoole, near Preston, Horrocks projected a 16 cm solar image onto a graduated sheet of paper. The horizontal diameter was

[30] Tables of planetary motions were the first to make use of Kepler's newly formulated laws on planetary motions. Printed in Ulm in 1627, they were dedicated to Tycho Brahe and Rudolph II.

[31] At that time the Julian Calendar was still in use in England and the Gregorian 4 December corresponds to the Julian 24 November.

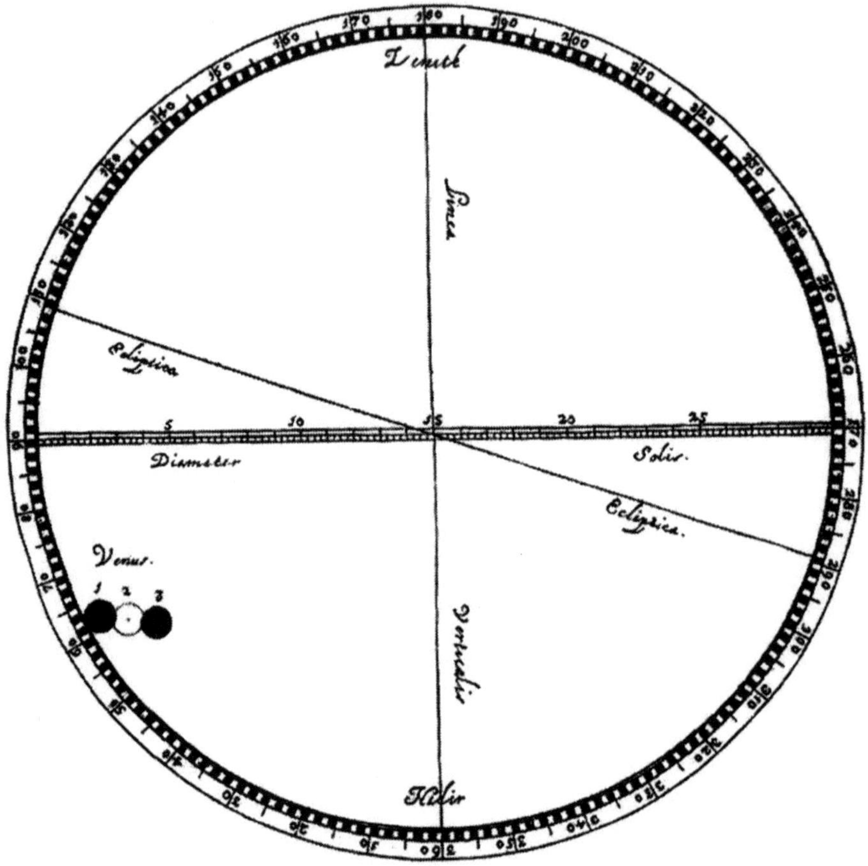

Fig. 5.27. Observations of Venus in transit by J. Horrocks, as was published in *Venus in Sole Visa* by J. Hevelius.

divided into 30 equal parts and in these units he recorded the measurements (Figure 5.27). 4 December was a Sunday and Horrocks was at the telescope until noon. Although the sky was overcast, he was able to see the Sun for some brief moments, seeing nothing unusual. At 13 hours in the afternoon he interrupted the observation by "business of the highest importance, which for these ornamental pursuits I could not with propriety neglect" (Maor, 2004). He reassumed the observation around 15 h 15 m noting "a spot of unusual magnitude and of a perfectly circular shape" had already fully entered upon the solar disk.[32] At Much Hoole the exterior ingress was at 14 h 59 m 42.7 s with the interior ingress occurring at 15 h 18 m 03 s and the sunset at 15 h

[32] Whatton (1859) commented that the reason of his absence was to conduct divine services at the church. Chapman is of the opinion that it may have been simply to take the children under his care to church.

53 m. Therefore during the remaining hours of the transit this was invisible to observers in western Europe.

Although missing the ingress, he was fortunate to see Venus crossing the solar disk before the sight was lost in the sunset (Chapman, 1990), measuring carefully the size and position of the planetary spot. Crabtree observed the transit from Salford, near Manchester, but the sky was mostly cloudy and he was unable to make any measurements except of the last phase of the event (Kollerstrom, 2005).

These observations were first published in Gdansk with the title *Venus in Sole Visa* under the auspices of J. Hevelius, who added his own explanations and appended it to his own parallel work on Mercury. However, the annotations of Hevelius were very voluminous, actually much longer than the original text. Horrocks describes the appearance of the planet as follows: ... *a spot of unusual magnitude and of a perfectly circular shape, which had already fully entered upon the Sun's disk on the left, so that the limbs of the Sun and Venus precisely coincided, forming an angle of contact. Not doubting that this was really the shadow of the planet, I immediately applied myself to sedulously observe it.*. See Whatton (1859) for an English translation.[33]

After a long process, John Wallis (1616–1703) proceeded to publish, in 1672, a compendium called *Opera Posthuma* where he was able to include all he could find of the Horrock's work, including forty letters to Crabtree, and the contributions of J. Flamsteed and W. Crabtree. A copy of the Latin original is in the Liverpool Central Reference Library.

Hughes (2005) suggests reasons why the event was not observed in other places of Europe, although it was visible over Italy, France, Spain and Portugal. He proposes that it was probably due to the drop in interest in solar phenomena after the discovery of sunspots, and that the event took place in the afternoon of a Sunday played a role. Scheiner was living at that time in Vienna, outside the visibility zone, and Galileo had already gone blind by 1637.

Mercury: 3 November 1651

Jeremiah Shakerley (1626–1670?) was born in Halifax, and became an important mathematician. Recognizing the importance of the observations of J. Horrocks, he went to Surat (India) with the purpose of observing the Mercury transit. However, he was unable to time either the ingress nor the egress.

[33] The Horrocks manuscript was published in Latin for better international diffusion and the originals were lost. They now appear translated again into English for the same reason.

Mercury: 3 May 1661

Published calculations predicted a transit between 1 May and 11 May. Hevelius planned 11 days of continuous observation, if need be. The third day was partly cloudy, and at 14:00 the Sun appeared for just a few seconds, with a spot; then at 16:30, the Sun reappeared and the spot had moved. Clouds parted again at 17:00 and 19:30. The image of the Sun was 80 mm on the easel, and Mercury was about 1 mm in diameter. Hevelius used his drawings to estimate times of first and last contact and deduced the elements of Mercury's orbit. He measured Mercury's apparent diameter at 12 arcseconds – close to the actual value of almost 13 arc seconds and much more accurate than Gassendi and Riccioli's contemporary measurements of 28 and 25 arcseconds (Volkoff, 1971). The event was also seen by Christiaan Huygens in London. No records seem to exist of the Mercury transit of 4 November 1664, visible in Europe.

Mercury: 7 November 1677

During his 13-month sojourn on the island of St. Helena in the southern Atlantic Ocean, the English astronomer Edmond Halley (1656–1742) succeeded in observing a complete transit of the planet Mercury over the solar disk. He used for this purpose a 24-foot telescope.

He made careful measurements and commented he had *very accurately obtained the very moment in which Mercury, entering the Sun's limb, seemed to touch it internally [...] Hence I discovered the precise quantity of time [...] and that without an error of a single second.* It was during these observations when Halley realised that transits could be used to measure the astronomical unit.[34]

Clouds covered most of Europe during the event, with some exceptions. In Lancashire, R. Towneley was able to observe the Sun through "flying clouds" during the last part of the event, timing Mercury's exit. Charles Gallet and M. de Beauchamps observed the transit from a church in Avignon using a lens of 23 feet (Gallet, 1677). Figure 5.29 shows the passage of the planet across the solar disk.

Mercury: 5 May 1707

It was the longest transit in historical times (7 h 56 m), but the weather conditions were rather unfavourable in central Europe (Van Biesbroeck, 1912). Only O.C. Roemer communicated some observations to E. Halley made in

[34] He wrote to Jonas Moore on 2 December: *I have notwithstanding had the opportunity of observing the ingress and egress of Mercury on the Sun, which compared with the like observations made in England, will give a demonstration of the Sun's parallax, which hitherto was never proved, but by probable arguments.* Cited in E.F. Mac Pike (ed.), 1932, Correspondence and Papers of Edmond Halley, Clarendon Press, Oxford.

MERCURII TRANSITUS
Sub Solis diſco, *Octob.* 28. Anno 1677,
cum tentamine pro Solis Parallaxi.

RArum iſtud, & a mortalibus non niſi ter, (quod mihi ſcire contigit,) hactenus obſervatum Phænomenon tranſitus Mercurii ſub Solis diſco, mihi, in Inſula Sanctæ *Helenæ* commoranti, felicius obſervare, quam cuivis alio Aſtronomo, contigit: *Gaſſendus* enim in tranſitu Anni 1631, & in hoc noſtro Clariſſimus *Gallet*, exitum ſolum ſpectaverunt; ingreſſu, huic ſub denſa nubium compagine, illi ſub terra Orientali, latente; Atque imperfectius adhuc Anno 1661, inclytus ille *Hevelius Gedani*, & noſtrates *Londini*, qui ſolo ſitu intra faciem Solarem ſumpto contenti erant: Mihi primo & ingreſſus & egreſſus momenta accuratiſſime conſpecta ſunt, idque peculiari & inſolito Cœli favore; erat enim nocte præcedente *Octobris* 28vum Cœli facies triſtiſſima, cum vento valido, interdumque deſcendentibus Nubibus denſa Nebula Inſulæ ſummitates obvelavit; luce reverſa, vento licet paulo remiſſiore, idem manſit Cœli vultus; Juxta Solis ortum, ad inſtrumenta me contuli, languente jam omne ſpe obſervationis habendæ, tuboque 24 pedum in plagam Solis verſo, patienter expectavi, an per Nubium aliquem hiatum conſpici poſſit deſideratiſſimus Phœbus: Juxta horam octavam Nubes rareſcere ceperunt, ita ut 8 *h.* 36 *m.* Sole clare conſpecto, Mercurium nondum intraſſe pronunciavi; inde brevibus intervallis ſæpius eluxit, ac ſequentem habui obſervationem.

A *In.*

Fig. 5.28. First page of the description of the 1677 Mercury transit by E. Halley. Included in the Catalogus Stellarum Australium (1679). Courtesy: Smithsonian Institution Libraries.

the morning of 6 May, but the quality was poor. Baily (1835) discussed the reliability of an observation by A. Sharp (1651–1742), whose description was found amongst the documents of J. Flamsteed.

Mercury: 6 November 1710

In 1742, Johann Gabriel Doppelmayr (1677–1750) published the *Atlas Coelestis*. At the lower right corner of its plate 7, a drawing of the Mercury transit is included (Figure 5.30).

Mercury: 11 November 1736

Vaquero et al. (2007) describe the existing literature on this event.

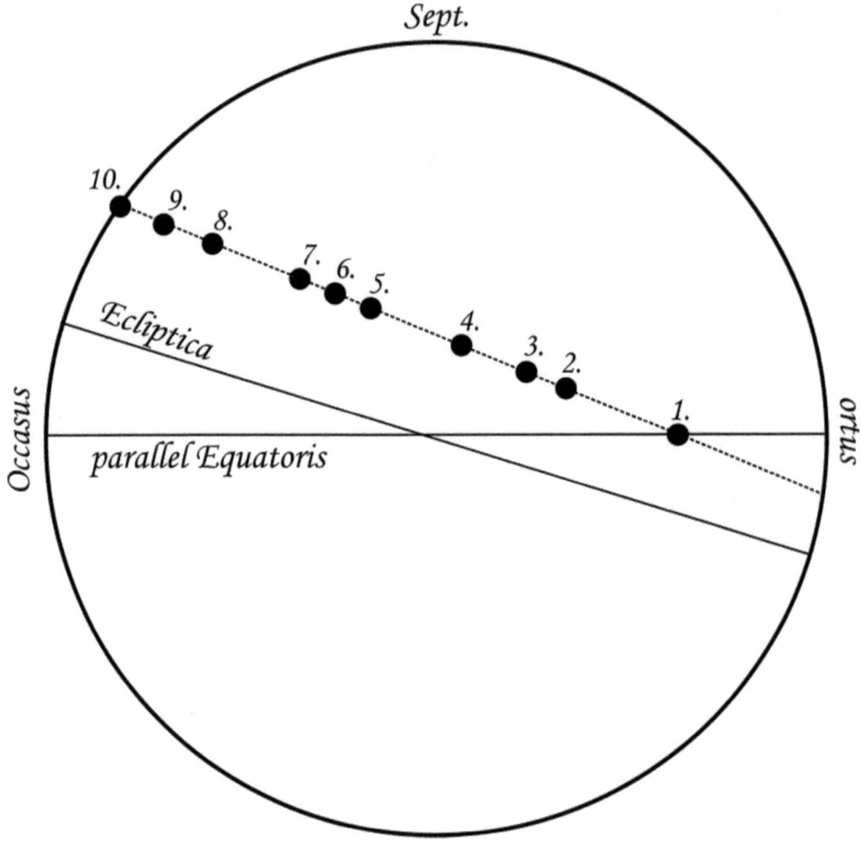

Fig. 5.29. The 1677 Mercury transit observed by Charles Gallet. Available from Gallica.

Mercury: 21 April 1740

John Winthrop (1714–1779) reported to the secretary of the Royal Society his observations made in Cambridge (New England) with a 24-foot telescope.[35] He used a pocket-watch with a claimed accuracy of a quarter of a minute (Winthrop, 1742).

Mercury: 5 November 1743

John Winthrop again observed this Mercury transit in Cambridge, New England. His description of the event illustrates the difficulties of solar observations produced by turbulence of the Earth's atmosphere: *I could not be so certain of the moment when he (Mercury) left the Sun, as of his interior*

[35] He was Professor of Mathematics and Natural Philosophy in Harvard College.

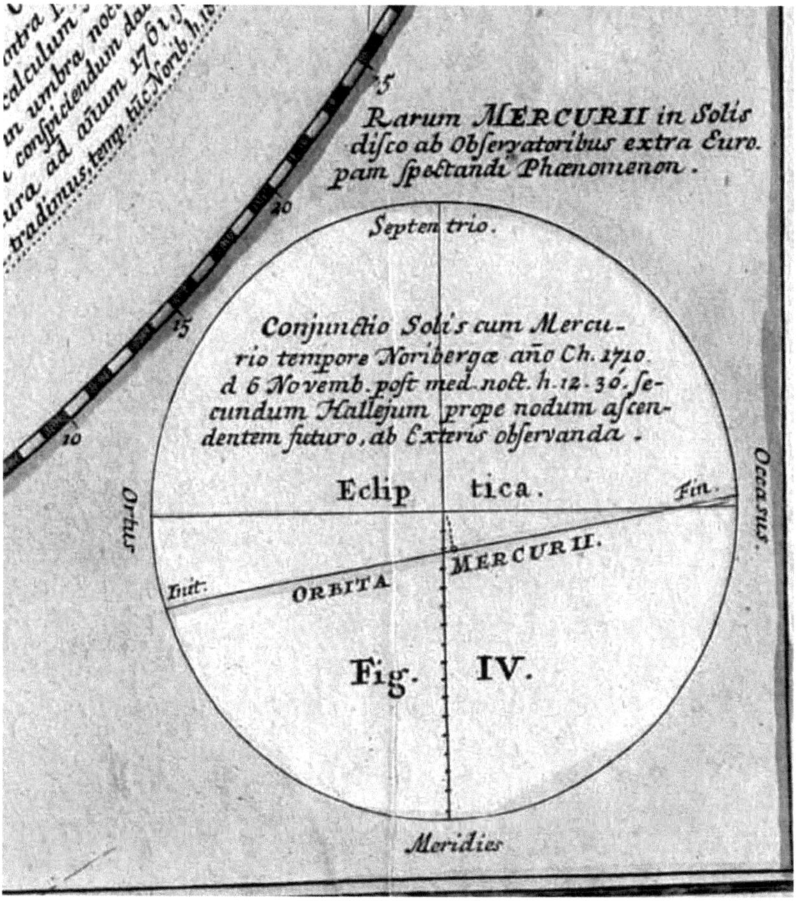

Fig. 5.30. Drawing of the 1710 Mercury transit included in the *Atlas Coelestis* (Plate 7) of J.G. Doppelmayr. Available from Gallica (Bibliothèque Nationale de France).

contact. For the Sun's limb, undulating in the vapours of the horizon, made it somewhat difficult to judge when the indenture, formed by the Planet upon it, entirely ceased (Winthrop, 1763). Joseph Delisle coordinated observations worldwide. Nicolas Louis de Lacaille (1713–1762) report that the discrepancies among different observers of the transit times were of 40 seconds, too much to obtain reliable results.[36]

Mercury: 6 May 1753

J. N. Delisle observed the transit from Paris and part of his drawings are preserved at Gallica, the National Library of France. Probably this event was

[36] He is mainly known for his observations of the southern skies.

observed by King Louis XV at Meudon. Benjamin Franklin (1706–1790) also observed the event and described it in a four-page pamphlet (Cohen, 1950). Figure 5.31 summarizes the different predictions of Mercury's path across the Sun.

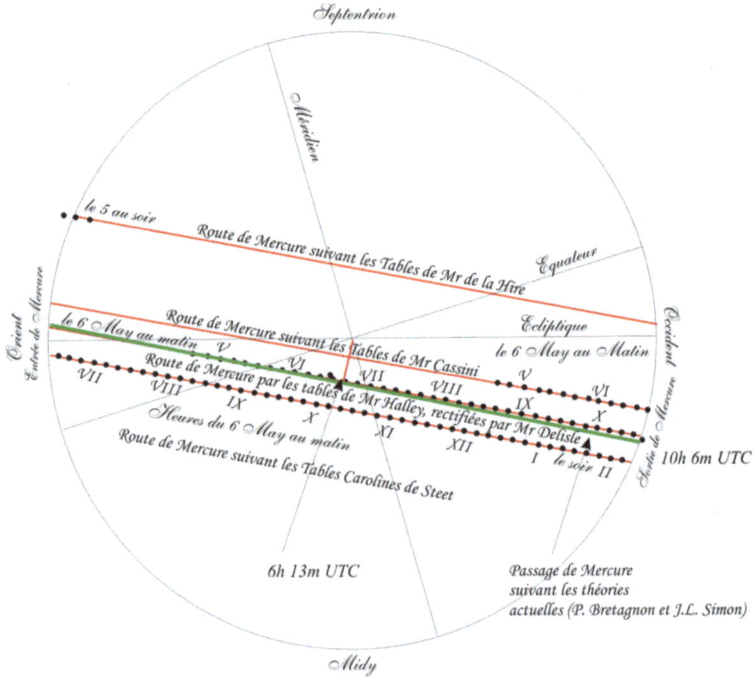

Fig. 5.31. Transit of 1753: drawing of J.N. Delisle with the various possible transits. In green is included the transit of Mercury calculated using the current theories. Courtesy: P. Rocher (Institut de Mécanique Céleste et de calcul des Ephémerides, Observatoire de Paris).

Venus: 6 June 1761

By this time, the middle of the 18th century, advancements in navigation had made it easier to organize scientific expeditions to remote places. However, ongoing wars between the superpowers of that time made any scientific enterprise more difficult.

Hornsby (1763) summarized numerous observations made in the visibility zone of the transit. The values of solar parallax ranged from $9''.101$ to $10''.145$ with a mean of $9''.695$. His paper finishes with a list of recommendations for the next transit in 1769. From 62 observing stations, the French Academy received about 120 reports reporting values of solar parallax between $8''.6$ and $10''.6$ (Débarbat, 2005).

James Ferguson (1710–1776) commented on the whole series of observations:[37] *Whoever compares the times of the internal contacts, as given by different observers, will find such difference among them, even those which were taken from the same spot, as will show that the instant of either contact could not be so accurately perceived by the observers as Dr. Halley thought it could, which probably arises from the difference of people's eyes and the different magnifying powers of those telescopes through which the contacts were seen.* Once more the influence of the personal and instrumental profiles on the observations is evidenced. Figure 5.32 shows a detail of a plate included in his book *Astronomy explained upon Sir Isaac Newton's Principles.*

Mikhail V. Lomonosov (1711–1765) was born in Arkhangelsk, in the northern part of European Russia. Founder of the Moscow University, in 1755, he was recognized in different fields of natural sciences. A discussion on the accuracy of the transit ephemerides brought his attention to the transit of 1761. Under his direction, observations were made at St Petersburg Observatory by the astronomers A.D. Krasinilkov and N.G. Kurganov using a 6-foot long focus telescope. Figure 5.33 shows drawings obtained during the event.

Lomosonov himself used a telescope of 4.5 feet with two-lens tubes covered by a smoky glass. These observations led him to discover the existence of an atmosphere around Venus. In his own words, *Before the Venus ingress, when*

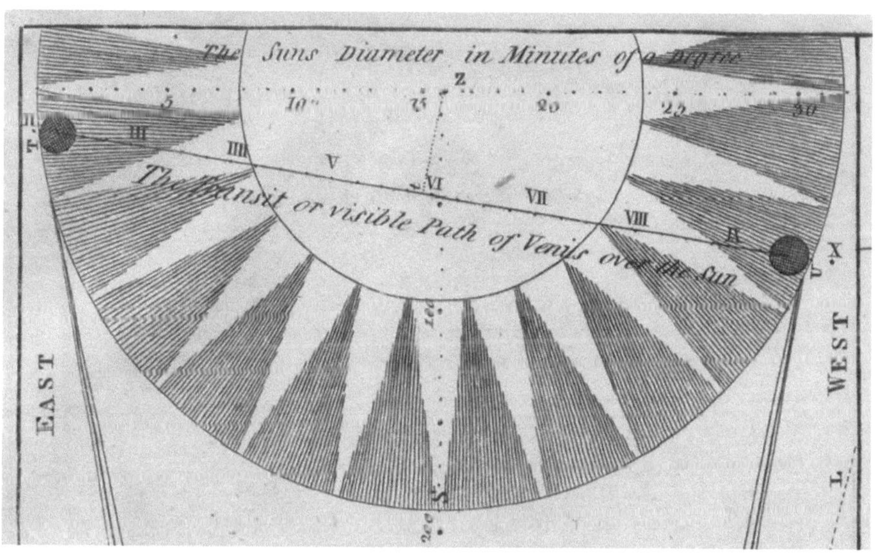

Fig. 5.32. Venus at both ends of its path across the solar disk, at the two moments of internal contact with the limb of the Sun. Courtesy: Museum of the History of Science, Oxford.

[37] Cited in Proctor (1882) p. 53.

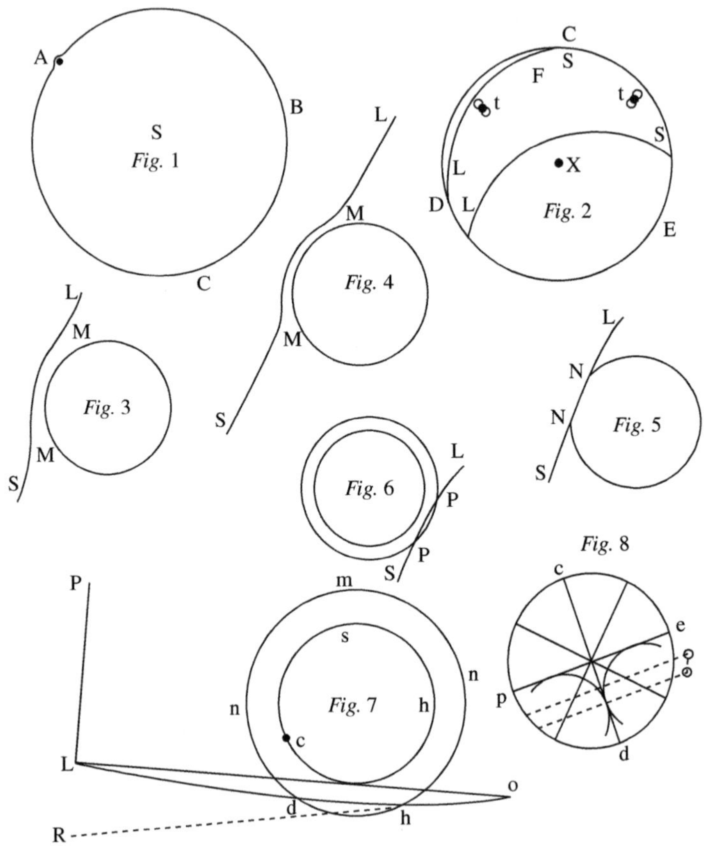

Fig. 5.33. Drawings of the 1761 Venus transit made by M. V. Lomosonov.

its front side approached the solar edge at about one tenth of the planet's diameter, a bulge set up which progressively became more pronounced as Venus came to leave the Sun. Soon after that the bulge disappeared and instead Venus appeared with no edge [...] Based on these observations I conclude that the planet Venus is surrounded by a distinguished air atmosphere similar (or even possibly larger) than that is poured over our Earth. (Marov, 2005).

Jean Baptiste Chappe d'Auteroche (1722–1769) was a French astronomer of Moroccan origin. In 1761, he travelled to Tobolsk (Siberia) to observe the transit. The details of the journey are described in his book *Voyage en Sibérie*, whereas the observations are included in *Memoirs du passage de Venus sur le soleil* published by the Imperial Academy of Sciences at St. Petersburg (see Figure 5.34). Using data of the transit he determined the solar radius to be 15′ 19″ and observed that the radius of Venus increased by using larger telescopes.

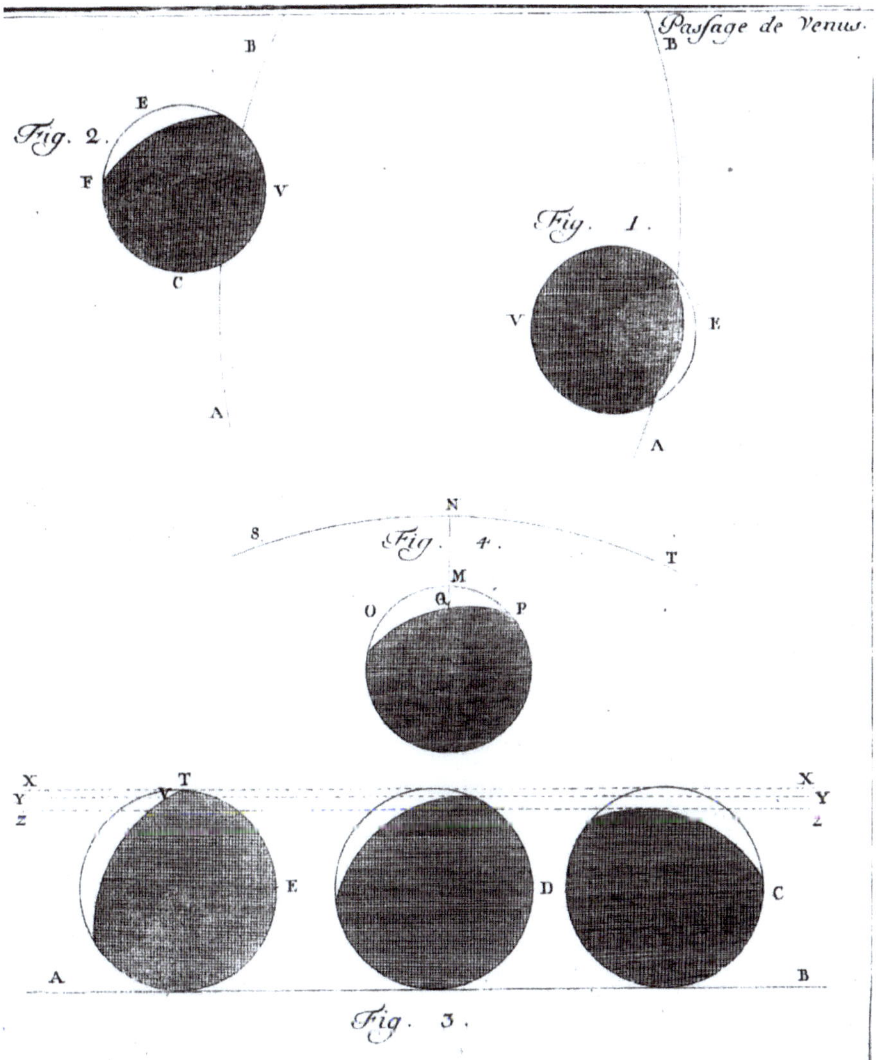

Fig. 5.34. Drawings of the Venus transit made by J. Chappe. Courtesy: Niedersächsische Staats- und Universitätsbibliothek Göttingen, Digitalisierungzentrum.

His equipment for the observations consisted of a quadrant with a telescope and micrometer. Table 5.8 gives the most relevant data of these instruments.

Alexander-Gui Pingrè (1711–1796) headed a French expedition to Madagascar, but unfortunately due to cloud cover he was only able to observe the middle of the transit.

At the Royal Observatory Paris, Jean D. Maraldi (1709–1788) remarked that at the egress, the disk of Venus was encircled by a reddish glow extending

Table 5.8. Equipment of Chappe in the observations in Tobolsk (Siberia)

Focal objective	Telescope diameter	Focal eyepiece	Magn.	Micrometer parts	Corresponding angles
19pied 7po.	2po 0lig.	1po. 9lig,	134	3303	10′ 18″ 21‴
10. 3.	1. 6.		70	3414	20′ 28″ 4‴
5. 10	−10		40	1919	20′ 28″ 40‴

over the planetary radius, that he attributed to the fatigue of his eyes, since he had been observing the Galilean satellites the previous night.[38]

Several British campaigns were organized for this event by the Royal Society. Neville Maskelyne (1732–1811) was sent to the island of St. Helena in the mid-Atlantic. Unfortunately, he got clouded out and missed the end of the transit, making his data mostly useless. Jeremiah Dixon (1733–1779) and Charles Mason (1730–1786) were originally dispatched to Bencoolen (modern Bengkulu) in Sumatra, but their ship (HMS *Seahorse*) was attacked by a French frigate only a few hours after sailing from Portsmouth England, forcing them to limp back into port. After much delay repairing and refitting they reluctantly set sail again (after threats from the Society if they decided to bale out), but only made it as far as Cape Town, South Africa before the transit. Undeterred, they changed their plans, set up their observing station in time, and got excellent data. On the passage home, they stopped at St Helena in October and, after discussion with Maskelyne, who had observed the transit there, Dixon returned temporarily to the Cape with Maskelyne's clock to carry out gravity experiments.

At the Royal Observatory in Greenwich the transit was observed by Nathaniel Bliss (1700–1764). The sky was very cloudy in the morning, till the end of the transit, when it was tolerably clear. Bliss timed both the interior and exterior contacts at egress. For this purpose he used a reflecting telescope, two feet in focal length, to which was fitted a Dollond micrometer, both executed by J. Short (Bliss, 1762). Hornsby was more fortunate with the weather at Shirburn Castle.

Italian observations of this transit have been summarized by Pigatto (2005), where the work of the Dominican G.B. Audiffredi (1714–1794) during this event is discussed in detail.[39] Eustachio Zanotti observed the transit from the tower of the observatory in Bologna together with two assistants.

In the Netherlands, the 1761 transit of Venus was observed by several people. Johan Lulofs (1711–1768) was professor of mathematics and philosophy at Leiden University and made his observations from the simple university

[38] The report was published at the "Histoire de l'Academie Royale des Sciences" in 1763 in the pages 76–77.

[39] The observations were carried out at the roman church of Santa Maria sopra Minerva and published, around 1762, in the report *Transitus Veneris ante Solem observati Romae 6 junii 1761 expositio.*

observatory which was located on the roof of the Academy Building. Dirk Klinkenberg (1709–1799) watched the sunrise with a Gregorian telescope from a dune west of Zorgvliet. Later on he went back to Zorgvliet and observed the remaining part of the transit with a Cassegrain telescope. Afterwards, he discovered that his time piece was in error of at least ten seconds, making his observations useless. Jan de Munck constructed his own observatory next to his house in Middelburg. From this observatory he observed the transit of Venus with his 25 cm refractor, providing a projected image of the Sun for a couple of friends, whom he had invited to watch the phenomenon. The observatory tower was torn down in 1775, but the house is still present.

Venus: 3 June 1769 and the Black Drop Effect

The Royal Society presented a memorial to King George III, requesting a vessel to convey a group of astronomers to a suitable site in the Southern Pacific. The petition was granted and the ship *Endeavour*[40] was placed under the command of Captain James Cook (1728–1779). Also on board were the astronomer Charles Green (1735–1771) and the botanists D.C. Solander (1733–1782) and J. Banks (1743–1820), who also helped with the observations. Setting sail from Deptford on 30 July 1768, the expedition arrived at Tahiti, the selected site, on 10 April 1769, where in one month "Fort Venus", the observing station for the event, was built.[41] They also built two auxiliary stations on the islands of Moorea and Puaru. The drawings of the expedition are preserved at the Armagh Observatory.

A remarkable side result of these observations was the descriptions and drawings of a phenomenon showing the influence of the instrument and the atmosphere on the astronomical observations. Figure 5.35 shows clearly the black drop effect. Modern observations from space show that this effect is due to the PSF (Point Spread Function, see Chapter 1) of the instrument and atmospheric seeing convolved with solar limb darkening (Wittmann, 1974; Schneider et al., 2004; Licchelli, 2005; Pasachoff et al., 2005) and is better evaluated using Mercury transits. The worse the atmospheric seeing is, the better the black drop is observed.[42]

The observatory of the Earl of Macclesfield (George Parker 1695–1764), at Shirburn Castle, participated in the observations. The Earl himself made use of a refracting telescope of $3\frac{1}{2}$ feet, made by J. Dollond, magnifying 150 times, and a stop-watch to calculate the timing of the transit. On the north side, Mr. Barlett observed with a 14 feet refractor, hearing the seconds counted out by Mr. Phelps, another assistant. The sky was free of clouds but the solar limbs suffered constant undulations indicating the presence of strong air turbulence.

[40] A barque of 370 tons, which had been built for the coal trade.

[41] Of the 94 people who set off on the *Endeavour*, only 54 came back in August 1771. Charles Green died on 29 January 1771 and Zachary Hicks on 25 May 1771.

[42] J. Lalande had already proposed in 1771 that the terrestrial atmosphere would blur the image such that a meniscus (the black drop) appears.

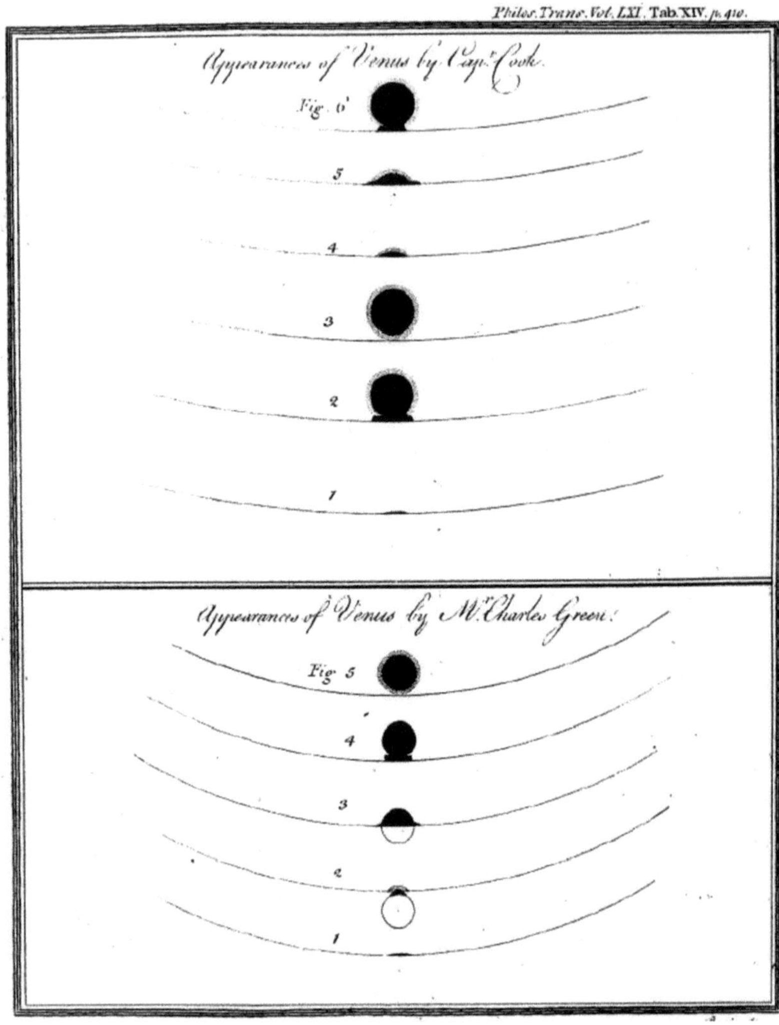

Fig. 5.35. Observation of the Venus transit by James Cook and Charles Green from Fort Venus (Tahiti). *Philosophical Transactions of the Royal Society* Vol. 61, p. 410 (1771). Available from Armagh Observatory.

Thomas Hornsby (1733–1810), at that time professor of Astronomy in Oxford, described the observations and made his own observations in the upper room of the tower of the schools. He observed with a refractor of 12 feet, furnished with a system of eye-glasses and magnifying 68 times (Hornsby, 1769). Observations were also carried out at Norriton in Pennsylvania.

Hornsby (1771) used the data of Green and Cook (1771) in Tahiti and Wales and Dymond (1769) in Hudson bay to calculate the solar parallax using his own method. Evaluation of these data by Browne (2005), applying the method of Halley, led to a solar parallax of 8.61″, very close to the modern value of 8.65″.

Dunn (1770) reveals why accurate measurements of the exact moment of contact were impossible. His drawings show how the edge of Venus appeared to smear instead of remaining sharply defined as it approached the Sun. This phenomenon played havoc with the transit observer's measurements (Figure 5.36).

For this transit, J.B. Chappe led a French-Spanish expedition to the southern tip of Baja California.[43] He observed the transit under ideal conditions, describing the black drop as follows: *The edge of the disk of Venus lengthened*

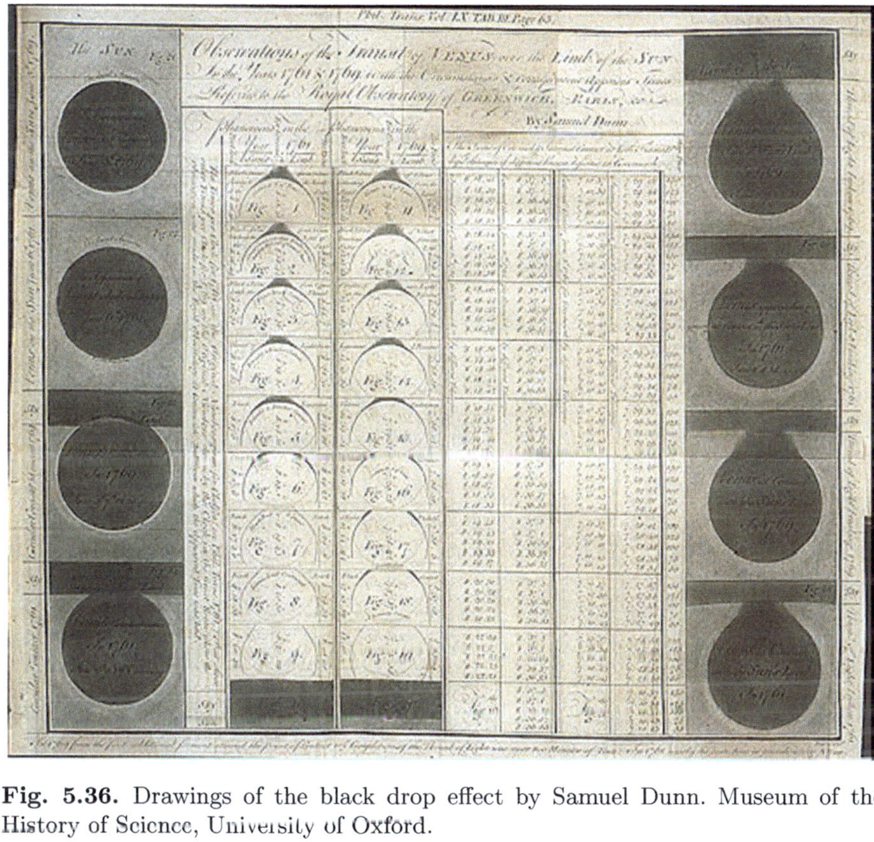

Fig. 5.36. Drawings of the black drop effect by Samuel Dunn. Museum of the History of Science, University of Oxford.

---

[43] The French group was led by Chappe and composed of Jean Pauly, Alexander Noel and Dubois, the Spanish were Vicente de Doz and Salvador Medina. A Mexican group headed by Joaquín Velázquez Cárdenas was observing the transit a few kilometers south from the present town of San Antonio. This was the first permission given by Spain to a foreign scientific expedition to enter in its American domains. A similar proposal from the Royal Society of London was denied. For more details see Piñera Ramírez (1982).

itself, as if it had been attracted by the Sun....but not being able to doubt that this black point was not part of the opaque body of Venus, I observed the moment when it ended in such a sort that the total ingress could not have occurred earlier.... The black point was a little less dark than the rest of Venus.. He died, together with Dubois and Medina, a month after the transit as a consequence of an epidemic but his colleague Jean Pauly was able to return to France with the results which were published in the book *A Voyage to California.*

David Rittenhouse (1732–1796), one of the most important early American astronomers, viewed the transit of Venus in 1769 from Norriton, Pennsylvania. For this purpose he had prepared a wooden observatory with a 2-foot Gregorian reflector and a Dollond micrometer. He describes that just before the internal contact, as Venus began to cross the Sun's edge, a faint halo appeared around the edge of the planet.

The Mexican José Antonio Alzate was an enthusiastic meteorology observer and a corresponding member of the French and Spanish Academy of Sciences. In 1770 he described observations of the Venus transit made together with Bartoleche in Mexico City. Figure 5.37 shows the solar disk during the event.

Vicente Tofiño (1732–1795) observed the 1769 transit from the observatory of Cádiz, when the Sun was setting. The observatory, founded in 1753, was located at the Castillo de la Villa. This castle no longer exists.

Benjamin Martin (1705–1782) was a popularizer of British astronomy. In his *Institutions of Astronomical Calculation containing a survey of the Solar System*, published in 1773, he included an artistic view of the part of the Venus transit that could be see from London (Figure 5.38).

Jeremiah Dixon (1733–1779) sailed to Norway in 1769 with William Bayly to observe another transit of Venus. The two split up, with Dixon at Hammerfest Island and Bayly at North Cape, in order to minimize the possibility of inclement weather obstructing their measurements (Robinson, 1950).

In total, 200 memoirs of the transit were sent to the French Academy of Sciences and more than 400 to the Royal Society.

Mercury: 9 November 1769

After having observed the Venus transit in Tahiti, Cook sailed to New Zealand and observed the transit of Mercury in November 1769 from an area now called Mercury Bay. In Cook's logbook, he states: *Thursday, 9 th: Variable light breezes and clear weather. At 8 Mr. Green and I went on shore with our instruments to observe the transit of Mercury, which came on at 7h20'58" apparent time, and was observed by Mr. Green only. I at this time was taking the sun's Altitude, in order to ascertain the time.*

J.A. Alzate mentioned that 20 spots were visible during this transit, much more than during the previous transit of Venus (see Figure 5.39).

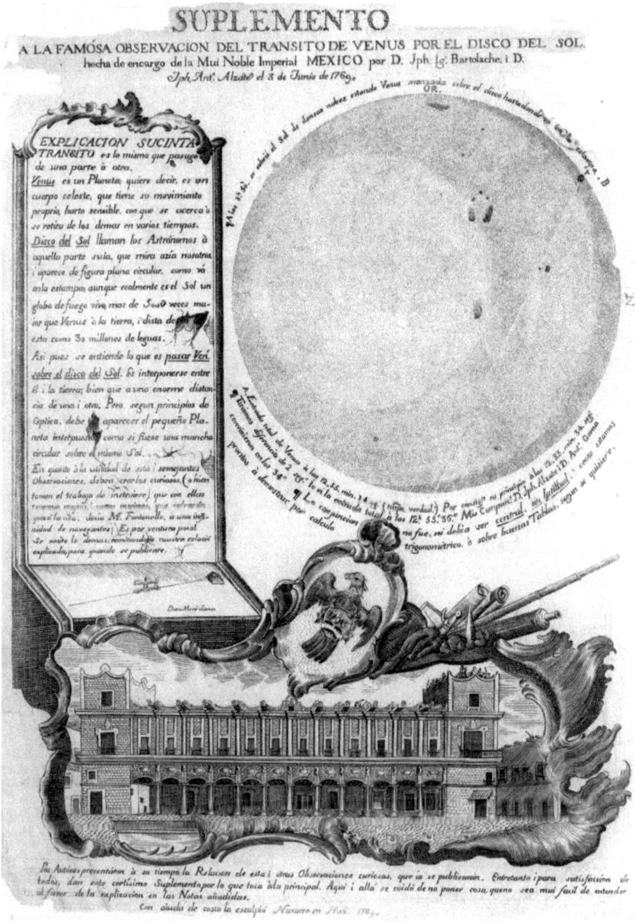

Fig. 5.37. The solar disk observed during the Venus transit by J.A. Alzate.

Mercury: 12 November 1782

Observed by Charles Messier (1730–1817) in Paris where the weather was fine (Messier, 1782). J.A. Hamilton (1748–1815) also reported observations of the event to the Astronomer Royal N. Maskelyne (1732–1811), made at the Irish village of Cookstown (Hamilton, 1783; Butler, 2005).

Mercury: 4 May 1786

Edward Pigott (1753–1825) was able to observe the transit at Louvain (Belgium) using a two-feet achromatic telescope, magnifying about 70 times. Although the there were some clouds during the event, he calculated the contact times with the solar limbs (see McConnell and Brech, 1999). Similar

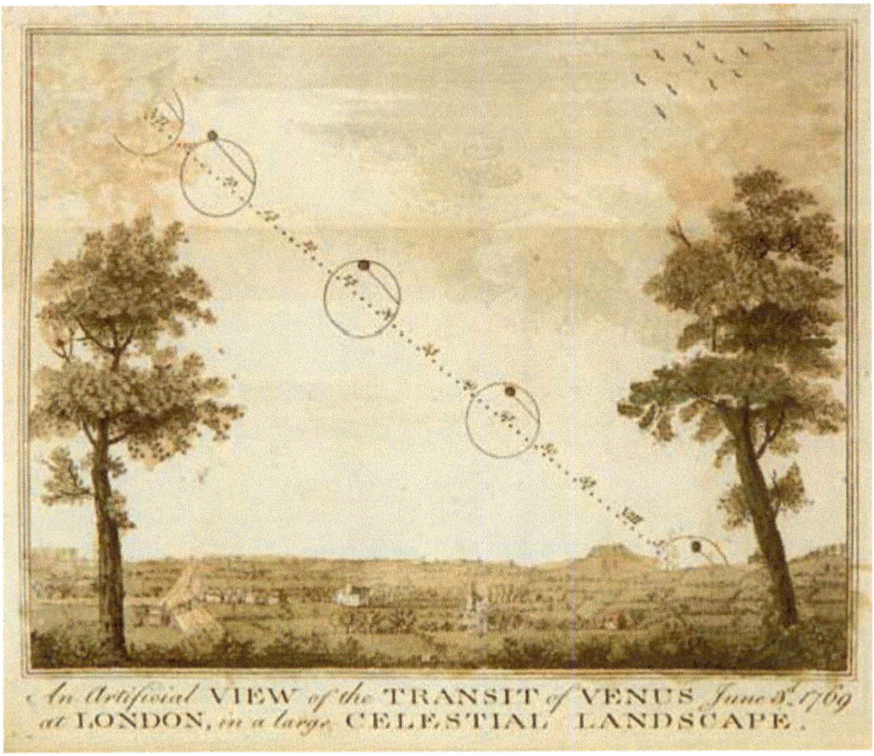

Fig. 5.38. Reproduction of the 1769 Venus transit as observed from London after a working demonstration by B. Martin. The scene itself is to the north of the city, looking westwards from Islington towards Hamstead. Courtesy: Museum of the History of Science, Oxford.

observations were made by M. Rumovski at Saint Petersburg (Russia) determining the diameter of Mercury and the characteristic distances with a micrometer (Rumovski, 1787).

Mercury: 9 November 1802

The transit of 9 November 1802 was viewed by W. Herschel at 7 a.m., using a telescope with a glass mirror 6.3 inches in diameter and 7 feet focus. He also used a 10 feet reflector at 130 power; the same telescope he used at 1000 power at night. *With two small double convex lenses, both made of dark green glass, and one of them having the side which is nearest the eye thinly smoked....power was about 300. With a single eye-glass, smoked on the side towards the eye, and magnifying 460 times...the appearance of the planet...remained well defined.* Describing the event he noticed: *It (Mercury) was easily distinguished from the openings in the luminous clouds generally*

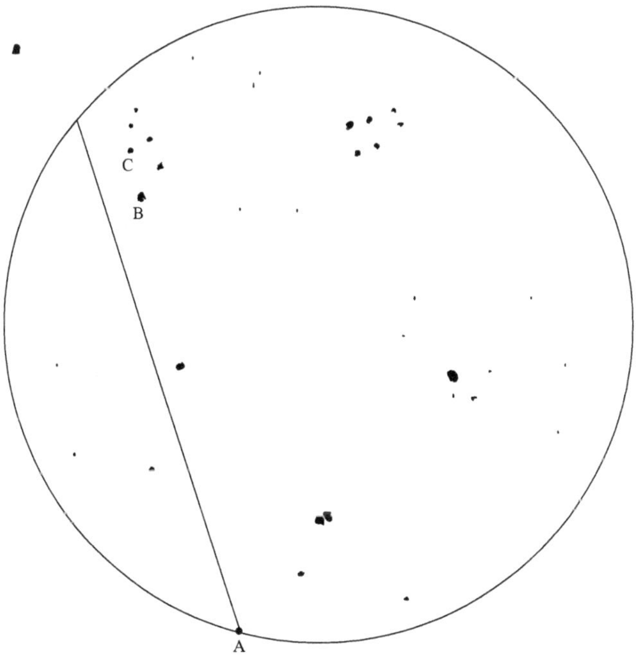

Fig. 5.39. Engraving showing the 1769 Mercury transit by J.A. Alzate. The path of the planet is marked by a line.

called spots, of which there were more than forty in number. Its perfect roundness would have been sufficient to point it out, had I not already known where to look for it.. Finally he commented: ... *the corrugations of the luminous surface of the Sun were visible up to the very edge of the planet.* (Herschel, 1803).

The Venus transits of 1874 and 1882 were mainly observed with photographic techniques and are therefore beyond the scope of this book.

5.6.5 A Message from the Past Toward the Future

Richard Proctor (1837–1888) finished his 1882 book on transits with some phrases addressed to the astronomers of our times: *We cannot doubt that when the transits of 2004 and 2012 are approaching, astronomers will look back with interest on the operations conducted during the present "transit–season" and although in those times in all probability the determination of the Sun's distance by other methods [...] will far surpass in accuracy those now obtained by such methods, yet we may reasonably believe that great weight will even then be attached to the determinations obtained during the transits of the present century. The astronomers of the first years of the twenty–first century, looking back over the long transitless period which will then have passed, will understand the anxiety of astronomers in our own time to utilize to the full whatever opportunities the coming transits may afford; and I venture to hope that should there then be found, among old volumes on their book–stalls, the essays and charts by which I have endeavoured to aid in securing that end, they will not be disposed to judge overharshly what some in our own day may have regarded as an excess of zeal..*

Our book demonstrates that solar observations carried out in the past are still a valuable tool for understanding the present and forecasting the future. The next Venus transit will take place on 6 June 2012, after which there will not be another until 2117. Probably on 10 November 2084, terrestrial astronomers will be able to see our blue planet, the Earth, crossing the solar disk as seen from the surface of Mars. They will no doubt experience the same feelings as we do today.

References

Aughton, P. *The Transit of Venus: The Brief, Brilliant Life of Jeremiah Horrocks, Father of the British Astronomy.* Windrush Press, 2004.

Azcárate, T. Eclipse total de Sol del 17 de Abril de 1912. *Revista Sociedad Astronómica de España y América*, 33, 1914.

Badache-Damiani, C. and Rozelot, J. P. Solar apparent radius variability: a new statistical approach to astrolabe multisite observations. *Monthly Notices Royal Astronomical Society*, 369:83–88, 2006.

Baily, F. Account of Mr. Abraham Sharp's observations of the transit of Mercury in April 1707, and Flamsteed's correspondence thereon. *Monthly Notices Royal Astronomical Society*, 3:105–107, 1835.

Balmforth, N. J., Gough, D. O., and Merryfield, W. J. Structural changes to the Sun through the solar cycle. *Monthly Notices Royal Astronomical Society*, 278: 437–448, 1996.

Bartolini, S. *I fori gnomonici di Egnazio Dante in Santa Maria Novella.* Edizione Polistampa, 2006.

Baum, R. and Sheehan, W. *In Search of Planet Vulcan: The Ghost in Newton's Clockwork Universe.* Basic Books, 2003.

Bliss, C. Observations on the Transit of Venus over the Sun, on the 6th of June 1761. *Philosophical Transactions of the Royal Society*, 52:173–178, 1762.

Bònoli, F. Parmeggiani, G., and Poppi, F. Il Sole nella Chiesa: Cassini e le grandi meridiane come strumenti di indagine scientifica. *Giornale di Astronomia, Atti del Convegno Il Sole nella Chiesa: Bologna*, 32:1–128, 2006.

Bouguer, P. De la mesure des diametres des plus grandes Planètes: description d'un novel instrument qu'on peu nombrer Héliometre, prope a les déterminer; et observations sur le soleil. *Mémoires de l'Academie Royale des Sciences, année 1748*, pp. 11–34, 1748.

Browne, L. W. B. Halley's method for calculating the Earth-Sun distance. *Archive for History of Exact Sciences*, 59:251–266, 2005.

Burton, E. *The History of Clocks and Watches.* London, Little, Brown & Co., 1992.

Butler, C. J. Observations of planetary transits made in Ireland in the 18th Century and the development of astronomy in Ireland. In D. W. Kurtz (ed.), *IAU Colloq. 196: Transits of Venus: New Views of the Solar System and Galaxy*, pp. 87–99, 2005.

Butrica, A. J. In conjunction with Venus. *IEEE Spectrum*, 34:31–38, 1997.

Chapman, A. Jeremiah Horrocks, the transit of Venus, and the 'New Astronomy' in early seventeenth-century England. *Quartely Journal of the Royal Astronomical Society*, 31:333–357, 1990.

Chapman, A. *Dividing the Circle: The Development of Critical Angular Measurement in Astronomy 1500-1850.* John Wiley and Sons, 2nd. edition, 1996.

Cohen, I. B. Benjamin Franklin and the Transit of Mercury in 1753. *Proceedings of the American Philosophical Society*, 94:222–232, 1950.

Cox, A. N. *Allen's Astrophysical Quantities*, 4th ed. Publisher: New York: AIP Press; Springer, 2000.

Danjon, A. The contribution of the impersonal astrolabe to fundamental astronomy. *Monthly Notices Royal Astronomical Society*, 118:411–431, 1958.

Dearborn, D. S. P. and Blake, J. B. Magnetic fields and the solar constant. *Nature*, 287:365–368, 1980.

Débarbat, S. Venus transits – A French view. In D. W. Kurtz (ed.), *IAU Colloq. 196: Transits of Venus: New Views of the Solar System and Galaxy*, pp. 41–51, 2005.

Débarbat, S. and Guinot, B. *La méthode des hauters égales en astronomie.* Gordon and Breach, Paris, 1970.

Delisle, J. N. Observations du passage de Mercure sur le disque du Soleil, le 6 Novembre 1756; avec des réflexions qui peuvent servir a perfectionner les calculs de ces passages, et les élements de la theorie de Mercure, deduits des observations. *Mémoires Acad. Roy. Sci. Paris*, pp. 134–154, 1763.

Dorham, W. Extracts from Mr. Gascoigne's and Mr. Crabtrie's Letters, probing Mr. Gascoigne to Have Been the Inventor of the Telecopic Sights of Mathematical Instruments, and Not the French. *Philosophical Transactions*, 30:603–610, 1717.

Dollond, J. A Letter from Mr. John Dollond to Mr. James Short, F. R. S. concerning an Improvement of Refracting Telescopes. *Philosophical Transactions of the Royal Society of London, part I*, 48:103–107, 1753.

Dollond, J. An explanation of an Instrument for Measuring Small Angles. *Philosophical Transactions*, 48:551–564, 1754.

Dunham, D. W., Sofia, S., Fiala, A. D., Muller, P. M., and Herald, D. Observations of a probable change in the solar radius between 1715 and 1979. *Science*, 210: 1243–1245, 1980.

Dunn, S. A Determination of the exact Moments of Time when the Planet Venus was at external and internal contact with the Sun's Limb in the Transits of June 6, 1761 and June 3,1769. *Philosophical Transactions Royal Society London*, 60:65–73, 1770.

Encke, J. F. *Der Venusdurchgang von 1769.* Gotha, In der Beckerschen Buchhandlung, 1824.

Encke, J. F. Der wahre u. scheinbare Sonnenhalbmesser, nach Herrn Professor Encke's neuesten Untersuchungen. *Astronomische Nachrichten*, 3:287–288, 1825.

Endal, A. S., Sofia, S., and Twigg, L. W. Changes of solar luminosity and radius following secular perturbations in the convective envelope. *Astrophysical Journal*, 290:748–757, 1985.

Fiala, A. D., Dunham, D. W., and Sofia, S. Variation of the solar diameter from solar eclipse observations, 1715-1991. *Solar Physics*, 152:97–104, 1994.

Forbes, E. G. *Tobias Mayer's Opera Inedita.* Mac Millan, London, 1971.

Forbes, E. G. *The unpublished writings of Tobias Meyer.* Vandenhoech and Ruprecht, Göttingen, 1972.

Fuhg, A. Über den Sonnendurchmesser. *Astronomische Nachrichten*, 85:375–376, 1875.

Galindo Trejo, J. and Allen, C. Maya observations of 13th century transits of Venus? In D. W. Kurtz (ed.), *IAU Colloq. 196: Transits of Venus: New Views of the Solar System and Galaxy*, pp. 124–137, 2005.

Galle, J. G. *Ueber eine bestimmung der sonnen-parallaxe aus correspondirenden beobachtungen des planeten Flora auf mehreren sternwarten der nordlichen und sudlichen halbkugel im october und november 1873.* Maruschke & Berendt, Breslau, 1875.

Gallet, C. Mercurius sub Sole visus avenione die 7. Novembris 1677. *Journal des Scavans*, pp. 241–246, 1677.

Geyer, E. H. The heliometer principle and some modern applications. *Astrophysics and Space Sciences*, 110:183–192, 1985.

Gill, D. On the solar parallax derived from observations of Mars at Asccension (abstract). *Monthly Notices Royal Astronomical Society*, 41:317, 1881.

Gilliland, R. L. Solar Luminosity Variations – Models of Solar Convection Zone. *Nature*, 286:838–839, 1980.

Gilliland, R. L. Solar radius variations over the past 265 years. *Astrophysical Journal*, 248:1144–1155, 1981.

Goldstein, B. R. Some Medieval Reports of Venus and Mercury Transits. *Centaurus*, 14:49–59, 1969.

Goldstein, B. R. *Astronomy of Levi Ben Gerson.* Studies in the History of Mathematics and Physical Science, Vol. 11, Springer, 1985.

Gough, D. On the seat of the solar cycle. In S. Sofia (ed.), *Variations on the Solar Constant*, pp. 185–206, 1981.

Gough, D. Solar physics: Sizing up the Sun. *Nature*, 410:313–314, 2001.

Green, R. M. *Spherical Astronomy.* Cambridge University Press, Cambridge, 1985.

Hamilton, J. A. Extract from a letetr from Rev. James August Hamilton giving an account of his observation of the transit of Mercury over the Sun, of Nov. 12, 1782, observed of Cook's Town, near Dungannon, in Ireland. *Philosophical Transactions Royal Society London*, 73:453–455, 1783.

Harkness, W. Erratum [On the masses of Mercury, Venus and the earth, and on the solar parallax]. *Astronomical Journal*, 9:31–31, 1889a.

Harkness, W. On the masses of Mercury, Venus and the earth, and on the solar parallax. *Astronomical Journal*, 9:9–15, 1889b.

Heath, T. L. *Aristarchus of Samos, the Ancient Copernichus.* Dover Publications, 1981.

Heilbron, J. L. *The Sun in the Church.* Harvard University Press, 1999.

Herald, D. Correcting predictions of solar eclipse contact times for the effects of lunar limb irregularities. *Journal of the British Astronomical Association*, 93:241–246, 1983.

Herschel, W. Observations of the transit of Mercury over the disk of the Sun; to which is added an investigation of the causes which often prevent the proper action of mirrors. *Philosophical Transactions Royal Society London*, 93:214–232, 1803.

Hilfiker, J. *Über die bestimmung der constante der sonnenparallaxe mit besonderer berucksichtigung der oppositionsbeobachtungen.* Bern, Buchdruckerei B. F. Haller, 1878.

Hinks, A. R. Reduction of 295 Photographs of Eros made at Nine Observatories during the period 1900 November 7–15, with a determination of the Solar Parallax. *Monthly Notices Royal Astronomical Society*, 64:701, 1904.

Hooke, R. A description of an Instrument for Dividing a Foot into many thousand parts, and thereby measuring the Diameters of the Planets to great exactness. *Philosophical Transactions*, 29:541–544, 1667.

Hornsby, T. A Discourse on the Parallax of the Sun. *Philosophical Transactions Series I*, 53:467–495, 1763.

Hornsby, T. An Account of the Observations of the Transit of Venus and of the Eclipse of the Sun, Made at Shirburn Castle and at Oxford. By the Reverend Thomas Hornsby, M. A. F. R. S. and Savilian Professor of Astronomy in the University of Oxford. *Philosophical Transactions Series I*, 59:172–182, 1769.

Hornsby, T. The Quantity of the Sun's Parallax, as Deduced from the Observations of the Transit of Venus, on June 3, 1769. *Philosophical Transactions Series I*, 61:574–579, 1771.

Hughes, D. W. Why did other European astronomers not see the December 1639 transit of Venus? In D. W. Kurtz (ed.), *IAU Colloq. 196: Transits of Venus: New Views of the Solar System and Galaxy*, pp. 146–153, 2005.

Humbert, P. *L'oeuvre astronomique de Gassendi.* Paris, 1936.

Johnson, S. J. On a probable Assyrian transit of Venus. *Monthly Notices Royal Astronomical Society*, 43:41–42, 1882.

Kollerstrom, N. William Crabtree's Venus transit observation. In D. W. Kurtz, editor, *IAU Colloq. 196. Transits of Venus: New Views of the Solar System and Galaxy*, pp. 34–40, April 2005.

Kovalevsky, J. *Modern Astrometry*, 2nd edition, Springer, 2002.

Kubo, Y. Position and radius of the Sun determined by solar eclipses in combination with lunar occultations. *Publications Astronomical Society of Japan*, 45:819–829, 1993.

Kurtz, D. W. (ed.). *IAU Colloq. 196: Transits of Venus: New Views of the Solar System and Galaxy.* Cambridge University Press, 2005.

Laclare, F., Delmas, C., Coin, J. P., and Irbah, A. Measurements and variations of the solar diameter. *Solar Physics*, 166:211–229, 1996.

Lalande, J. J. *L' Astronomie.* 2nd. ed. Editions Desaimp, Paris, 1771.

Lefebvre, S. and Rozelot, J.-P. The PICARD space mission: Ground-space synergy. In Societe Francaise d'Astronomie et d'Astrophysique, editor, SF2A-2001: *Semaine de l'Astrophysique Francaise*, pp. 61–62, 2001.

Licchelli, D. The transit of Venus and the Black Drop Effect. *Memorie della Societa Astronomica Italiana Supplement*, 6:17–21, 2005.

Lisicki, A. Johannes Hevelius as an observer. In Glebocki & Zbierski (eds.). *On the 300th Anniversary of the Death of Johannes Hevelius*, The Polish Academy of Science, pp. 23–42, 1992.

Maeyama, Y. Ancient Stellar Observations: Timocharis, Aristyllus, Hipparcus, Ptolomy – the Dates and Accuracies. *Centaurus*, 27:280–310, 1984.

Mancha, J. L. Egidius of Baisiu's theory of pinhole images. *Archive for History of Exact Sciences*, 40:1–35, 1989.

Mancha, J. L. Astronomical Use of Pinhole Images in William of Saint Cloud's Alamanach Planetarum. *Archive for History of Exact Sciences*, 43:275–298, 1993.

Mancinelli, F. and Casanovas, J. *La Torre dei Venti in Vaticano*. Libreria Editrice Vaticana, 1980.

Maor, E. *Venus in Transit*. Princeton University Press, 2004.

Marov, M. Y. Mikhail Lomonosov and the discovery of the atmosphere of Venus during the 1761 transit. In D. W. Kurtz (ed.), *IAU Colloq. 196: Transits of Venus: New Views of the Solar System and Galaxy*, pp. 209–219, 2005.

Maunder, M. *Transit: When Planets Cross the Sun*. Practical Astronomy Series, Springer, 2000.

McConnell, A. and Brech, A. Natahaniel and Edward Pigott, Itinerant Astronomers. *Notes and Records of the Royal Society London*, 53:305–318, 1999.

Meeus, J. *Transits*. Willmann-Bell Inc, Richmond, 1989.

Meeus, J. *Astronomical Tables of the Sun, Moon and Planets*. Willmann-Bell Inc., Richmond, p. 450, 1995.

Messier, C. Observation du passage de Mercure sur le disque du soleil. *Memoires Academie Royale des Sciences*, 1782:652–657, 1782.

Mignard, F. *The Solar Parallax with the Transit of Venus*. Observatoire de la Cote d'Azur, 2004.

Mouton, G. *Observationes Diametrorum Solis et Lunae apparentium*. Lyon, 1670.

Neugebauer, O. *A History of Ancient Mathematical Astronomy. Part 1; Part 2; Part 3*. Studies in the History of Mathematics and Physical Sciences, Berlin; Springer, 1975.

Newcomb, S. *Discussion of Observations of the Transits of Venus in 1761 and 1769, Astronomical Papers prepared for the use of the Americal Ephemeris and Nautical Almanac*. U. S. Government Printing Office, part V, 259; Washington, 1890.

O'dell, C. R. and Van Helden, A. How accurate were seventeenth-century measurements of solar diameter? *Nature*, 330:629–631, 1987.

Pap, J., Rozelot, J. P., Godier, S., and Varadi, F. On the relation between total irradiance and radius variations. *Astronomy and Astrophysics*, 372, 2001.

Parkinson, J. H., Morrison, L. V., and Stephenson, F. R. The constancy of the solar diameter over the past 250 years. *Nature*, 288:548–551, 1980.

Pasachoff, J. M., Schneider, G., and Golub, L. The black-drop effect explained. In D. W. Kurtz (ed.), *IAU Colloq. 196: Transits of Venus: New Views of the Solar System and Galaxy*, pp. 242–253, 2005.

Piñera Ramírez, D. Sondeo historiografico sobre Astronomia en Baja California. In M. A. Moreno Corral (ed.), *Historia de la Astronomía en Mexico, UNAM*, pp. 193–214, 1982.

Pigatto, L. The 1761 transit of Venus dispute between Audiffredi and Pingré. In D. W. Kurtz, editor, *IAU Colloq. 196: Transits of Venus: New Views of the Solar System and Galaxy*, pp. 74–86, 2005.

Proctor, R. A. *Transits of Venus*. Longmans, Green and Co., 1882.

Ribes, E., Beardsley, B., Brown, T. M., Delache, P., Laclare, F., Kuhn, J. R., and Leister, N. V. The variability of the solar diameter. In C. P. Sonnett, M. S. Giampapa, and M. S. Matthews (eds.), *The Sun in Time*, pp. 59–97, 1991.

Ribes, E., Merlin, P., Ribes, J.-C., and Barthalot, R. Absolute periodicities in the solar diameter, derived from historical and modern data. *Annales Geophysicae*, 7:321–329, 1989.

Ribes, J. C., and Nesme-Ribes, E. The solar sunspot cycle in the Maunder minimum AD1645 to AD1715, *Astronomy and Astrophysics*, 276:549–563, 1993.

Ribes, E., Ribes, J. C., and Barthalot, R. Evidence for a larger sun with a slower rotation during the seventeenth century. *Nature*, 326:52–55, 1987.

Robinson, H. W. Jeremiah Dixon (1733–1779): A Biographical Note. *Proceedings of the American Philosophical Society*, 94:272–274, 1950.

Rosa, P. *Studii intorno al diametri solari.* Roma, A. Befani., 1873.

Rozelot, J. P. Possible links between the solar radius variations and the Earth's climate evolution over the past four centuries. *Journal of Atmospheric and Terrestrial Physics*, 63:375–386, 2001.

Rumovski, M. Observations of the Transit of Mercury at St. Petersburg. In a Letter from M. Rumovski, Astronomer in the Imperial Academy, to Mr. J. H. de Magellan, F. R. S. *Philosophical Transactions Series I*, 77:48–49, 1787.

Savery, S. A New Way of Measuring the Difference between the Apparent Diameter of the Sun at the Times of the Earth's Perihelion and Aphelion, with a Micrometer placed in a Telescope Invented for that purpose. *Philosophical Transactions*, 48:167–178, 1754.

Schneider, G., Pasachoff, J. M., and Golub, L. TRACE observations of the 15 November 1999 transit of Mercury and the Black Drop effect: Considerations for the 2004 transit of Venus. *Icarus*, 168:249–256, 2004.

Sellers, D. *The Transit of Venus: The Quest to Find the Distance of the Sun.* MagaVelda Press, 2001.

Shapiro, I. I. Is the sun shrinking. *Science*, 208:51–53, 1980.

Sheenan, W. *The Transits of Venus.* Prometheus Books, 2004.

Sigismondi, C. and Fraschetti, F. Measurements of the solar diameter in Kepler's time. *The Observatory*, 121:380–385, 2001.

Smart, W. M. *Textbook on Spherical Astronomy, sixth edition.* Cambridge University Press, 1977.

Sofia, S., Basu, S., Demarque, P., Li, L., and Thuillier, G. The Nonhomologous Nature of Solar Diameter Variations. *Astrophysical Journal*, 632:L147–L150, 2005.

Spencer Jones, H. Eros, On the suitability of, for the accurate determination of the solar parallax. *Monthly Notices Royal Astronomical Society*, 100:422–434, 1940.

Spiegel, E. A. and Weiss, N. O. Magnetic activity and variations in solar luminosity. *Nature*, 287:616–617, 1980.

Spruit, H. C. Theory of luminosity and radius variations. In C. P. Sonett, M. S. Giampapa, and M. S. Matthews (eds.), *The Sun in Time*, pp. 118–158, 1991.

Stephenson, F. R. Historical Evidence concerning the Sun: Interpretation of Sunspot Records during the Telescopic and Pretelescopic Eras. *Royal Society of London Philosophical Transactions Series A*, 330:499–512, 1990.

Stephenson, F. R., Jones, J. E., and Morrison, L. V. The solar eclipse observed by Clavius in A.D. 1567. *Astronomy and Astrophysics*, 322:347–351, 1997.

Sterken, C. *Astronomical heritages: Astronomical archives and Historic transits of Venus.* Vrije Universiteit Brussel, 2005.

Sveshnikov, M. L. Solar-radius variations from transits of mercury across the solar disk. *Astronomy Letters*, 28:115–120, 2002.

Taft, L. G. *Computational Spherical Astronomy.* Kriger Pub. Co., 1991.

Thomas, J. H. Variations of the sun's radius and temperature due to magnetic buoyancy. *Nature*, 280:662–663, 1979.

Toulmonde, M. The diameter of the Sun over the past three centuries. *Astronomy and Astrophysics*, 325:1174–1178, 1997.

Van Biesbroeck, G. Mercury, transit of, 1707. *Monthly Notices Royal Astronomical Society*, 73:34–36, 1912.

Van Brummelen, G. *The Mathematics of the Heavens and the Earth: The Early History of Trigonometry.* Princeton University Press, 2009.

Van Helden, A. The importance of the transit of Mercury of 1631. *Journal for the History of Astronomy*, 7:1–9, 1976.

Van Helden, A. *Measuring the Universe.* University of Chicago Press, 1985.

Vaquero, J. M., Gallego, M. C., and Trigo, R. M. Sunspot numbers during 1736–1739 revisited. *Advances in Space Research*, 40:1895–1903, 2007.

Volkoff, I. *J. Hevelius and his Catalog of Stars.* Brigham Young University Press, 1971.

Wales, W. and Dymond, J. Astronomical observations made by order of the Royal Society, at Prince of Wales's Fort, on the North-West Coast of Hudson's Bay. *Philosophical Transactions Series I*, 59:467–488, 1769.

Watts, C. B. *The Marginal Zone of the Moon.* Washington, US Government Printing Service, 1963.

Waugh, A. E. *Sundials: Their Theory and Construction.* Dover, 1973.

Whatton, A. B. *Memoir of the Life and Labours of Jeremiah Horrocks, To Which Is Appended A Translation of His Celebrated Discourse The Transit of Venus across the Sun*, London, 1859.

Winthrop, J. A Letter from Mr. John Winthrop to C. Mortimer, M. 'D. Sec. R. S. concerning the Transit of Mercury Over the Sun, April 21. 1740. and of an Eclipse of the Moon, Dec. 21. 1740. *Philosophical Transactions Series I*, 42:572–578, 1742.

Wittmann, A. Numerical simulation of the Mercury transit black drop phenomenon. *Astronomy and Astrophysics*, 31:239–243, 1974.

Wittmann, A. Tobias Mayer's observations of the sun – Evidence against a secular decrease of the solar diameter. *Solar Physics*, 66:223–231, 1980.

Wittmann, A. D. The solar diameter derived from Tobias Mayer's observations 1756-1761. *Acta Historica Astronomiae*, 3:49–51, 1998.

Wittmann, A. D. and Debarbat, S. The solar diameter and its variability. *Sterne und Weltraum*, 29:420–426, 1990.

Wittmann, A. D. and Neckel, H. On the Relation between the Sun's (Geocentric) Angular Diameter and the Observed (Topocentric) 'Drift Time'. *Solar Physics*, 163:1–5, 1996.

Wlodarczyk, J. Solar eclipse observations in the time of Copernicus: tradition or novelty? *Journal for the History of Astronomy*, 38:351–364, 2007.

Woolf, H. *The Transits of Venus: A Study in Eighteen-Century Science.* Princeton University Press, 1959.

6

Terrestrial Aurorae and Solar–Terrestrial Relations

Few phenomena have made as much of an impression on human beings as the aurora borealis, also known as the northern lights. The aurora is one of the most spectacular and earliest-known manifestations of the links between the Sun and the Earth. The northern and southern lights appear in the night sky with a great variety of colours and forms (Figure 6.1). The auroral activity is not rare, although few aurorae can be observed from low-latitude sites. Magnificent aurorae have been observed by humans since the beginnings of civilization. As the sky was the heaven where the gods of many civilizations were seated, the relationship between systems of belief and omen was rapidly established. Only in recent times was the scientific method applied, which soon revealed the cause: the variability of our Sun. The basic concepts of solar–terrestrial relations were laid down in Chapter 1.

6.1 Auroral Physics in Brief

6.1.1 Geomagnetism

William Gilbert (1544–1603), an English physician and a natural philosopher, concluded in his *On the Magnet and Magnetic Bodies, and on the Great Magnet the Earth* (1600) that the Earth was itself magnetic and that this was the reason compasses pointed north. We can imagine a gigantic bar magnet at the centre of our planet. Thus, the magnetic field lines emerge from near the geographic south pole and re-enter at the north polar region (Figure 6.2). We must notice two facts. First, the magnetic axis is tilted at a slight angle (11.5° approximately) with respect to the Earth's rotation axis and, therefore, compasses do not point exactly to geographic north. And second, the magnetic and geographic poles are inverted. The north magnetic pole corresponds to the south geographical pole and *vice versa*.

The geomagnetic field vector, **B**, varies over the terrestrial surface. We could measure its orthogonal components (Figure 6.3): X (northerly intensity),

J.M. Vaquero, M. Vázquez, *The Sun Recorded Through History*,
Astrophysics and Space Science Library 361, DOI 10.1007/978-0-387-92790-9_6,

Fig. 6.1. Sketch of the auroral display observed in Paris (13 May 1869).

Y (easterly intensity) and Z (vertical intensity, positive downwards). Other important magnitudes are the total intensity F and the horizontal intensity H. However, the first magnitudes related with the geomagnetic field vector, **B**,

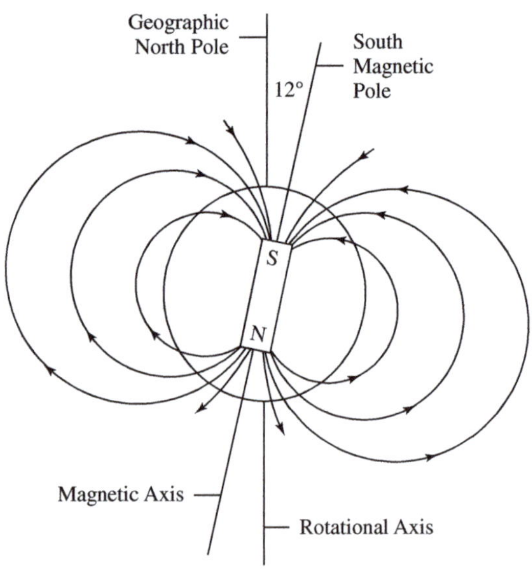

Fig. 6.2. Earth's Magnetic dipole from K. R. Lang (2006).

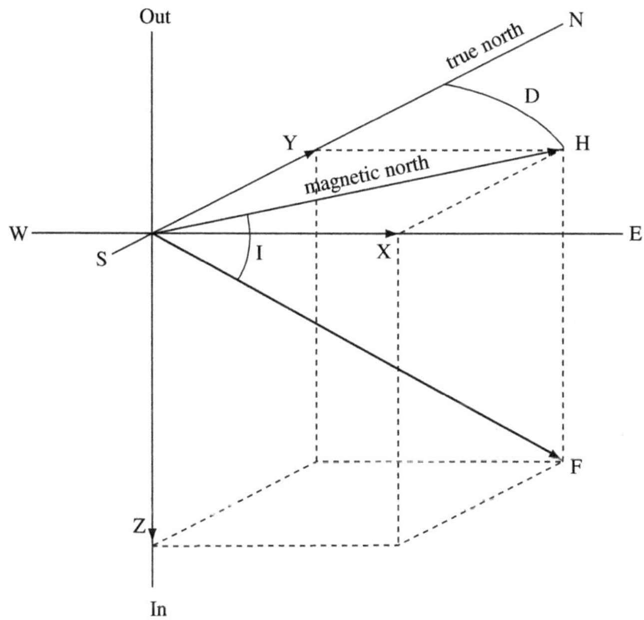

Fig. 6.3. Geomagnetic components in a local reference frame.

that were measured were the angles of inclination (or dip) I and declination (or magnetic variation) D. The geomagnetic inclination is the angle between the horizontal plane and the field vector, measured positive downwards. The geomagnetic declination is the horizontal angle between true north and the field vector, measured positive eastwards.

All these magnitudes are related by the equations

$$\mathrm{D} = \arctan \frac{\mathrm{X}}{\mathrm{Y}}$$

$$\mathrm{I} = \arctan \frac{\mathrm{Z}}{\mathrm{H}}$$

$$\mathrm{F} = \sqrt{\mathrm{H}^2 + \mathrm{Z}^2}$$

and

$$\mathrm{H} = \sqrt{\mathrm{X}^2 + \mathrm{Y}^2}$$

The tesla is the unit most commonly used in geomagnetism for the magnetic field intensity (strictly flux density). The total intensity varies from 24 000 nanotesla[1] (nT) to 66 000 nT at the Earth's surface. Moreover, other units likely to be encountered are the gauss (1 gauss = 100 000 nT), the gamma

[1] The tesla (symbol T) is the SI derived unit of magnetic flux density and it is equal to one weber per square metre.

(1 gamma = 1 nT) and the oersted[2] (1 Oe = 10^5 nT). The equatorial geomagnetic field strength is 31 000 nT approximately on the surface.

The greatest difficulty when studying long-term changes in the magnetic field of our planet arises from the lack of reliable long time series data. Since the early European voyages around the globe (16th century), the Earth's magnetic field has been monitored constantly (Jackson et al., 2000; Jonkers et al., 2003). However, these observations correspond mostly to on–board ship measurements (Jonkers et al., 2003) and few data of fixed positions are available for long periods.

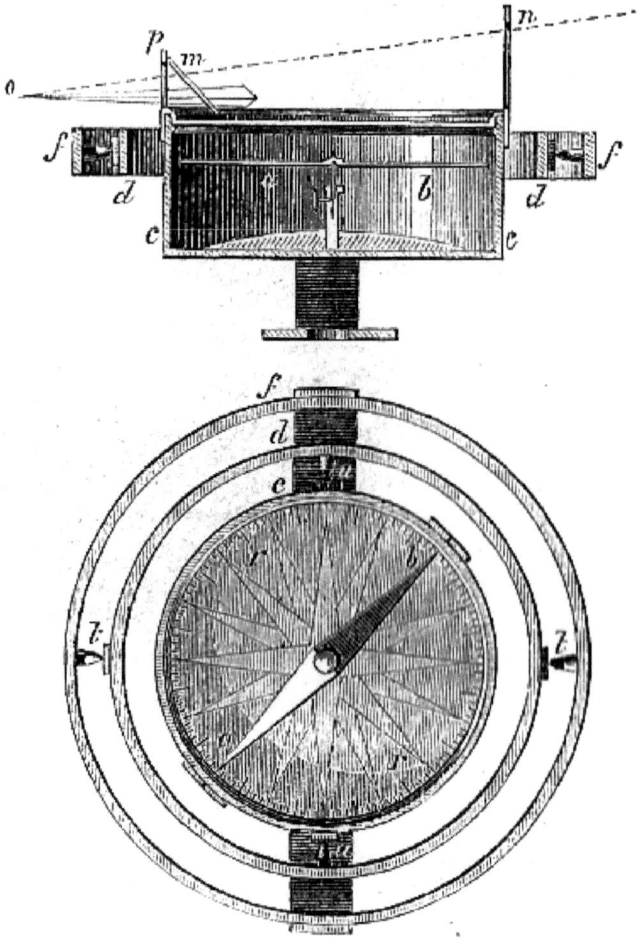

Fig. 6.4. Geomagnetic declination compass from Jamin (1870).

[2] Oersted (abbreviated as Oe) is the unit of magnetic field strength or intensity in the CGS system of units.

In the early 19th century, a considerable number of geomagnetic observatories were set up in many countries, mostly European, partially owing to the influence of Alexander von Humboldt (Malin and Barraclough, 1991). This was a consequence of the fact that the study of the geomagnetic field was fostered in the early 19th century. Two inventions at the end of the 18th century allowed the geomagnetic measurements to be performed more accurately. First, C. A. de Coulomb (1736–1806) developed the suspended magnet (or torsion balance) and, second, J. C. Poggendorf (1796–1877) built a variometer based on the mirror-and-scale method. Moreover, H. C. Oersted (1777–1851), M. Faraday (1791–1877) and A. M. Ampére (1775–1836) discovered electromagnetic interaction in the early 19th century. Finally, C. F. Gauss (1777–1855) established in 1839 the separation of the magnetic fields into their internal and external sources through the use of spherical harmonical analysis (Gauss, 1839; Jackson et al., 2000). Figures 6.4 and 6.5 show typical 19th century instruments used to measure geomagnetic declination and inclination.

There is a range of literature dealing with geomagnetism, for example several treatises which introduce the subject with a succinct historical perspective (e.g. Chapman and Bartels, 1940; Malin, 1987; Merril et al., 1996).

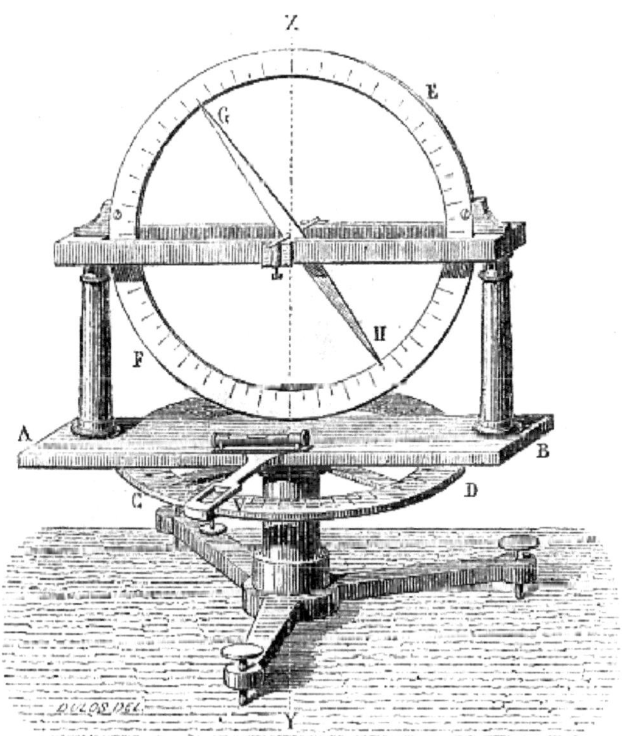

Fig. 6.5. Geomagnetic inclination compass from Jamin (1870).

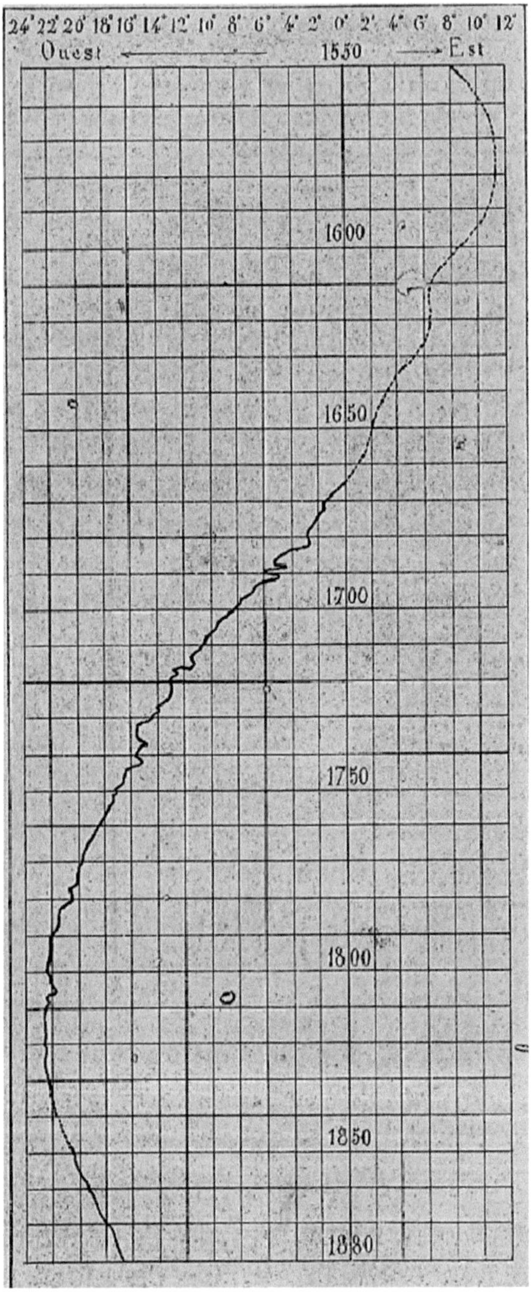

Fig. 6.6. Long-term evolution of the geomagnetic declination in Paris during the period 1550–1880 from Guillemin (1887).

Moreover, there is an abundant bibliography about the history of geomagnetism, for example the books by Aczel (2001) and Jonkers et al. (2003) about the origin of geomagnetism. Several review papers about related aspects have also been published (e.g. Smith, 1992; Schröder and Wiederkehr, 2000; Schröder, 2001; Stern, 2002; Courtillot and Le Mouël, 2007). Some biographies of relevant figures who have made contributions to geomagnetism can be consulted in Gubbins and Herrero-Bervera (2007), and there are several texts devoted to pioneers of geomagnetism (e.g. Smith, 1970; Radelet de Grave and Speiser, 1975; Frankel, 1987; Sarda, 2002; Mandea and Mayaud, 2004; Nevanlinna, 2005; Soffel, 2006; Trigo and Vaquero, 2008). Finally, papers about scientific communities or institutions related to geomagnetism can be found as well, such as Green (1972), Good (1985, 1988), and Schröder and Wiederkehr (2000).

Long time series of direct observations provide useful information, which is intermediate between the actual data from magnetic observatories and archaeomagnetic and paleomagnetic data. Long-term time series of geomagnetic measurements are available from Paris (Alexandrescu et al., 1996), London (Malin and Bullard, 1981), Rome (Cafarella et al., 1992) and Edinburgh (Barraclough, 1995). Figure 6.6 shows the time evolution of the geomagnetic declination in Paris during the period 1550–1880 from Guillemin (1887).

During the second half of the 19th century, the use of magnetographs to obtain a continuous record of geomagnetic variation is generalized. With these magnetographs, geomagnetic storms could be studied more easily. Figure 6.7 shows magnetograms of the H, D and Z components, recorded at the Magnetic Observatory of Coimbra. These records show the magnetic storms that

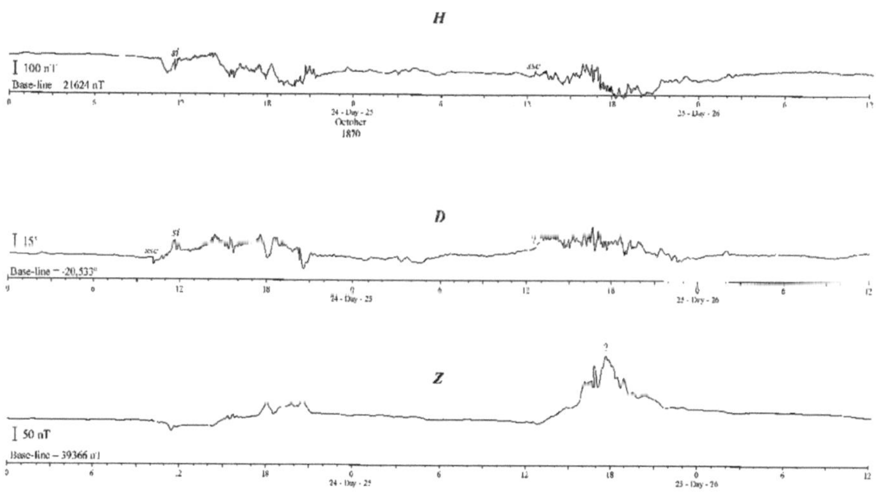

Fig. 6.7. Magnetograms of H, D and Z components recorded during 24–25 October 1870 at the Magnetic Observatory of Coimbra (courtesy of Paulo Ribeiro, University of Coimbra, Portugal).

occurred on 24 and 25 October 1870, some of the greatest storms during the 19th century (Vaquero et al., 2008).

6.1.2 Magnetosphere and Solar Wind

In Chapter 1, we briefly discussed the solar wind. The dipolar configuration of the geomagnetic field is distorted by solar wind when it is far from the surface of the Earth, providing an asymmetric shape like a "drop falling to the Sun" (see Figure 6.8). Fortunately for life on our planet, a protective cavity is thus formed, protecting life on the ground from lethal energetic solar particles. Thomas Gold (1920–2004) coined the term magnetosphere for this protective cavity in 1959. Therefore, the magnetosphere can be defined by the area of

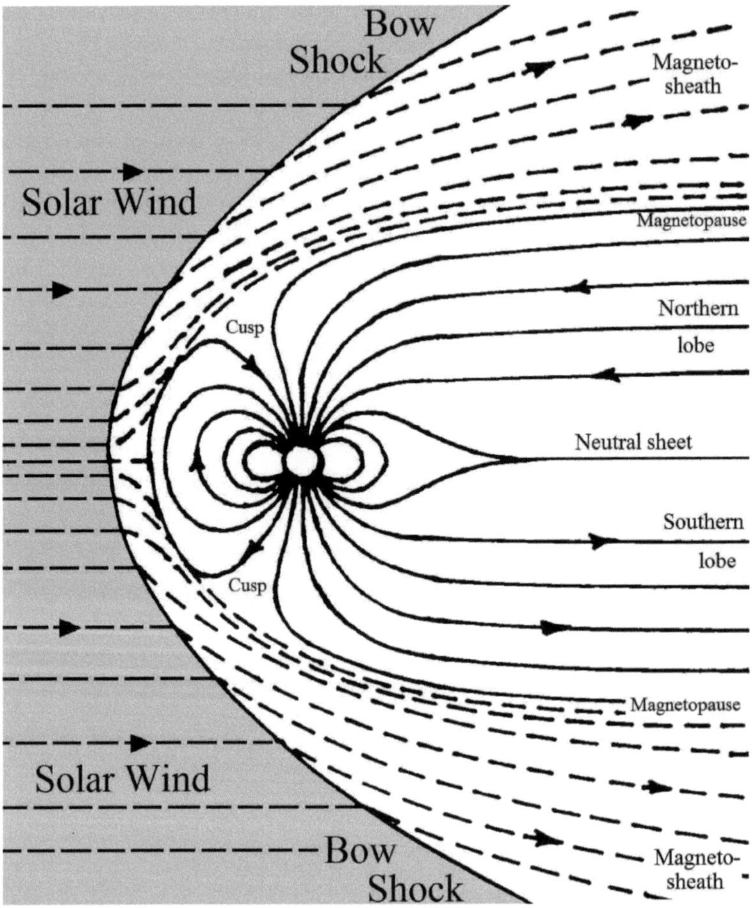

Fig. 6.8. The Earth's magnetosphere, showing its typical asymmetrical shape produced by the solar wind.

space around the Earth that is controlled by the Earth's magnetic field. In the past few years, magnetospheres have been detected in other planets of the Solar System with instruments aboard spacecraft.

The space enclosed by the magnetosphere is filled with trapped particles, namely ions and electrons. These particles are controlled by magnetic forces because they are much stronger than gravity forces. In Figure 6.8 one can see the *magnetopause*, the real shape of the boundary of the magnetosphere. This shape is strongly modified by the solar wind . The distance of the magnetopause is 10–12 r_E, 15 r_E, and $> 100r_E$ on the day side, over the poles and on the night side, respectively, where r_E is the radius of the Earth (6 378 km). These distances are measured from the Earth's centre. In Figure 6.8 other parts of the magnetosphere are also shown. The *bow shock* is the front region where solar wind particles hit the magnetosphere. The *magnetosheath*, a region where the plasma is highly turbulent, is comprised between the bow shock and the magnetopause. The terrestrial magnetic field is dragged and stretched by the solar wind creating a very long *magnetotail* on the night side.

The magnetic fields of the Earth and the solar wind can point in opposite directions in some places. The merging between both magnetic fields is then most effective and they become linked, allowing access of the solar wind particles to the magnetosphere. Figure 6.9 shows this magnetic reconnection process at the night side. The solar wind and terrestrial magnetic field come together in the magnetotail. The thin arrows indicate the direction of the magnetic fields. Particles in the plasma sheet are accelerated (thick arrows). Thus, some of the plasma is ejected down the magnetotail and other charged particles follow magnetic field lines back towards Earth.

Carl Stömer (1874–1957) predicted from theoretical studies the existence of trapping regions in the magnetosphere in the early 20th century. Later, this hypothesis was confirmed from the measurements recorded by detectors aboard Explorer I in 1958 and these regions were called Van Allen belts. Particles trapped in these regions were once thought to play a primary role in governing auroral activity. However, the plasma sheet is the main magnetospheric

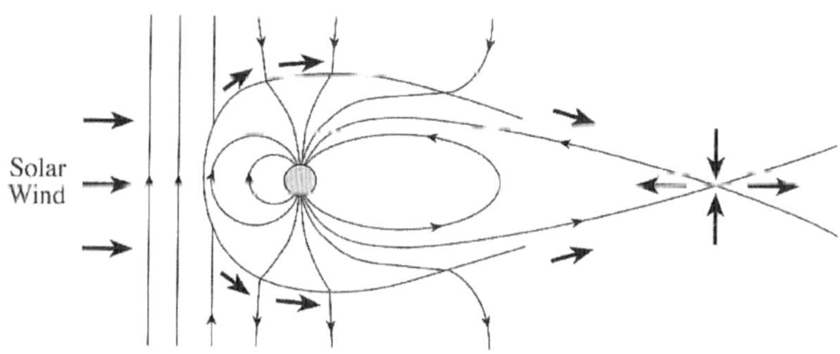

Fig. 6.9. Magnetic reconnection at the night side.

reservoir of the particles that cause the aurorae according to the most widely accepted theories. The upper layer of our atmosphere (ionosphere) is another source of particles. Some particles have sufficient energy to escape from this layer, forming the plasmasphere when the Earth's magnetic field traps oxygen ions, protons and electrons from the ionosphere.

The ionosphere is very important for modern telecommunications due to its influence on radio waves. It is divided into the lowest D region (between 50 and 90 km), the E region (between 90 and 150 km) and the F region that contain the F1 and F2 layers (Figure 6.10). The top of the ionosphere is 1000 km approximately. The most important ions are O_2^+ and NO^+ in the E region and O^+ in the F region. The peak in the electron concentration is located in the F2 layer at about 400 km. The variability of the ionosphere is related mainly with the Sun from different mechanisms. Thus, the ionospheric variation occurs in a 24 h period (night-day time, see Figure 6.10) and 11-year period (solar cycle).

Disturbances in the surface of the Earth's magnetic field are measured when charged particles are guided down the field lines into the upper atmo-

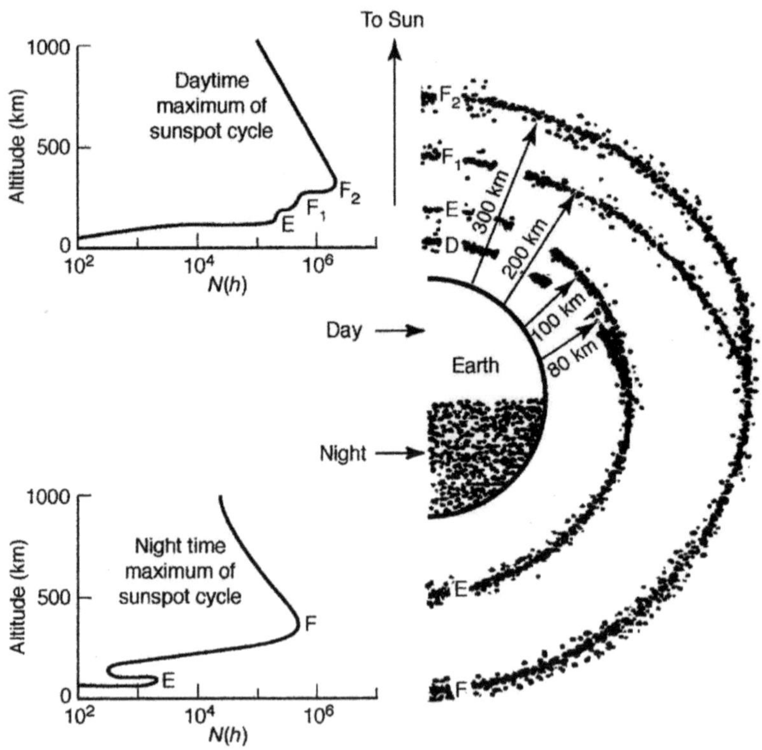

Fig. 6.10. Ionospheric layers for middle latitudes including the day-to-night variation (from Goodman, 2005).

sphere, producing large horizontal currents that flow in the D and E regions of the auroral ionosphere (east-west currents). The intense magnetic field associated with these auroral electrojets causes the disturbances in the geomagnetic field and produces the aurora.

The interplanetary origin of geomagnetic storms depends strongly on the stage of the 11-year solar cycle. Around solar maximum, the interplanetary manifestations of coronal mass ejections (CMEs) cause intense geomagnetic storms ($Dst \leq -100\,\mathrm{nT}$). These interplanetary manifestations are the sheath region just behind the forward sock and the CME ejecta itself (Gonzalez et al., 2002). However, highspeed streams from coronal holes control the interplanetary medium activity (Gonzalez et al., 1999) during solar minimum. The interaction between the fast stream from a coronal hole and the heliospheric current sheet (HCS) creates high-field regions called Corotating Interaction Regions (CIRs). Consequently, continuous auroral activity can take place on Earth. This phenomenon is called "High Intensity Long Duration Continuous AE Activity" or HILDCAAs (Tsurutani et al., 2006).

6.1.3 Geomagnetic Indices

There are a number of indices that describe magnetic activity in the Earth's environment. These indices are calculated using ground based magnetometer measurements. The changes in solar radiation produce daily regular variations of magnetic field measured on the Earth's surface. However, other irregular current systems produce magnetic field changes. These can be caused by the interaction of the solar wind with the magnetosphere, the magnetosphere itself, the interactions between the magnetosphere and ionosphere or the ionosphere itself. Different magnetic activity indices were designed to describe variations in the geomagnetic field caused by these irregular current systems.

The aa index is a simple 3-hourly global geomagnetic activity index (Mayaud, 1973) which is produced from two approximately antipodal observatories (i.e. Hartland Observatory in the UK and Canberra Observatory in Australia). The main advantage in using aa indices for long-term solar-terrestrial research purposes is that the time series spans further back (to 1868). A daily average of aa can be obtained and it is called the Aa index. The daily Ak index was defined by Nevanlinna and Kataja (1993), and it has been adjusted with the aa index. Thus, the longest uniform index of global geomagnetic activity was obtained (Nevanlinna and Kataja, 1993), extending from 1844 until now. The AE index is an auroral electrojet index that provides a measure of the overall horizontal current strength. Anomalies in this index indicate magnetospheric substorms with durations from tens of minutes to hours (Davis and Sugiura, 1966). Another famous index is Kp defined by Bartels et al. (1939). This index is calculated from magnetometer stations at mid-latitudes where these stations are rarely under an intense horizontal current and, therefore, magnetic perturbations are dominant in the H or D

Table 6.1. Summary of geomagnetic indices (from Hansleimer, 2007)

Index	Description
aa	3-hour range index, derived from two antipodal stations
AE, AU, AL	1- , 2.5-minute, or hourly auroral electrojet indices
am, an, as	3-hour range (mundial, northern, southern) indices
Ap	3-hour range planetary index derived from Kp
C, Ci, C9	Daily local (C) or international (Ci) magnetic character; C9 was first derived from Ci, then from Cp
Cp	Daily magnetic character derived from Kp
Dst	Hourly index mainly related to the ring current
K	3-hour local quasi-logarithmic index
Km	3-hour mean index derived from an average of K indices (not to be confused with the Km of the next item)
Km, Kn, Ks	3-hour quasi-logarithmic (mundial, northern, southern) indices derived from am, an, as
Kp, Ks	3-hour quasi-logarithmic planetary index and the intermediate standardized indices from which Kp is derived (not to be confused with the Ks of the preceding item)
Kw, Kr	3-hour quasi-logarithmic worldwide index and the intermediate from which Kw is derived
Q	Quarter hourly index
R	1-hour range index
RX, RY, RZ	Daily ranges in the field components
sn, ss	3-hour indices associated with an and as
U, u	Daily and monthly indices mainly related to the ring current
W	Monthly wave radiation index

component. The Kp index utilizes both these perturbations by taking the logarithm of the largest excursion in H or D over a 3-h period. Finally, the index is scaled from 0 to 9. Table 6.1 lists a brief summary of useful geomagnetic indices.

There is a semiannual variation of geomagnetic (and auroral) activity characterized by stronger and more frequent storms in spring/fall vs. summer/winter. The cause of this variation has long been a subject of discussion (e.g. Sabine, 1856). Three mechanisms have been proposed that could explain the semiannual variation: the equinoctial hypothesis, the axial hypothesis and the Russell–McPherron effect.

According to the equinoctial hypothesis (Bartels, 1925, 1932; McIntosh, 1959), activity maximizes for as yet unknown reasons at the equinoxes when the angle between the solar wind flow direction and the Earth's dipole axis is 90°. The axial hypothesis (Cortie, 1912) is based on the fact that in early March and September the Earth is at its maximum angular distance from the solar equatorial plane. Thus, the Earth is more closely aligned with both the sunspot zones and also with coronal holes that extend down from the solar poles, as can be seen in Figure 6.11. The last

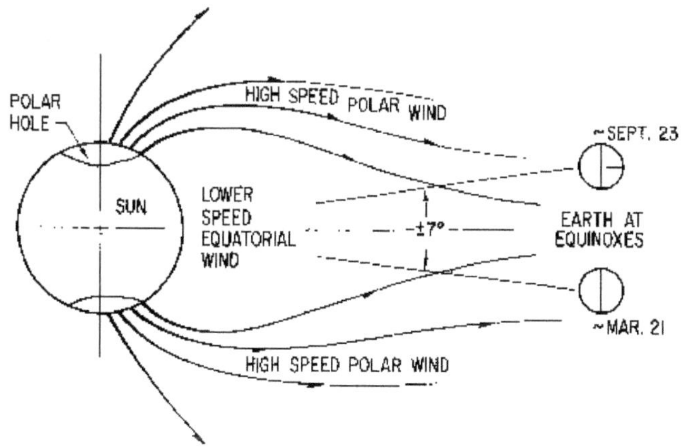

Fig. 6.11. The axial effect: high-speed streams from polar coronal holes can contribute to the semiannual variation of geomagnetic (and auroral) activity (from Bohlin, 1977).

mechanism proposed is the Russell–McPherron effect, which is based in the magnetic fields in the solar equatorial plane that have a peak southward component at the Earth in Geocentric Solar Magnetospheric coordinates in early April or October, depending on their polarity (Russell and McPherron, 1973). The latest studies have shown that the equinoctial hypothesis plays a more important role.

6.1.4 Observing the Aurora

Although the forms of the aurorae can seem chaotic, there are several patterns which are observed regularly. It is very typical to observe a *homogeneous arc* spanning across the sky from east to west. Moreover, rays following the Earth's magnetic field can be observed (*arc with ray structure*). Also, *homogeneous bands* and *bands with structure* are frequently seen. The appearance of *rays* is typical of periods with high solar activity. *Rays* line up along the Earth's magnetic field and variations occur rapidly. Also *coronas* with rapid movements and variations appear in periods of high solar activity. A type of magnificent auroral pattern in which the widths of the bands and the lengths of the rays fill most of the sky is named a *curtain*. Other shapes, such as *diffuse surfaces* or *spiral structures*, and the *pulsating* aurorae are visually impressive.

Generally, the aurorae can be classified as active or quiet. Homogeneous arcs and bands, and diffuse surfaces are classified as quiet auroral forms. Active aurorae have rapid variations and ray structures such as coronas, curtains, arcs with rays and spirals.

Observations of aurorae borealis at low latitudes are rare. A recent case is described by Alexeff and Parameswaran (2005). These aurorae are clearly associated with strong geomagnetic storms. Morphologically, they are characterized by a diffuse red colour (produced by the 630 nm oxygen emission) with no rapid motions. Silverman (1995) established a threshold magnetic latitude of about 15 degrees for a visual auroral event. Rassoul et al. (1993) proposed a classification of low-latitude aurorae based on OI 630/558 ratios. Moreover, Stable Auroral Red (SAR) arcs are also often observed at subauroral latitudes, predominantly during the recovery phase of magnetic storms (Rees and Roble, 1975) lasting for more than ten hours. They are produced by electron heating in the upper F-region of the ionosphere. Modern monitoring of airglow lines can be used to study the physical characteristics of low-latitude aurorae and SARs. Shiokawa, Ogawa and Kamide (2005) conducted such a study in Japan during the period 1999–2004. Some historical observations of SARs can be consulted in Zhang (1985).

Surprisingly, low-latitude aurorae have been observed also during periods of weak-to-moderate geomagnetic activity. Silverman (2003) has shown from US auroral data from 1880 to 1940 that some auroral phenomenon occurred under conditions of quiet or moderate magnetic activity and at low latitudes. He used the term "sporadic aurora" for this type of auroral phenomenon. Previously, this expression had been adopted in the papers by Abbe (1895) and Botley (1963). Sporadic aurorae have been described for North America (Silverman, 2003), East Asia (Willis et al., 2007) and Europe (Vaquero et al., 2007).

Willis et al. (2007) compiled 42 Chinese and Japanese auroral observations during the period 1840–1911 and found that at least 29 out of the 42 observations (i.e. 69 per cent) occurred at times of weak-to-moderate geomagnetic activity. According to Willis et al. (2007) the term "sporadic aurorae" may be used to refer to localised auroral displays that occur:

1. at low geomagnetic latitudes (approximately, $-45° <$ geomagnetic latitude $< +45°$);
2. during intervals of weak-to-moderate geomagnetic activity (aa index<50).

Aurorae also occur on other planets just like on Earth. These bodies need to have an atmosphere and a magnetic field (i.e. magnetosphere). The configuration, colours and the rapidly varying displays of the aurora on other planets or bodies of the Solar System are different from what we see on our planet. As an example, on both Saturn and Jupiter, large ovals at higher magnetic latitudes circle the magnetic poles, similar to the Earth, showing large transient events especially in ultraviolet light. Aurorae have been observed on several planets and satellites such as Jupiter, Saturn, Titan, Triton, Io, Uranus or Neptune.

There is a long list of monographs and popular books on the aurora. Some books that may be of interest for readers are Akasofu and Kamide (1987) Akasofu (2007), Bone (1991, 2007), Brekke and Eggeland (1994), Bryson (2001), Davis (1992), Eather (1980), Savage (2001) and Ytter (1999). Carlowicz and Lopez (2002), Freeman (2001) and Pope and Jorden (2006) are focused on space weather.

6.2 Folklore, Omen and Myths

The footprint that both the northern and southern lights have left on the legends and folklore of native cultures in the auroral regions can easily be seen. Holzworth (1975) related a variety of accounts in several parts of the world: North American Eskimos, Scandinavian, Scottish, Siberian, Australian and Maori people. Bone (1991) also provides other interesting examples.

Several native North Americans have a mythology that explains the aurora. The Eskimos of Alaska, the west coast of the Hudson Bay, and west Greenland think that the aurorae mean the struggles of the spirits when they play ball, the Eskimo's favourite game. On the other hand, aurorae are the torches of spirits to the Eskimos near the Hudson Strait and they are a manifestation of the spirits that brought good weather to the Eskimos of the Coronation Gulf. Other explanations of aurorae by Eskimos can be consulted in Weyer (1969) and Holzworth (1975). The aurorae were also explained through myths by several native tribes in North America (Holzworth, 1975; Hamilton, 1903; Bell, 1903; Jones, 1911–1912).

As for European mythologies, the aurora is described in Norwegian folklore as a harbinger of harsh weather, although another Norwegian legend explains the aurora as a dance in the sky by the souls of dead maidens. The bridge "Bifrost" is used by the gods to travel from Heaven to Earth in Norse mythology; Bone (1991) suggested that the aurora was the inspiration for this bridge. Moreover, the vivid red colours in intense aurorae could be associated with the Viking "vigrod" or war-reddening (Bone, 1991; Egeland and Brekke, 1984).

In the Scottish city of Aberdeen, the aurora is not a rare phenomenon and it is associated to "Heavenly Dancers" in a popular song. Bone (1991) and Livesey (2005) have published some poetic descriptions of aurora made by Robert Burns (1759–1796), the national poet of Scotland. Recently, Farquharson (2006) published another mention of aurora by Burns in his ballad 'As I stood by yon roofless tower' (1794):

The cauld blae North was streaming forth
Her lights, wi hissing, eerie din:
Athort the lift they star and shift,
Like Fortune's favours, tint as win.[3]

Also, Maycock (2006) cite the mention of an aurora made by the Cumberland poet Susanna Blamire (1747–1794) in the verses *Once, when the sky was up in arms, with norther lights at war ...* (Maycock, 2003, p. 124). Maycock (2006) indicates that Blaimere probably observed the great aurora of August 1768.

---

[3] According to A. Mackay James, the editor of the complete poetical works of Burns, we can understand the significance of the following words and expressions: cauld: cold; blae: blue; Athort: athwart, across; lift: sky, horizon; tint as win: lost as gained (Burns, 1993).

In the Mediterranean region, where the aurora is not frequent, the Roman Pliny the Elder (AD 23–79) declared in his *Historia Naturalis* (Chapter 27): "*There is no presage of woe more calamitous to the human race than a flame in the sky, which seems to descend to the earth on showers of blood*". On 25 January 1938, the heavens over south-western Europe "were filled with a strange and terrible crimson fire" that presage the start of the Second World War.

There are also other examples from Northern Asia. The Chuvash people in Siberia believed in the "Surutan-Tura", a birth-giving heaven. According to the Cuvash tribe, the heavens gave birth to a son during the aurora. In north-east Asia, the aurora is a dwelling chiefly for those who died a violent death for the Chukchee tribe. In the surroundings of the Baltic sea, the Estonians explain the aurora as a heavenly war.

Dragon legends could be related to auroral observations (Bone, 1991), especially in the East. Active auroral bands could have reminded the ancients of the shape of a snake. Thus, it is easy to find "celestial serpents" and "dragons" in ancient chronicles. One example is the "dragon" seen at Dehly during the night of 22 May 1753 reported by the Portuguese Carlos de Bivar de Aragam (1753). Another example is the aurora represented as a dragon seen in Pressburg, Hungary, on 10 February 1681 (see Figure 6.12).

The examples presented so far have only discussed aurorae in folklore from the North Hemisphere, because the only land mass in the Southern Hemisphere where the aurora is a common phenomenon is uninhabited Antarctica. However, Holzworth (1975) found two examples from New Zealand, where the Maori believed that the auroral light was a great fire made by their ancestors.

Fig. 6.12. Aurora seen in Pressburg, Hungary, 10 February 1681 represented as a dragon in the sky.

Also, other tribes believed that Kangaroo Island was the home of the dead because on occasion they could see the campfires of the spirits of the dead.

6.2.1 Babylonia and the Bible

Which was the first aurora dated and registered in a document? According to Stephenson et al. (2004), a late Babylonian astronomical text probably contains the earliest account of a datable aurora borealis. This text is in the form of a sun-dried clay tablet inscribed with a cuneiform syllabic script. In the Vorderasiatisches Museum, Berlin, a clay tablet named VAT 4956 from the year 568–567 BC is conserved. This tablet contains a fragment of an astronomical diary where one can read: "*Night of the 29th* [11th month], *red glow flared up in the west; two-*[hours...]" (Sachs and Hunger, 1989). The date of this red glow phenomenon corresponds with the night between the days 12 and 13 March in 567 BC because the Babylonian day began at sunset. It should also be noted that (1) the geomagnetic latitude of Babylon was about 41°N in 567 BC (instead of the current 27.5°N), and (2) the geomagnetic storm activity has a maximum near the equinoxes (as in this case). Thus, the most reliable interpretation of this description is an aurora (Stephenson et al., 2004).

Another possible dated auroral observation is the first of the visions by the Old Testament prophet Ezekiel (I: 1–28). Eather (1980), Silverman (1998) and Siscoe et al. (2002) have studied this case in detail. The date of the event is recorded but, unfortunately, the interpretation of the year is questionable (probably 593 BC, see Greenberg, 1983). This passage is the clearest description of an auroral display among the four fragments of *The Old Testament*, the first part of the Bible (see Table 6.2), that could describe this phenomenon. In fact, several passages in the Bible may be descriptions of the aurora (Silverman, 1998; 2006). Ezekiel's fragment can be interpreted as different phases of an auroral report.[4]

[4] "As I looked, behold, a stormy wind came out of the north, and a great cloud, with brightness around it, and fire flashing forth continually, and in the midst of the fire, as it were gleaming metal. And from the midst of it came the likeness of four living creatures. And this was their appearance: they had a human likeness, but each had four faces, and each of them had four wings. Their legs were straight, and the soles of their feet were like the sole of a calf's foot. And they sparkled like burnished bronze. Under their wings on their four sides they had human hands. And the four had their faces and their wings thus: their wings touched one another. Each one of them went straight forward, without turning as they went. As for the likeness of their faces, each had a human face. The four had the face of a lion on the right side, the four had the face of an ox on the left side, and the four had the face of an eagle. Such were their faces. And their wings were spread out above. Each creature had two wings, each of which touched the wing of another, while two covered their bodies. And each went straight forward. Wherever the spirit would go, they went, without turning as they went. As for

Table 6.2. Possible aurora descriptions according to some interpretations of four passages from *The Old Testament*

Cite	Date	Text
Genesis 15: 17	2000 BC	When the sun had gone down and it was dark, behold, a smoking fire pot and a flaming torch passed between these pieces.
Jeremiah 1: 13	626 BC	The word of the Lord came to me a second time, saying, "What do you see?" And I said, "I see a boiling pot, facing away from the north."
Zechariah 1: 8	518 BC	I saw in the night, and behold, a man riding on a red horse! He was standing among the myrtle trees in the glen, and behind him were red, sorrel, and white horses.
II Maccabees 5: 1–5	170 BC	About this time Antiochus made his second invasion of Egypt. And it happened that, for almost forty days, there appeared over all the city golden-clad cavalry charging through the air, in companies fully armed with lances and drawn swords troops of cavalry drawn up, attacks and counter-attacks made on this side and on that, brandishing of shields, massing of spears, hurling of missiles, the flash of golden trappings, and armour of all kinds. Therefore everyone prayed that the apparition might prove to have been a good omen.

---

the likeness of the living creatures, their appearance was like burning coals of fire, like the appearance of torches moving to and fro among the living creatures. And the fire was bright, and out of the fire went forth lightning. And the living creatures darted to and fro, like the appearance of a flash of lightning.

Now as I looked at the living creatures, I saw a wheel on the earth beside the living creatures, one for each of the four of them. As for the appearance of the wheels and their construction: their appearance was like the gleaming of beryl. And the four had the same likeness, their appearance and construction being as it were a wheel within a wheel. When they went, they went in any of their four directions without turning as they went. And their rims were tall and awesome, and the rims of all four were full of eyes all around. And when the living creatures went, the wheels went beside them; and when the living creatures rose from the

Auroral physicists believe that Ezekiel's account was inspired by a very strong magnetic storm accompanied by coronal aurorae at low latitudes. The site of the observation was about 100 km south of Babylon (approximately 32°N, 45°E) and the date of the vision was around 593 BC. The geomagnetic dipole moment increased to a peak nearly 50% greater than at present, ca. 2500 years ago according to several paleomagnetic studies. Siscoe and Siebert (2002) used a global magnetohydrodynamic simulation to calculate, for a 50% increased dipole moment, the correspondingly increased auroral-zone potential and its extension to low latitudes. These results can explain the auroral observation by Ezekiel. Moreover, Raspopov et al. (2003) hypothesized that the geomagnetic Sterno-Etrussia excursion, which occurred about 2200–2800 years ago, was the cause of the coronal aurorae seen by Ezekiel at low latitudes.

As for the first printed auroral description, Hellman (1921) indicates a pamphlet published in 1490 describing "lightning" seen from Constantinople on 13 July. The description may be interpreted as an auroral phenomena. Other pamphlets were published over the next few years describing probable aurorae. However, the first definite printed auroral observation was a pamphlet printed in 1527 describing an aurora which occurred on 11 October. According to Eather (1980), the first scientific illustration depicting the aurora is contained in the treatise *De Naturae Divinis Characterismis* by Cornelius Gemma in 1575. Schröder (2006) and Silverman (2007) have pointed out that the first accurate description of an aurora appears in the work *Das Buch der Natur*

---

earth, the wheels rose. Wherever the spirit wanted to go, they went, and the wheels rose along with them, for the spirit of the living creatures was in the wheels. When those went, these went; and when those stood, these stood; and when those rose from the earth, the wheels rose along with them, for the spirit of the living creatures was in the wheels.

Over the heads of the living creatures there was the likeness of an expanse, shining like awe-inspiring crystal, spread out above their heads. And under the expanse their wings were stretched out straight, one toward another. And each creature had two wings covering its body. And when they went, I heard the sound of their wings like the sound of many waters, like the sound of the Almighty, a sound of tumult like the sound of an army. When they stood still, they let down their wings. And there came a voice from above the expanse over their heads. When they stood still, they let down their wings.

And above the expanse over their heads there was the likeness of a throne, in appearance like sapphire; and seated above the likeness of a throne was a likeness with a human appearance. And upward from what had the appearance of his waist I saw as it were gleaming metal, like the appearance of fire enclosed all around. And downward from what had the appearance of his waist I saw as it were the appearance of fire, and there was brightness around him. Like the appearance of the bow that is in the cloud on the day of rain, so was the appearance of the brightness all around.

Such was the appearance of the likeness of the glory of the Lord. And when I saw it, I fell on my face, and I heard the voice of one speaking."

[The Book of Nature] written by Konrad von Megenberg (1309–1374) between 1348 and 1350. This eight-volume work was the first German encyclopedia of natural phenomena.

6.2.2 The Classical Period

During the Classical period, the enquiry into the aurora corresponded to philosophers interested in meteorology. Aristotle (384–322 BC) wrote his treatise *Meteorologica* in 340 BC approximately. It is the oldest comprehensive work on meteorology and has dominated the development of this science until the 17th century (Frisinger, 1973). Aristotle's ideas on meteorological phenomena were based on two theories that come from other classical philosophers. The first theory is that the Universe has a spherical form, accepting Eudoxus' ideas. The second is the four-element theory of Empedocles (Frisinger, 1972).

In Chapter 5 of *Meteorologica*, we can find the first Western rational description of auroral phenomena and an explanation of their causes. Aristotle used the name "chasmata" to describe the aurora (Silverman, 1962). According to Aristotle, vapour rises from the surface of Earth and collides with the element fire. The vapour is caused by the heat from the Sun. Finally, the fire element bursts into flames and the aurora is produced. The observed colours are produced due to the reflection and attenuation of light through the dense air layer and the sublunar character is confirmed by its wide angular extension and speed. The Roman philosopher Seneca (c. 4 BC-AD 65), born in Corduba (now Córdoba, Spain), also wrote about aurorae in his book *Naturales Quaestiones* [Natural Questions] where aurorae are described as natural phenomena (Senecae, 1979).

Stothers (1979a,b) compiled an exhaustive catalogue of aurorae directly from the classical Greek literature to avoid errors, incorrect dates and omissions of previous catalogues. Table 6.3 lists a summary of Stothers' catalogue. A more complete catalogue can be found in Stothers (1979a), which includes exact bibliographical references for each auroral event. According to Stothers (1979a), the aurorae described in classical literature can be separated into the following nine divisions:

1. X. Chasm (*hiatus* or *discessus*).
2. SF. Sky fire (*caelum ardens*).
3. NS. Night Sun (*sol* or *lux notu*).
4. BR. Blood rain (*pluvia sanguinea*).
5. MR. Milk rain (*pluvia lactea*).
6. B. Beam (*trabs*).
7. P. Pilar (*columna*).
8. T. Aurora-like torch (*fax*).
9. K. Aurora-like comet (*stella crinita*).

Stothers (1979a,b) selected a set of sky phenomena which the ancient authors had failed to connect with the same physical origin. Thus, the last

Table 6.3. Catalogue of aurorae mentioned in classical literature. Date and typology (see text) for each auroral event are available

Year	Category	Year	Category
BC 467	SF, B	BC 117	MR
373	B, T	114	BR, MR
349	X, BR, SF	113	SF, NS
344	X, SF, T	111	MR
223	SF, NS	108	MR
217	X, SF	106	BR, MR
214	B, R	104	BR, MR
209	MR	102	NS, BR
206	NS	95	MR
204	NS	94	SF
200	SF	93	X, SF
198	SF	92	MR
197	NS	91	X, BR
183	BR	63	SF, B, K
181	BR	49	SF, BR
172	BR	48	B, P
169	SF	42	SF, NS
166	NS, BR	32	T
163	SF, NS, MR	BC 17	T
162	SF	AD 9	SF, P, K
147	SF	14	SF, BR, K
134	NS, BR	50	SF
130	MR	54	BR
128	MR	76	K
125	MR	196	SF
124	MR	333	SF
118	MR		

six categories were practically ignored in previous works. Moreover, Stothers rejected most of the reported "comets" and "torch" when specific auroral properties were not described.

The catalogue of classical aurorae is geographically very homogeneous, except for only three questionable cases not included in Table 6.3: 30 BC (Egypt), AD 30 (Judea) and AD 212 (Carthage); the places where the aurorae are observed are Greece, Italy and southern Gaul. Also the catalogue is homogeneous respect to the sources; although 30 classical authors were used by Stothers, there are three main sources: Livy (*Ad urbe condita*), his exceptor Obsequens (*Prodigiorum liber*) and Cassius Dio (*Roman History*). Chronological accuracy is usually limited to the year of the event. The possible error in the date is ±1 year, especially for the dates previous to the Julian calendar reform which took place in 45 BC.

Stothers found some important gaps in the auroral records that could be identified with an aurorally quiet period. The first gap in the Roman record occurs between 459 and 223 BC. The second gap in the auroral records occurs during the period 91–49 BC with only one aurora in the year 63 BC. There is also a series of gaps after AD 76; however, all these gaps can be explained using historical reasons and Stothers was not able to associate the gaps with aurorally quiet periods.

Stothers (1979b) developed a mathematical method to detect periodicities in a series of dates as seen in Table 6.3. He had sufficient data in the more reliable interval 223–91 BC to analyse separately the categories of "Sky fire" (SF), "night suns" (NS), "blood rain" (BR), and "milk rain" (MR). The mean period was 11.5 years (associated with the 11-year period of the solar cycle). As for the categories, the most important periods were: SF (8.7 yr), NS (10.2 yr), BR (13.3 yr) and MR (11.6 yr). Moreover, Stothers (1979b) found a secondary cycle of 80–100 years for the well-documented time interval 223–91 BC. Stothers concluded that the solar activity two millenia ago was similar to the present time.

Solow (2005) revisited the well-documented time interval of the Stothers' catalogue using a statistical procedure based on a nonparametric estimation of periodic functions. The purpose of this analysis was to estimate the form of the 11-year cycle in the rate of auroral events. The result indicates a bimodality in the estimated form of the 11-year cycle. This form can be related to the bimodality of the geomagnetic activity recorded during the last century. This result lends striking support to Stothers' previous findings.

Figure 6.13 shows a comparison between secular solar activity and the dates of the "classical" aurorae provided by Stothers (1979b). It is difficult to

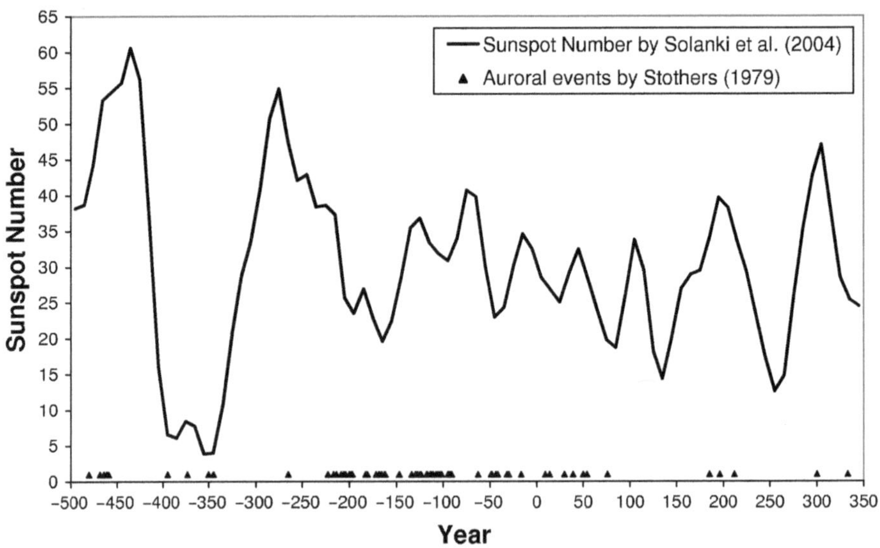

Fig. 6.13. Comparison between secular solar activity and the dates of the "classical" aurorae.

Table 6.4. Astronomical orientations of the Temples of Apollo from Ancient Greece and its colonies (Liritzis and Vassiliou, 2006)

Temple	Date	Azimuth
Apollo Epicurean	c.450–425 BC	2° 36'
Apollo Thermios	c.630–610 BC	12° 53'
Apollo	600 BC	137°
Apollo	c.366–320 BC	51°
Apollo	c.1000 BC	56° 25'
Apollo	c.600 BC	69° 10'
Apollo	350 BC	84° 45'
Apollo	600 BC	91° 45'
Apollo Pythius	c.600 BC	95°
Apollo	c.550 BC	96° 40'
Apollo Pythius	3rd–2nd century BC	96° 48'
Apollo Lyk	c.550 BC	126° 39'
Apollo Lyk	–	153° 30'
Apollo Erethimius	c.5th–4th century BC	242° 14'

interpret the aurora observations according to the levels of solar activity in this epoch. This implies that the appearance of aurorae in the classical literature does not so much depend on the level of solar activity but on several other factors. It is interesting to point out that the data of the aurorae observed by classical authors are not a good proxy of the solar activity. However, these records show us the existence of an 11-year cycle with two maxima by using a statistically appropriate analysis.

Finally, an interesting fact was discovered by Liritzis and Vassiliou (2006). They examined the astronomical orientation of Greek temples dedicated to the god Apollo. The results appear in Table 6.4. Two temples (Apollo Epicurean and Apollo Thermios) have a rare and exceptional north-south orientation although most of them have an east-west orientation approximately. Liritzis and Vassiliou (2006) proposed from historical and mythological accounts that this northern orientation may be related to the appearance of northern lights. Following on ancient information, the Temple of Epicurean Apollo was built as a thanksgiving to Apollo for deliverance from a plague in 430 BC. Local people may have oriented the temple towards these lights because the occurrence of a great aurora coinciding with the cessation of the plague may have been taken as a sign of the god Apollo. Moreover, the level of solar activity was very high around 450 BC (see Figure 6.13).

6.3 Reports During the Last Two Millennia

6.3.1 Aurorae Borealis

Abundant documentation exists on the appearance of the aurora borealis during the last two millennia (especially during the past 400 years) in a number of geographical areas in the world. We can mention the Asian sources, especially Chinese, Korean and Japanese sources. Another major source of information comes from Europe. European scientists have been observing the aurora for the past four centuries. Also, there are several observations from popular literature and historical chronicles. The Arab sources are especially remarkable for their reports of great aurorae that were observed in low latitudes. Lastly, we can mention the American sources with reports of aurorae observed over the past few centuries.

Yau et al. (1995) have published a catalogue of auroral observations from China, Korea and Japan (193 BC–AD 1770) that contains more than 1100 events. Additional information can be found in the catalogue published (in Chinese) by the Beijing Observatory (1988), which extends the Chinese auroral observations back to at least 210 BC and up to AD 1911. Moreover, Osaki (1994) published (in Japanese) a list of Japanese auroral records since AD 1600 and Matsushita (1956) published some cases of ancient aurorae also seen in Japan. This information has been used in some papers to study several aspects of the solar–terrestrial relations (Willis et al., 2005; Willis et al., 2007).

Lee et al. (2004) examined historical sources of Korea to collect naked-eye sunpot and auroral observations for the period covering the 11–18th centuries. Figure 6.14 shows the distribution of the 788 auroral records compiled by Lee et al. (2004). Great minima of secular solar activity are clearly visible as the Spörer and Maunder minima with very low auroral activity. High auroral activity is also very marked around 1370.

Nakazawa et al. (2004) compiled 16 probable auroral events for the 12–19th centuries from Japanese historical literature. They are often referred

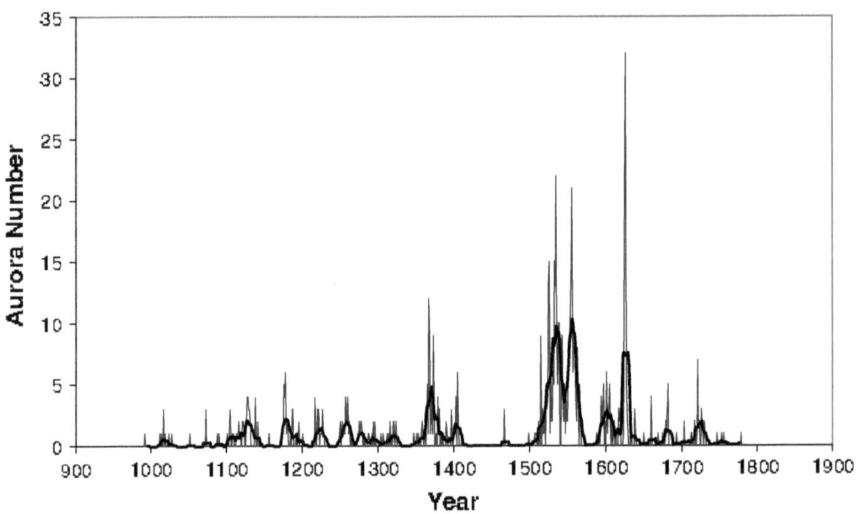

Fig. 6.14. Korean auroral records of 11–18th centuries from Lee et al. (2004).

Table 6.5. Dates of the aurorae reported in Japanese historical sources for 12–19th centuries

Event No.	Date	Event No.	Date
1	12 August 1150	9	October 1371
2	08 October 1150	10	02 March 1653
3	19 December 1202	11	17 September 1672
4	21 February 1204	12	15 February 1730
5	10 August 1247	13	17 September 1770
6	30 July 1363	14	25 September 1770
7	27 October 1370	15	06 January 1781
8	25 November 1370	16	02 September 1859

to in the historical sources as "SEKKI", a Japanese word that means "the red atmosphere". The dates of the compiled events are listed in Table 6.5. The best descriptions correspond to the events occurring on 21 February 1204 and on 17 September 1770. From these detailed descriptions, the authors conclude that the SEKKI event could be considered a large-scale appearance of low-latitude aurorae. In modern times, some reports and analysis of auroral observations from Japan have been published (Furuhata, 1958; Hikosaka, 1958; Kakioka Magnetic Observatory, 1969; Miyaoka et al., 1990; Shiokawa et al., 2005).

Several studies have been carried out on aurorae observed from Europe. In the north of Europe, the auroral frequency is high and many scientists have studied, registered and classified them (see, as examples, Dalton, 1793; Link, 1962, 1964; Dall'Olmo, 1979). In the south of Europe, however, the auroral frequency is low or very low, but nevertheless scientists have also studied these appearances over the past five centuries. Furthermore, simple people were fascinated when they observed an aurora and, hence, popular literature is an important source of information on historical aurorae. We can mention some Spanish and Portuguese examples such as *Phoenix of Tempests* [A Fenix das Tempestades] (Anonymous, 1732), *Physical, Astrological and Medical Dissertation* [Disertacion physica, astrologica, y medica] (Antonio Serrano, 1739) or *Physical Discourse on Aurora Borealis* [Discurso Physico sobre la Aurora Boreal] (Anonymous, 1769). Figure 6.15 shows a sketch of a low-latitude aurora observed in Valencia (Spain) published by Rosell (1764).

Basurah (2004, 2006) published records of aurora displays in Islamic historical texts during the 9–16th centuries. It should be stressed that this paper is the result of an extensive literature search through a wide variety of Arabic chronicles from the major Islamic cities of the time. Some examples are Baghdad (Iraq), Cairo (Egypt), Damascus (Syria), Fas (Morocco), and Córdoba (Spain). Table 6.6 lists the auroral events recorded. Vaquero and Gallego (2001) present evidence concerning the observation of aurorae recorded by

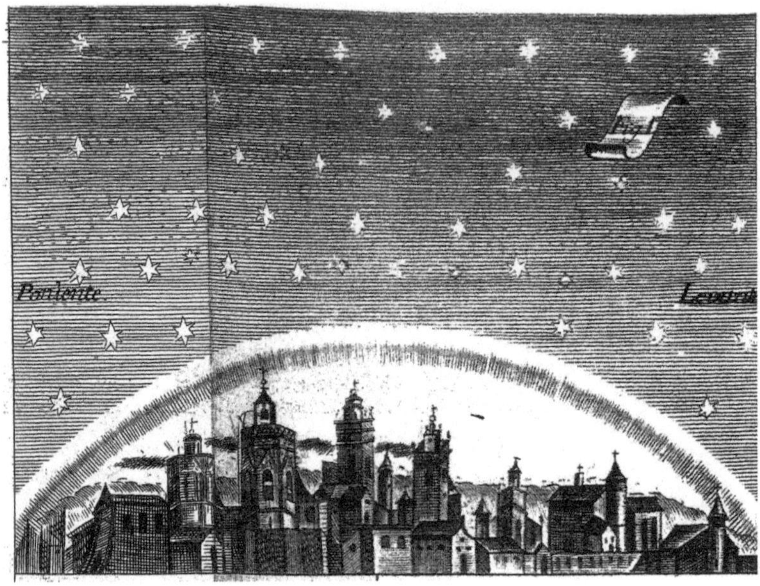

Fig. 6.15. Sketch of the auroral display observed on 5 March 1764 from Valencia (Spain).

Table 6.6. Aurorae reported in Islamic chronicles during the 9–16th centuries from Basurah (2006) including dates, country and category. The category classifications is defined as DA (definite sighting of aurora), SA (suspected aurora), and AP (atmospheric phenomena)

Event No.	Date	Country	Category
1	29 Oct 817	Iraq	DA
2	11 Oct 879	Morocco	DA
3	7 May–4 Jun 897	Egypt	AP
4	9 Nov 931	Iraq	AP
5	17 Oct 939	Syria	DA
6	6 May 941	Spain	DA
7	23 Aug 977	Egypt	DA
8	29 Sep 979	Morocco	DA
9	19 Mar 991–7 Mar 992	Egypt	SA
10	25 Apr–24 May 1050	Egypt	SA
11	9 Nov–7 Dec 1060	Egypt	DA
12	6 Sep–5 Oct 1176	Syria	SA
13	7 May 1179	Syria	DA
14	26 Oct 1223	Syria	SA
15	20–30 Jul 1264	Syria	SA
16	27 Nov 1370	Syria	DA
17	14 Dec 1422–3 Dec 1423	Egypt	DA
18	4 Jun 1570–24 May 1571	Spain	DA

the Arabs from the north of Africa in the year AD 880 and from the Iberian Peninsula in AD 942.

Reports of auroral observations are also available from North America. The first report comes from Ellesmere Island about AD 1255 (Silverman, 2002). However, the auroral event of 22 December 1719 in New England is considered to be the earliest North American aurora in standard auroral catalogues (Lovering, 1866). Silverman (2005) notes as well the observation of an aurora in 1611, in what is now New Brunswick, Canada. A general account of the auroral observations made in colonial North America is given by Mendillo and Keady (1976) and by Eather (1980) in his Chapter 6.

An important collection of auroral observations from North America was made by Wilson Bentley (1865–1931), a farmer in northern Vermont. He made auroral observations over a period of 49 years (1883–1931). His data set illustrates the usefulness and importance of self-consistent observations by single and careful observers over a long period. Silverman and Blanchard (1983) analysed Bentley's data (abstracted from the original manuscript). The auroral activity secular minima around 1900, comparable to the similar sunspot secular minima, is clearly evident.

6.3.2 Aurorae Australis

We can also find some records of observations of southern (or austral) aurorae in the South Hemisphere, although the appearances are not numerous because there is little land in southern high latitudes.

In several books, we can read that the first European that observed (and registered) southern aurorae was Captain Cook (see, as an example, the beautiful and recent book by Lang, 2006). We would like to point out here that a letter describing auroral observations during a voyage around Cape Horn in March and April of 1745 by the Spanish mariner and scientist Antonio de Ulloa (1716–1795) appears in the second edition of the *Traité* by Mairan (1754).

Ulloa was a member of the "Geodesic Mission" that the French Academy of Sciences was sending to Peru to measure a degree of the meridian at the equator (led by Pierre Bouguer) and co-discoverer of the element platinum. According to his account, he "saw a perceptible illumination [...] which had altogether the appearance of the polar lights so well known to him in the northern hemisphere".

There are obscure reports of early observations. Angot (1896) includes the observation of an aurora in 1640. It is interesting note that a "battle in the air" is described in a historical chronicle entitled *Historica relacion del reyno de Chile* [Historical account of the Kingdom of Chile] by the Jesuit Alonso de Ovalle (1601–1651) published in 1646. Figure 6.16 shows the "battle" and other portentous signs. However, the auroral character of this event needs further study.

Fig. 6.16. Battle in the air recorded by Alonso de Ovalle in Chile in 1640.

6.4 The Search for the Cause

6.4.1 Scientific Research on Aurorae

The beginning of the Scientific Revolution also marked the beginning of the scientific investigation of the auroral phenomena. The great aurora of 12

September 1621 was observed by many European scientists. The name "aurorae borealis" comes from this time (Siscoe, 1978) when scientists such as P. Gassendi and Galileo used these words systematically to name the phenomenon. These words had been used in an incidental way by Gregory of Tours (AD 538–594) one millennium before (see Tours, 1849).

Few aurorae were observed during the second half of the 17th century and the first years of the 18th century due to the Maunder minimum (1645–1715), when the solar activity was very low for approximately half a century. However, the aurora that all Europe observed on 17 March 1716 motivated a qualitative change in the scientific study of this phenomenon (Figure 6.17). Schröder (1984) provides some detailed descriptions of this aurora from Germany carried out by important scientists such as Christian Wolf (1679–1754) or Rudolph Christian Wagner. Edmund Halley, who was also interested in geomagnetic measurements, wrote two papers describing his observations of the aurorae seen in 1716 and in 1719 from London (Halley, 1717, 1720).

The most important work on aurora borealis in the 18th century was the *Traité physique et historique de l'Aurora Boréale* [Physical and historical treatise of the Dawn Borealis] written by Jean-Jacques d'Ortous of Mairan (1678–1771). The first edition of this work was published in 1733. Later on, a more complete edition was published in 1754. This is the first document where a connection is indicated between sunspots and aurorae (see Mairan, 1754,

Fig. 6.17. Aurora observed in 1716 according to the plates published in *Acta Societatis Hafniensis* in 1745.

p. 264), and the first attempt to show how the solar atmosphere might extend towards Earth and cause the aurora. Also, in this work Mairan offers the first northerly catalogue of auroral observations, which would be the origin of a subsequent abundant bibliography.

George Graham (1675–1751) discovered the daily variation in geomagnetic declination in 1722 (Graham, 1724). Graham from London and A. Celsius and O. P. Hiorter from Uppsala observed simultaneous perturbations of a magnetic needle and an aurora on 1 March 1741 (Hiorter, 1747; Stern, 2002; Courtillot and Le Mouël, 2007). Thus, the connection between aurora and geomagnetism was settled.

Another important task for the early auroral scientists was the measurement of the aurora height. During the 17th century, it was widely accepted that the aurora occurred in the lowest part of the atmosphere, very near to the Earth's surface. Halley (1717) is considered to be the first to have proposed a practical method of measuring the height of the aurora using simultaneous observations from two separate places. Other scientists tried using this technique, for example Chancelor Wolf (1716) estimated 45 km in the first scientific measurement ever made. Mairan also made some attempts, obtaining heights of 500–1500 km from observations dated between 1726 and 1730. These early measurements showed that it was very difficult to do all the operations accurately. At the late 18th century, heights about 80–160 km were obtained (Dalton, 1793; Cavendish, 1790).

The theory of the measuring method is very simple (Figure 6.18): Two auroral observers, who are separated by a distance d, measure the angular elevation of the aurora over the horizon (α and β, respectively). Then, simple trigonometry allows calculating the auroral height as

$$\mathrm{h} = \frac{\sin\alpha\sin\beta}{\sin(\beta-\alpha)}\cdot \mathrm{d} \tag{6.1}$$

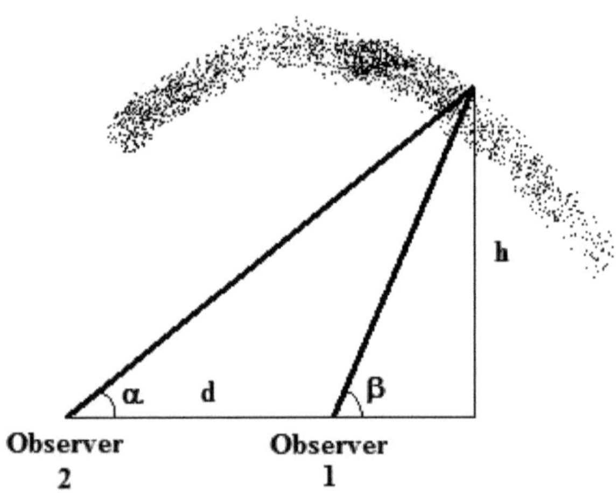

Fig. 6.18. Auroral height determined by triangulation.

Table 6.7. Hypotheses, theories and models on aurorae proposed and discussed in the scientific literature during the 19th century (modified from Eather, 1980, p. 64)

Hypotesis, Theory or Model	Author
Reflected sunlight from ice particles	R. Descartes
Reflected sunlight from clouds	M. Monge
Sulphurous vapours	P. Musschenbroek
Nitrous gases	A. Libes
Mixed mass of gaseous exhalations	Coates
Burning of gases from putrefaction, ignited by falling stars	G.F. Parrot
Luminous particles of Earth's atmosphere	L. Euler
Combustion of inflammable air	R. Kirwan
Magnetic effluvia	E. Halley
Luminous magnetic particles	J. Dalton
Meteoric dust ignited by friction with atmosphere	J.B. Biot
Atmospheric circulation patterns	B. Franklin, Rowell
Electrified molecular circulation	E. Edlund
Electric discharge between fine icy needles	L. Bradley
Electric fluid in vacuum	F. Hawksbee, J. Canton
Electric discharge in magnetic field	S. Lemström, M.A. de la Rive
Electric discharge between Earth's magnetic poles	B.V. Marsh
Electric currents in aqueous vapour	G. Planté
Condensation of vapours carrying latent electric fluid	A. Volta
Thin clouds illuminated by free electricity flow	A.P. Holden
Phosphoric electric light	A. Bertholen
Phase of certain thunderstorms	M. Silberman
Zodiacal light	J. Mairan
Cosmic dust	D. Olmsted
Cosmical particles	A. Humboldt
Currents generated by compressed cosmic ether	J. Unterweger
Solar particle streams	H. Becquerel

However, in practice actually taking this measurement is quite complicated. The observers must look at the same part of the aurora and at the same time. Precise auroral heights were not measured until the 20th century, when photography and the telephone allowed detailed analysis and communication between observers. Carl Stömer, using about 40 000 photographs and statistical analysis, established an average auroral height of 110 km.

During the 19th century, theories that tried to explain the appearance of aurorae proliferated in a spectacular way. Table 6.7 tries to show, though not exhaustively, the most important and representative theories.

An essential step came from the work of Norwegian Kristian Birkeland (1867–1917), one of the scientists who enthusiastically endeavoured to prove a theory, but never received any recognition in life. Educated in the French school of Henri Poincaré (1854–1912) and with a good knowledge of electromagnetism, he studied the variations in geomagnetic records and the appearance of aurorae that convinced him of the relationship of these phenomena with the Sun (Jago, 2001).

To simulate the process in the laboratory, he shot a beam of electrons into a vacuum camera in whose centre a sphere was located with an electric reel that simulated the geomagnetic field. The fluorescent painting of the sphere allowed visualizing the movements of the electrons, which impacted on the polar regions. Figure 6.19 shows a comparison between an image of the trajectory of the electrons in the experiments of Birkeland and a real image of an aurora observed from space. The places of penetration of the solar particles in the terrestrial atmosphere describe a circle known as the "auroral oval". This ring has its centre at the pole (north or south) of the geomagnetic field and it is usually located in latitudes between 60° and 70°. In the area between the oval and the poles, the caps, the lines of force are open to the exterior, while in the oval and in lower latitudes they are closed on the terrestrial surface.

To give a history of auroral physics is not our purpose. There is an abundant bibliography on the topic. There are excellent partial summaries in popular books such as Bone (1991), Soon and Yaskel (2003) and Clark (2007a). Robert H. Eather (1980) has written profusely on this topic, especially in Chapters 8, 9, 12 and 13, including a long list of references. Likewise the book by Brekke and Eggeland (1994) contains outstanding information. The book of Schröder (1984) can also be interesting to consult, as well as the more specific papers of Wilhelm Foerster (Schröder, 1999) and others

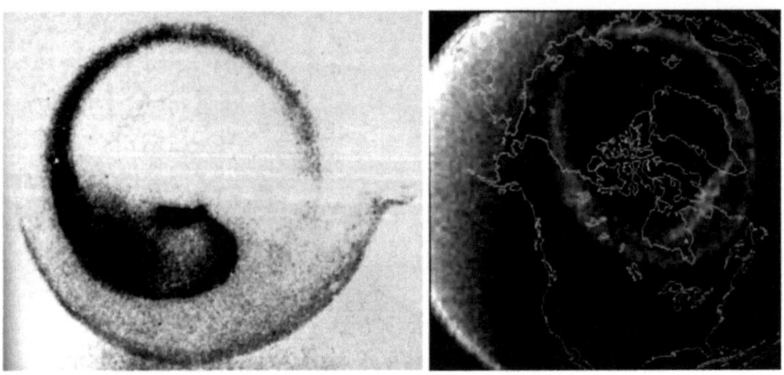

Fig. 6.19. (Left) Image with the polar vision of one of the experiments of Birkeland. (Right) Image from space of an aurora obtained in ultraviolet light by the satellite Dynamic Explorer, on which the contour of North America is superimposed (courtesy: University of Iowa).

(Schröder, 1979, 1997). Silverman (1989) wrote on auroral research in the 19th century, Brekke (1984) gave a history of the concept of oval auroral and Nygrén and Silén (1982) wrote on the contribution of Nordenskiöld to this concept. Finally, other papers that could be interesting for the reader are Meadows and Kennedy (1982), Feldstein (1986) and Rishbeth (2001).

6.4.2 The Discovery of Solar–Terrestrial Relations

The discovery of the Sun–Earth relationships had been initiated by the middle of the 19th century. However, we can find some previous dispersed ideas in the literature. In pre-classic Greece, the presence of (naked-eye) spots in the Sun was considered to be a sign of rain (Vaquero, 2007; Hardy, 1991). This idea was probably transmitted to Arab culture. This can be appreciated in an Arab text where a sunspot is described in AD 939 (Vaquero & Gallego, 2002). We have already mentioned the innovative ideas of Mairan on this topic in the 18th century. In the middle of the 19th century, three observational results allowed the initiation of studies on solar–terrestrial physics: the discovery of an 11-year cycle in sunspot number, and also in some simple geomagnetic indices, and the observation of an outstanding solar flare.

Samuel Heinrich Schwabe (1789–1875) discovered the famous 11-year cycle of sunspots (Schwabe, 1844). A decade later, four papers appeared related with the Schwabe discovery (Pomerantz, 1974).

Lamont (1851) analysed geomagnetic records taken during the previous 15 years. He found an approximate period of 10.33 years, in the amplitude of the daily variation but he did not relate it with the sunspot period discovered by S. Schwabe. Sabine (1852) also found a period of approximately 10 years in the occurrence of geomagnetic disturbances in Canada. However, he did point out the coincidence (casual or accidental) between the *solar* and *geomagnetic* cycles. Moreover, this relation was also independently identified by Wolf (1852) and Gautier (1852).

The third important observational result was the record of an intense solar white-light flare on 1st September 1859. White-light flares (WLFs) are major flares in which small parts of the Sun become visible in white light. Such flares are usually strong X-ray, radio, and particle emitters. The first observation of a WLF was made by Stephen Gray of Canterbury on 27 December 1705. It was found by Hoyt and Schatten (1996) in a manuscript of the Flamsteed collection (Cambridge University Library). Carrington (1859) and Hodgson (1860) observed the 1859 WLF and a number of scientists recorded brilliant aurorae and strong geomagnetic perturbations at the same time. There are some recent reviews of this flare observed on 1 September 1859. One can read the popular paper by Clark (2007b) and the technical papers by Tsurutani et al. (2003) and Cliver (2006). Moreover, there is a special issue of the journal *Advances in Space Research* devoted to "The great historical storm of 1859: a modern look" (vol. 38(2), pp. 115–388, 2006) and a brilliant popular science book on the context and consequences of this remarkable observation (Clark, 2007a).

Carrington's flare marked the beginning of an intense research effort on the relationships between the behaviour of the Sun and terrestrial phenomena. Over the past 150 years, relationships have been looked for between the solar variability and a wide variety of phenomena that happen on the Earth. Many of these phenomena are purely geophysical (geomagnetism, climatology, meteorology, etc.) but the relationship between the solar activity and a wide range of phenomena such as illnesses, market systems or the quality of crops has also been studied. The basic ideas on solar–terrestrial relationships can be consulted in the last section of Chapter 1. A complete revision of the *state of the art* can be found in Hoyt and Schatten (1997).

6.5 Catalogues of Aurorae Observations

6.5.1 Catalogues from 18th Until 20th Century

The systematic study of the seasonal and secular variations of the auroral appearances needed data with perfectly dated and correctly compiled records of aurorae. The catalogues of aurora borealis were born of this need. The first well-known catalogue appeared as a part of the already mentioned *Traité* of Mairan (1733). In the first edition, 229 aurorae were listed (many with descriptions) from the year 502 to 1731. This catalogue was expanded with many auroral records in the second edition (Mairan, 1754). Silverman (1998) provides the sources used by Mairan in his new compilation. Thus, the new version of the catalogue lists 1441 different aurorae (using 2137 observations).

Mairan's catalogue was followed by those of G. Kirch in 1735 (edited by Schröder, 1996) and Frobesius (1739). Frobesius's book is divided into two parts. The first is the catalogue itself and the second is a discussion about the auroral phenomenon. Pilgram (1788) compiled another catalogue of auroral observations from AD 394 to 1784. Moreover, he made an interesting study of the intervals between auroral observations. Catalogues of atmospheric phenomena that included aurorae were also made at this time. These catalogues were compiled to study the effect of weather on human illness. Short (1749, 1767) and Lowe (1870) are the best known although Short's catalogue is unreliable (Britton, 1937).

A great effort was made in the mid-19th century by J. Lovering (1813–1892), who compiled a catalogue using manuscripts and published material (also from the United States). The first version of the catalogue contains nearly 10 000 aurorae from 50 000 observations. Later, he added some 2000 aurorae. Additionally, Lovering analysed the annual and secular variation of the aurora. A paper on the biography of Lovering and his research can be consulted in Peirce (1909).

The best-known catalogue of the 19th century was that of Hermann Fritz (1830–1893) who drew on a great many sources including the Lovering

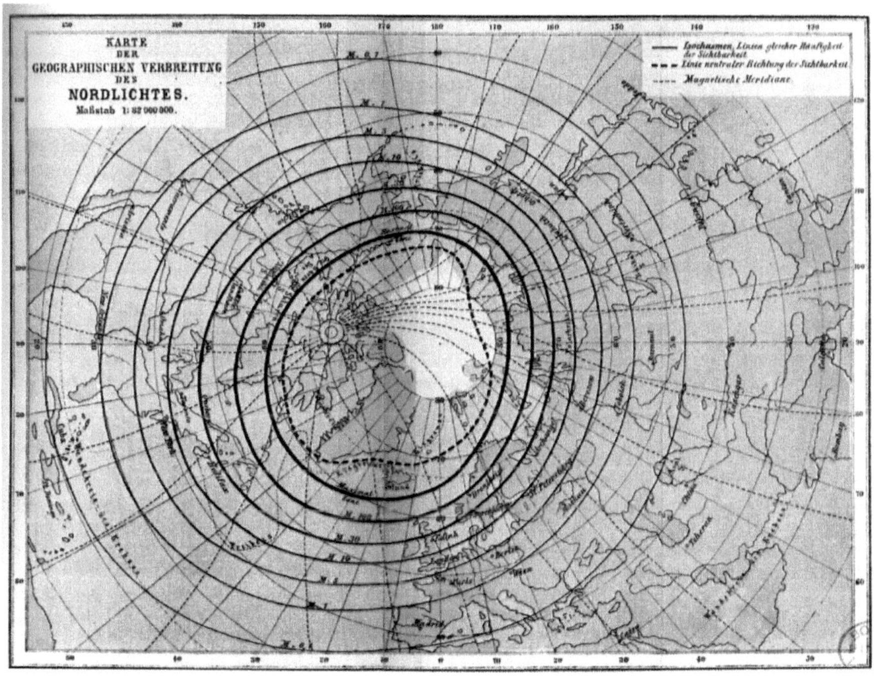

Fig. 6.20. Graphical representation of isochasms made by H. Fritz in 1881.

catalogue (Fritz, 1873). He was a student at the Technical College, Darmstadt. In 1872, he was appointed titular professor at the Technical College (Zürich, Switzerland). He taught mechanics and wrote some technical books. However, he was tremendously interested in solar–terrestrial physics and was influenced by J. R. Wolf (Schröder, 2004). Fritz proved a similar variation in sunspot number and auroral frequency. He also conceived the isochasms[5] (Figure 6.20) to describe the geographical distribution of auroral occurrence. Fritz's catalogue was complemented with the book *Das Polar Licht* (Fritz, 1881).

Lovering's and Fritz's catalogues have some errors. However, they are not numerous and the catalogues can be been used for statistical studies. Other 19th century regional catalogues were available for Sweden (Rubenson, 1879, 1882) and Norway (Schroeter, 1902). A short list of aurorae observed from the southern hemisphere appears in the last part of Fritz's catalogue (pp. 251–255). A more extensive catalogue of southern hemisphere aurorae was written by Boller (1898).

Of the 20th century catalogues, we would cite as of special interest that written by Link (1962, 1964). This is a carefully annotated catalogue for

[5] Isochasms are lines that connect geographical places where aurorae occur with the same frequency.

pre-1700 era covering Europe. Link (1962, 1964) included the original text and the date of the observations. Another useful catalogue was compiled by Křivský and Pejml (1988) for the interval 1000–1900. They give dates of observation only and are restricted to latitudes < 55°. Other 20th century regional catalogues are Loysha et al. (1989) for the former Soviet Union and Keimatsu (1976) for China. Lastly, we would mention the catalogue by Yau et al. (1995) for China, Korea and Japan.

6.5.2 Recent Catalogues

Some new datasets of auroral records have been published during recent years. Long-term series of auroral observations by a single observer or referring to the same geographical region are scarce. Therefore, they are important for the study of solar–terrestrial relationships. Here, Canadian (Broughton, 2002), Korean (Lee et al., 2004), English (Harrison, 2005), Spanish (Vaquero et al., 2003) and Portuguese (Vaquero and Trigo, 2005) records are briefly reviewed.

Using manuscript sources, Broughton (2002) studied over 1000 previously unused Canadian auroral observations during the period 1771–1819. He used meteorological records which simultaneously contain information on cloudiness. Therefore, the information is especially reliable. The data show an interval of low auroral activity around the period 1785–1815 (a decade earlier than the well-known Dalton minimun of solar activity). Broughton (2002) suggests that the auroral oval contracted more than usual during the Dalton minimum because the relationship between the auroral occurrence frequency and solar activity from 1795 to 1815 is similar to what is normally seen at lower latitudes. Moreover, it supports the type of relationship found by Ohl and Ohl (1979). The frequency of auroral occurrence at high latitudes during the declining portion of one solar cycle is related to the sunspot number at the following maximum and the rapidity with which the sunspot numbers return to a maximum.

Lee et al. (2004) collected auroral historical records of Korea during the period AD 1000–1800. They examined background data such as solar and lunar eclipses, and planetary motions, etc., in the historical sources used to estimate the astronomical observation frequency because auroral data in historical documents may be not homogeneous. To estimate the periods of auroral activity, they used a one-dimensional power spectrum method from inhomogeneous data. Lee et al. (2004) found 11.2 year and 88.4 year cycles using 542 auroral observations for the period 1397–1799.

Harrison (2005) published and analysed the auroral observations recorded in the 1771–1813 diary of Thomas Hughes from Stroud, Gloucestershire (51.75° N, 2.22° W). This diary has proved to be a rich source of historical meteorological data. It provided a list of 71 nights on which the aurora was seen, between 19 February 1771 and 13 October 1805 inclusive. About 90% of Hughes' aurora observations occurred between 1771 and 1789. These observations provide information on the incidence of the aurora in southern UK, before light pollution.

A catalogue of aurorae was compiled by the Spanish physicist Rico Sinobas (1821–1898) in 1855. He was one of the most important Spanish physicists of the 19th century, holding the Chair of Physics at the Central University in Madrid, and becoming vice-president of the Academy of Sciences in Madrid. The catalogue presents the aurorae observed in the Iberian Peninsula in the period 1700–1855, and is of special interest because observations were at low-latitude sites. Vaquero et al. (2003) analysed the reliability of the data, and found that this compilation confirms, for low latitudes, the absence of aurorae during the Dalton minimum.

Vaquero and Trigo (2005) presented a compilation of auroral observations made at Lisbon in the late 18th century by Jacob Praetorius and Henrique Schulze, two German artillery officers. Dates of 18 aurorae observed by Praetorius and Schulze were compared with those published in other catalogues for that period. The number of annual aurorae observed by the two Germans can be compared to some indices of solar activity. Figure 6.21 shows a comparison between the sunspot numbers and the number of aurorae observed in Portugal from Praetorius and Schulze observations and in Spain from Rico Sinobas' catalogue. This comparison shows a very good level of consistency between all time series.

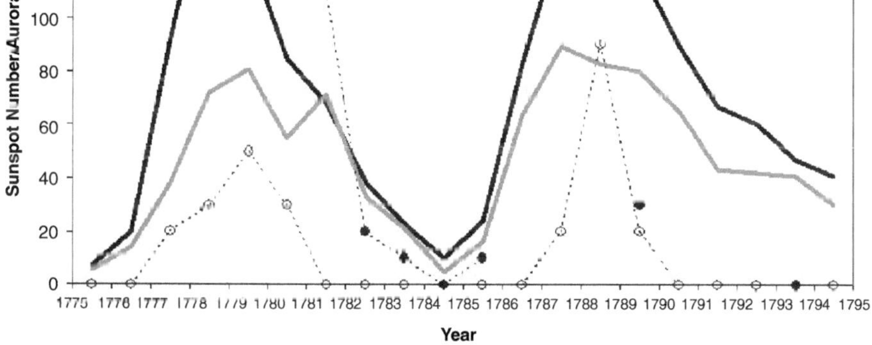

Fig. 6.21. A comparison between the sunspot numbers and the number of aurorae observed in Portugal from Praetorius and Schulze's observations and in Spain from Rico Sinobas' catalogue.

6.6 Aurorae and Secular Solar Activity

6.6.1 Aurorae as a "Proxy" of Solar Activity

Solar variability is manifested on different time-scales depending on the type of energy dissipated. The magnetic energy dominates on the scale of centuries. We can divide the effects on our planet into two different channels depending on the topology of the magnetic field.

Radiation domain: The magnetic flux of the active regions is characterized by magnetic configurations with closed field lines dominating the variations in the total irradiance and emission in the high-energy range of the solar spectrum (ultraviolet and X-rays). Most of the radiative losses from the outer layers occur in these regions.

Particles domain: Large-scale magnetic regions have field lines open toward the interplanetary medium. They are the main source of a continuous outward flow of charged particles (protons, electrons and He nuclei) known as the solar wind. The solar magnetic field of the open regions (OMFs) is frozen into this wind, configuring the interplanetary magnetic field (IMF), which produces a huge magnetic region, the heliosphere, which fills practically the whole Solar System. Galactic cosmic rays (GCRs) are high-energy particles (mainly protons with energies in the range 1–20 GeV), originating outside our planetary system and striking the Earth from all directions. Both the flux and energy spectrum of GCRs are modulated by the strength of the heliosphere, being stronger when the IMF is weaker. Variations in the solar wind are produced by transient events, such as coronal mass ejections, and the recurrent passage of equatorial coronal holes.

Aurorae are linked to the particle domain. Their brightness and large angular size have attracted the interest of many civilizations, with deep roots in mythology. Thus, numerous non-scientific documents have reported these events, and we can make use of them as a proxy of solar activity in past times.

During the 1960s and 1970s, several scientists decided to take up the data compiled in auroral catalogues to study the main characteristics of the secular solar activity. Link (1963) found 400-year and 80-year cycles in auroral activity using data from the period AD 400–1600. He also tried to improve on the knowledge of solar activity during the 17th century using auroral data where only fragmentary observations of sunspots exist. He obtained epochs that had 10 solar minima with mean periods of 11.3 years (Link, 1977) and explained the Maunder minimum as one of the minima of the 80-year cycle (or Gleisberg cycle) of solar activity (Link, 1978). D. J. Schove derived an auroral index from historical reports of aurora. A first version was published in Schove (1955) with subsequent improvements (Schove, 1962; 1980). Similarities have been found between this series and other proxy data of solar activity. The complete series covers the last two millennia (Figure 6.22).

After the outstanding contribution by Eddy (1976) that clearly established the Maunder minimum, a series of papers by different authors have

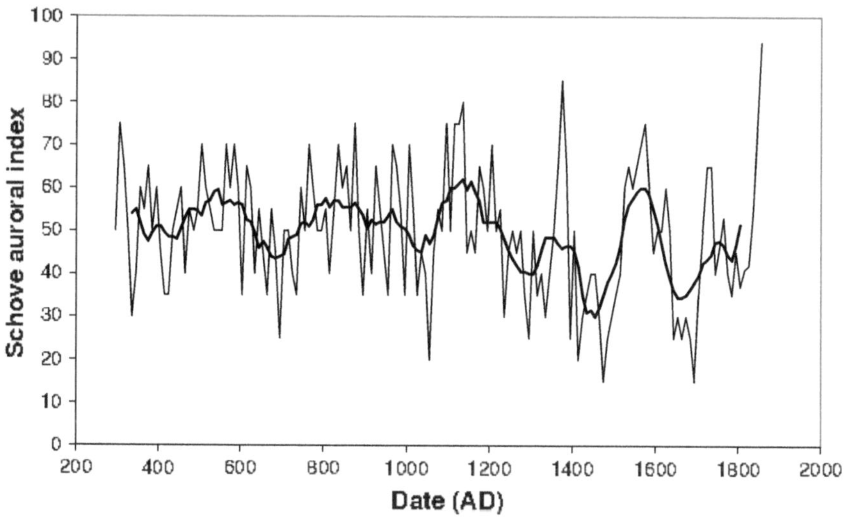

Fig. 6.22. Auroral index for the last two millennia approximately derived by Schove (1980). The thick line represents a smoothing using a 10-year moving average.

pointed out the apparent contradiction of great minima of solar activity and the appearance of aurorae in the same epoch. As an example, Botley (1981) noted that aurorae were observed from Hungary in the years 1648, 1652, 1664, 1668 and 1692 (see also Hedervari, 1981). Therefore, the number of aurorae observed in central Europe only fell to about 30% with respect to the previous years. Several papers (Schröder, 1994a,b) tried to demonstrate that the Maunder minimum was not confirmed by the auroral data. However, V. Letfus (2000) has indicated that the auroral events that appeared in years when no spots occurred could have been produced by proton flares. In fact, two-ribbon flares appeared in plages with only very small or no sunspots and Dodson and Hedeman (1970) have demonstrated that some of these flares are geoactive. Therefore, the auroral records during the Maunder minimum indicate the presence of magnetic activity in the upper layers of the solar atmosphere, probably related with the open magnetic field structure, even when no spots were visible at the photospheric level.

The auroral record from the fifth century BC to the 20th century was reviewed by Siscoe (1980), who found evidence of secular solar variability from his analysis. He found some features corresponding to the well-known Medieval minimum, Medieval maximum, and Spörer minimum in both European and oriental records. Some periods presented in the analysis (such as a quasi 80-year oscillation and a quasi-10-year periodicity during aurorally rich intervals) were attributed to solar activity cycles.

The study of long-term fluctuation of solar activity during the last millennium using ^{14}C data, historical observations of large sunspots by naked-eye and aurorae observed at latitudes <55° was the main purpose of Křivský (1984).

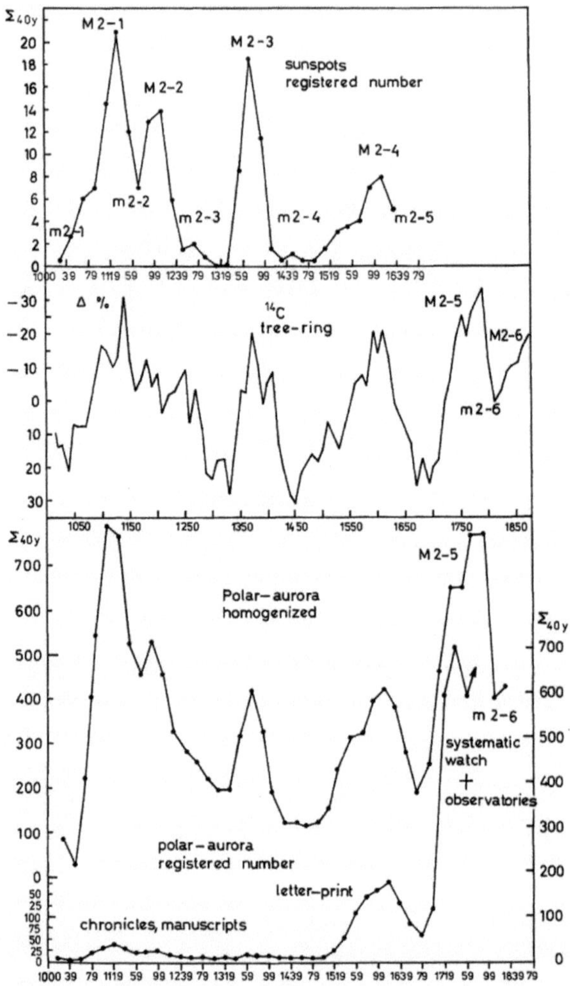

Fig. 6.23. Long-term fluctuation of three solar activity indexes: 40-year sums of sunspot observations by naked eye (statistically elaborated) where maxima (M) and minima (m) with numbering are indicated; ^{14}C deviation (%) from tree rings (Stuiver, 1980); 40-year sums of polar-aurora observations (statistically elaborated and homogenised) and 40-year sums of polar-aurora numbers without homogenization (from Křivský, 1984).

The series used for this study are shown in Figure 6.23. Křivský (1984) explains that the auroral series is considerably inhomogeneous due to different "civilization factors" although all the maxima and minima are similar to those in the naked-eye sunspots and ^{14}C data. Two intermediate series were constructed to obtain the final auroral curve. Series "a" and "b" consisted of 40-year sums of original data beginning in the year 1000 and 1020 respectively. For series "a", Křivský (1984) used as "civilization factors" the numbers 23 (for the period 100–1519), 6.3 (1520–1559), 2.4 (1560-1719) and 1.0 (1720-1880). For series "b", the "civilization factors" were 24 (1020–1499), 6.3 (1500–1539) 2.7 (1540–1699) and 1.0 for (1700-1799). The final auroral curve was obtained using an average of the series "a" and "b" (Figure 6.23). According to this study, secular maxima and minima using different proxies of solar activity are in good agreement.

The secular variation of the aurora during the last 500 years was reviewed by Silverman (1992). He used a database of visual observations for a period from 1450 to 1948, comprising about 45 000 observations. He compared the secular variation of the aurora to sunspot and magnetic activity data (where possible) and used Blackman–Turkey power spectra to determine periodicities. The well-known 11-year period is the strongest characteristic of the sunspot series spectrum. However, Silverman found other important shorter periods in auroral occurrence that induce a different temporal behaviour for sunspots and aurorae. Among his most interesting results, he confirmed that the 11-year cycle disappears during the Maunder minimum and at the end of the eighteenth and beginning of the nineteenth century. Moreover, the auroral data show the well-known Spörer, Maunder, Dalton and 1901–1913 minima including an unrecognized long-term minimum around 1765.

Figure 6.24 shows the number of days per year in which aurorae were recorded from the year AD 1000 to 1900 using the catalogue of Křivský and Pejml (1988). Very few aurorae before the Maunder minimum are recorded there, but the number increases considerably later. The number of auroral days per year during the first part of the total period is very low. During the 16th and 17th centuries, this number begins to increase, although it is still

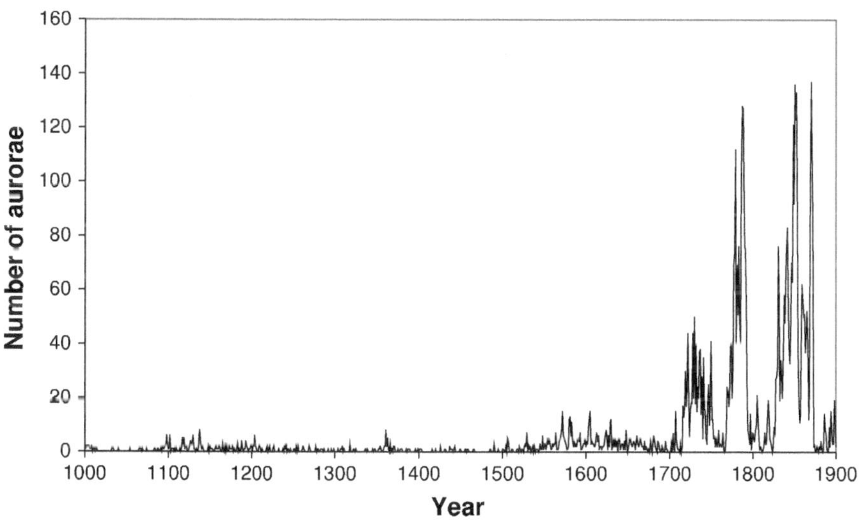

Fig. 6.24. Annual number of auroral days recorded in the catalogue of Křivský and Pejml (1988).

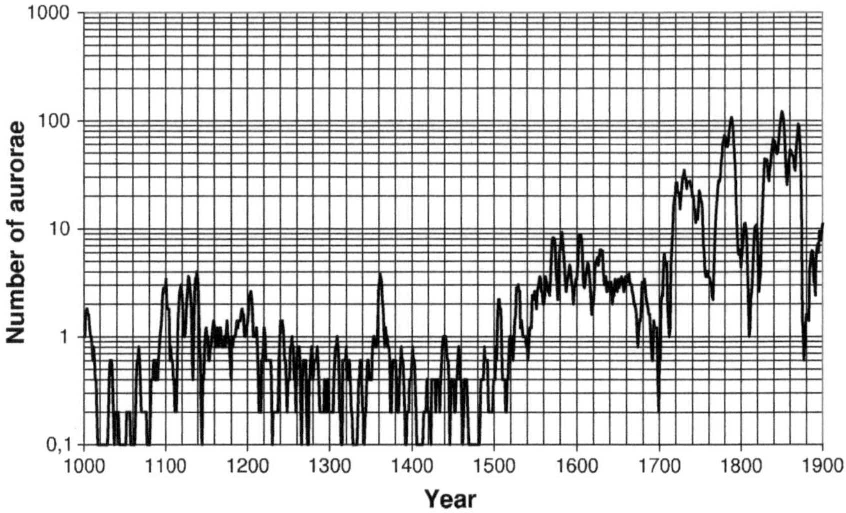

Fig. 6.25. Smoothed annual number of auroral days (5-year moving average) recorded in the catalogue of Křivský and Pejml (1988).

small compared with the number of recorded auroral days during the 18th and 19th centuries. We can improve our vision of the auroral frequency by using the logarithmic scale in Figure 6.24 due to the wide range of the numbers, and smoothing the series. Figure 6.25 has been made using a 5-year moving average and logarithmic scale for the annual number of auroral days recorded in the Křivský–Pejml catalogue.

Maunder (≈1700) and Dalton (≈1810) minima can be seen clearly in Figure 6.25. Several interesting periods are also revealed. There is a puzzling minimum in auroral activity around 1765 that was discovered by Silverman (1995). Another period of minimum auroral activity is located around 1880 (Figure 6.25), also cited by Silverman (1995). This striking minimum of auroral activity deserves more study.

6.6.2 Low-Latitude Aurorae

Low-latitude aurorae constitute a special and not frequent case. These aurorae can often be produced by massive solar flares, usually associated with huge Coronal Mass Ejection (CME). They are generally red and diffuse resulting primarily from an enhancement of the 630.0 nm [OI] emission due to the bombardment by soft electrons (<100 eV). The typical altitude for a low-latitude aurora is 250-400 km (Silverman, 1998). They can be confused with fires or twilights due to their colour. As an example we can cite the observation by A. Lang, Governor of the Saint Croix Island (17°44′32″ North, now a part of the Virgin Islands), who reported a red aurora on 17 November 1848 and commented that the red glare ascended high above the hills, leading several

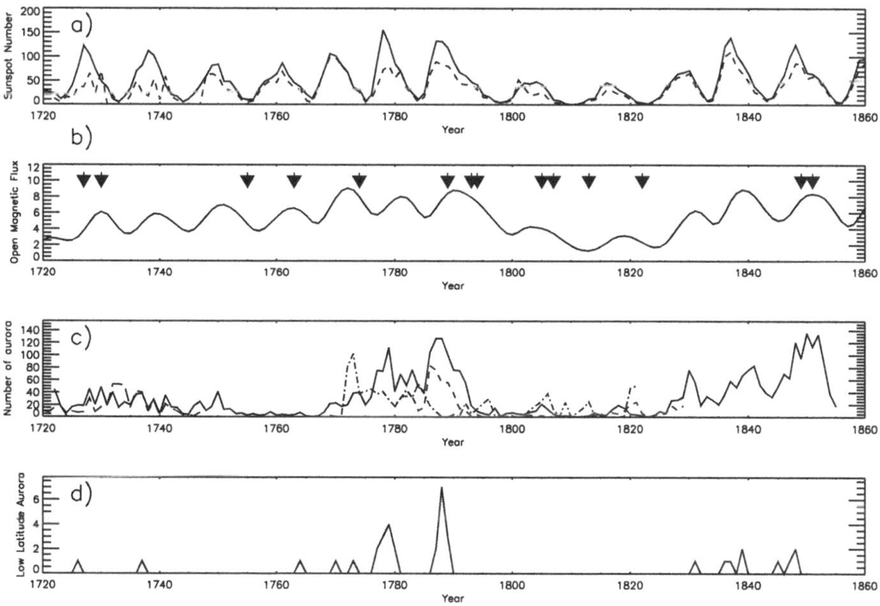

Fig. 6.26. (a) Variation of the closed magnetic field (CMF), indicated by the sunspot number. (b) Open magnetic flux calculated by Usoskin et al. (2002) (data courtesy of M. Schüssler). Arrows indicate the solar proton events reported by McCracken et al. (2004). (c) Yearly number of aurorae in the catalogue of Křivský and Pejml (1988) (solid line), Schröder (2003) data during the Dalton minimum (dashed), Canadian observations by Broughton (2002) (dashed-dotted), and Mairan (1733) complemented by Viera and Clavijo (1770) (long dashes). (d) A summary of low-latitude ($< 42°$) aurora, observed at the Iberian Peninsula. Adapted from Vazquez et al. (2006).

persons to believe that a tremendous fire had occurred (Lang, 1849). This confusion is also described by J. Viera y Clavijo (1731–1813), writing about the aurora observed in Tenerife on 18 January 1770 (Viera, 1770).

Vazquez et al. (2006) studied the relationship between the appearance of aurorae in mid- and low-latitudes and various indices describing the solar activity. They selected a sample that covered the period 1715–1860. This epoch is not very well covered in the existing literature, which includes the Dalton minimum. They studied the number of low-latitude aurorae visible in this period and their relationship with different parameters describing the physical state of the heliosphere. Aurorae occur mainly in the decaying phase of the solar cycle (as defined by the sunspot number) according to this study. However, auroral appearances coincide with the maximum of the Open Magnetic Field (OMF), a better descriptor of the physical state of the heliosphere. They also calculated the variation over time of the magnetic latitude of the places of observations. Figure 6.26 compares two parameters of solar activity (the variation of the closed magnetic field (CMF), indicated by the sunspot number, and the OMF calculated by Usoskin et al., 2002) with different sources of auroral observations from 1720 to 1855. Several phases can be distinguished in this figure:

(i) Recovery from the Maunder minimum (1715–1765): no clear correspondence is found between the number of aurorae and the solar activity indices.
(ii) First period with enhanced activity (1765–1795): all parameters show high values but curiously the maximum number of aurorae is not coincident with any maximum of CMF or OMF. The 1771–1813 diary of Thomas Hughes from Stroud, Gloucestershire (51.75°N, 2.22°W), lists 71 nights on which the aurorae were seen, between 19 February 1771 and 13 October 1805 inclusive. Ninety percent of Hughes aurora observations occurred between 1771 and 1789 (Harrison, 2005).
(iii) The Dalton minimum (1795–1823): a period of reduced activity. Usoskin et al. (2003) have studied in detail the beginning of this period, proposing the existence of a weak additional period from 1792 to 1794. This seems also to be supported by the auroral frequency during this interval (Usoskin et al., 2002).
(iv) Second period with enhanced activity (1823–1860): this was the start of a continuous increase of solar activity lasting until the present. Again we find good correlation between aurorae and activity indices.

6.6.3 Rieger Periodicity

Figure 6.27 shows the available information of the number of sunspot groups and the aurorae recorded in the catalogues of Rico Sinobas and Hughes during the period 1777–1789 (which includes the maxima of the solar cycles 3 and 4). A close inspection of the dates of auroral events reveals several time intervals of 155 days approximately (Rieger periodicity, painted with arrows in Figure 6.27).

Rieger et al. (1984) and Dennis (1985) detected a periodicity of approximately 158 days in γ and X-ray flare data taken by the SMM and GOES satellites during solar cycle 21. Later, this solar periodicity was also detected in several indexes of solar activity in different epochs. Moreover, Krivova and Solanki (2002) proposed that the Rieger period is the third harmonic of the 1.3 year periodicity, present in helioseismic data. Nevertheless, the majority of the results have been obtained using data sets of the last cycles due to the shortage of representative solar indexes for previous times. Silverman (1990) and Ballester et al. (1999) demonstrated that Rieger periodicity was present in the Sun in historical epochs. On the one hand, Silverman (1990) managed to show different historical auroral series with the above mentioned periodicity. On the other hand, Ballester et al. (1999) used the Group Sunspot Number (Hoyt and Schatten, 1998a,b). Unfortunately, the analysis made by Ballester et al. (1999) is not valid for all of the 18th century since the quality of the

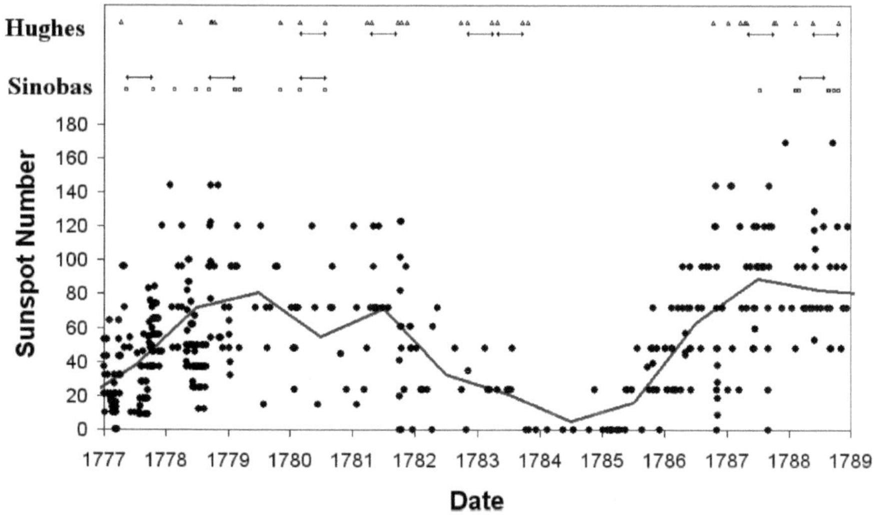

Fig. 6.27. Sunspot numbers during the time interval 1777–1789 (black points for daily data and grey line for monthly data) and auroral events recorded by Rico Sinobas and Hughes. The arrows point out auroral events separated by the Rieger period (155 days approximately).

above mentioned index is inadequate for any periods of the 18th century and, especially during the last quarter, corresponding to the solar cycles number 2 and 3. However, Rieger periodicity could be present during these cycles according to the Rico Sinobas and Hughes catalogues (Figure 6.27), although further study is necessary to establish the presence of this periodicity during these cycles.

6.7 Aurora and Great Space Weather Events

The appearance of aurorae at lower latitudes can be associated with intense solar storms, even during periods of relatively low solar activity (Silverman, 2003). The study of the largest geomagnetic solar storms that have occurred in the last two centuries is of relevance today. This kind of exercise provides additional information on extremely damaging space weather events that can strike in the near future (Lanzerotti, 2007).

Some important space weather events have been studied with some detail in recent years. However, several cases are still unexplored. Some studies are available about the historical events that have occurred in the "pre-satellite" era. We cite here the storms which occurred on 2 September 1859 (Cliver, 2006; Tsurutani et al., 2003), 25 September 1909 (Silverman, 1995), 14/15 May 1921 (Silverman and Cliver, 2001) and 25 January 1938 (Barlow, 1938; Botley, 1938; Bernhard, 1938; Hess et al., 1938).

Other, more recent, space weather events have already occurred with scientific satellites orbiting the Earth, providing a wealth of information impossible to attain prior to the 1960s. These include the episodes in 1972 August 4/5 (McKinnon et al., 1972; McIntosh, 1972; Rust, 1972), 1989 March 13/14 (Allen et al., 1989; Livesey, 1990) and the recent 2003 October–November (Veselovsky et al., 2004) event. The exact dates for other important geomagnetic/solar storms that occurred between 1850 and 1950 are relatively well known. However, most of these "pre-satellite" episodes are poorly understood and have not been analysed in detail, either because we have only partial information on them, or because, despite the existence of data, no research group has been able to merge and provide a good description of the events.

The 1859 and 1989 space weather extreme events are recognized as exceptionally large. However, one may ask how large or how exceptional. Space engineers, solar physicists and geophysicists are interested in the "worst case" examples for the various categories of space weather. Jones (1955) was a pioneer showing the 56 great events recorded in the magnetometers installed at Greenwich/Abinger Observatory from 1859 to 1954 in a supplement of his catalogue of sunspots and geomagnetic storms. Table 6.8 lists the date of outstanding geomagnetic storms recorded by Jones (1955) and the ranges of variation of different components of the measured geomagnetic field.

Cliver and Svalgaard (2004) compiled rank order lists of some parameters related with space weather events for the period 1859–2003. They found that the 1859 event has close rivals (and even superior) in each category of space weather activity used. However, the Carrington event is the only great storm of the last 150 years that is included at or near the top of all of the lists. As an example, Table 6.9 lists large solar energetic proton events, 1859–2000 from McCracken et al. (2001a,b). They used nitrate composition in ice cores to

Table 6.8. Outstanding geomagnetic storms recorded at Greenwich/Abinger, 1859–1954 from Jones (1955)

Date	Declination (′)	Horizontal force (nT)	Vertical force (nT)
01 Sep 1859	>>92	>>625	1500
04 Feb 1872	125	800	>950
17 Nov 1882	115	>1090	>1060
31 Oct 1903	119	1175	1440
25 Sep 1909	193	1710	>1080
14 May 1921	110	>>740	>>460
25 Jan 1938	126	1055	570
16 Apr 1938	307	1375	500
24 Mar 1940	131	1370	1000
01 Mar 1941	186	1650	1310
18 Sep 1941	123	1250	1115
28 Mar 1946	162	1660	920
21 Sep 1946	136	925	450

Table 6.9. Large solar energetic proton events, 1859–2000 from McCracken et al. (2001a) (after Cliver and Svalgaard, 2004). Unit for fluence is 10^9 protons cm^2. Only the year is given for events without identified candidate solar source

Date	> 30 MeV SEP Fluence
Aug–Sep 1859	18.8
1895	11.1
Nov 1960	9.7
1896	8.0
1894	7.7
1864	7.0
Jul 2000	6.3
1878	5.0
Aug 1972	≈ 5

find out about the great events of solar energetic protons (SEP) that occurred during the period 1561–1950, extending the period to 1994 with ionospheric and satellite data.

References

Abbe, C. An aurora in South Carolina and Kentucky. *Monthly Weather Review*, 23:297–298, 1895.

Aczel, A. D. *The Riddle of the Compass: The Invention that Changeg the World.* Harcourt, New York, 2001.

Akasofu, S.-I. *Exploring the Secrets of the Aurora.* Springer, 2 edition, 2007.

Akasofu, S.-I. and Kamide, Y. (eds.). *The Solar Wind and the Earth.* D. Reidel, 1987.

Alexandrescu, M., Courtillot, V., and Le Mouël, J. Geomagnetic field direction in Paris since the mid-sixteenth century. *Physics of the Earth and Planetary Interiors*, 98:321–360, 1996.

Alexeff, I. and Parameswaran, S. M. The Aurora Borealis in Southern USA. *IEEE Transactions on Plasma Science*, 33:500–501, 2005.

Allen, J., Sauer, H., Frank, L., and Reiff, P. Effects of March 1989 solar activity. *EOS Transactions AGU*, 70:1479, 1486–1488, 1989.

Angot, A. *The Aurora Borealis.* Kegan, Paul, Trench, Trubner, and Company, London, 1896.

Anonymous. *A Fenix das Tempestades.* Joseph Antonio da Sylva, Lisboa, 1732.

Anonymous. *Discurso Physico sobre la Aurora Boreal, observada en Madrid la noche del da 24 de Octubre de este año.* Madrid, 1769.

Antonio Serrano, G. *Disertacion physica, astrologica, y medica sobre las causas, y presagios del cometa, que se observo en Febrero de 1737. y del Phenomeno Igneo, que se vio en 16. de Diciembre del mismo año.* Fernando de Ros, Córdoba, 1739.

Aragam, C. B. *Relaam de un prodigioso, e estupendo phenomeno visto na Cidade de Dehly Corte do Gram Mogor, Imperador do Indostan.* Officina de Pedro Ferreira, Lisboa, 1753.

Ballester, J. L., Oliver, R., and Baudin, F. Discovery of the Near 158 Day Periodicity in Group Sunspot Numbers during the Eighteenth Century. *The Astrophysical Journal*, 522:L153–L156, 1999.

Barlow, E. W. The auroral display of January 2526, 1938. *Quarterly Journal of the Royal Meteorological Society*, 63:215–219, 1938.

Barraclough, D. R. Observations of the Earths magnetic field in Edinburgh, from 1670 to the present day. *Transactions of the Royal Society of Edinburgh Earth Sciences*, 85:239–252, 1995.

Bartels, J. Eine universelle Tagsperiode der erdmagnetischen Aktivität. *Meteorologische Zeitschrift*, 42:147, 1925.

Bartels, J. Terrestrial-magnetic activity and its relations to solar phenomena. *Terrestrial Magnetism and Atmospheric Electricity*, 37:1–52, 1932.

Bartels, J., Heck, N. H., and Johnston, H. F. The three-hour index measuring geomagnetic activity. *Journal of Geophysical Research*, 44:411–454, 1939.

Basurah, H. M. Auroral evidence for early high solar activities. *Solar Physics*, 225:209–212, 2004.

Basurah, H. M. Records of aurora in the Islamic chronicles during 9th–16th centuries. *Journal of Atmospheric and Solar-Terrestrial Physics*, 68:937–941, 2006.

Beijing Observatory. Zhongguo gudai tianxiang jilu zongji (a union table of ancient chinese records of celestial phenomena) (in chinese). In *Jiangsu Kexue Jishu Chubanshe*, pp. 27–60. Nanjing, 1988.

Bell, J. M. The fireside stories of the Chippwyans. *Journal of American Folklore*, 16:73, 1903.

Bernhard, H. J. Northern lights come south. *The Sky*, 2:10, 1938.

Bohlin, J. D. Extreme-ultraviolet observations of coronal holes. I - Locations, sizes and evolution of coronal holes, June 1973-January 1974. *Solar Physics*, 51:377–398, 1977.

Boller, W. Das Südlicht. *Gerlands Beiträge zur Geophysik*, 3:56–130, 550–609, 1898.

Bone, N. *The Aurora. Sun-Earth Interactions.* Ellis Horwood, London, 1991.

Bone, N. *Aurora. Observing and Recording Nature's Spectacular Light Show.* Patrick Moore's Practical Astronomy Series, Springer, 2007.

Botley, C. M. Some human reactions to the great aurora of January 2526, 1938. *Quarterly Journal of the Royal Meteorological Society*, 63:449–450, 1938.

Botley, C. M. Sporadic aurora. *Planetary and Space Sciences*, 11:723–724, 1963.

Botley, C. M. Hungarian auroral observations and the Maunder Minimum. *The Observatory*, 101:123–123, 1981.

Brekke, A. On the evolution in history of the concept of the auroral oval. *EOS Transactions AGU*, 65:705–707, 1984.

Brekke, A. and Eggeland, A. *The Northern Lights. Their Heritage and Science.* D. Reidel, 1994.

Britton, C. E. *A Meteorological Chronology to A.D. 1450.* HMSO, London, 1937.

Broughton, P. Auroral records from Canada 1769-1821. *Journal of Geophysical Research (Space Physics)*, 107:1152, 2002.

Bryson, G. *Northern Lights: The Science, Myth and Wonder of Aurora Borealis.* Sasquatch Books, 2001.

Burns, R. *Robert Burns, the Complete Poetical Works.* Edited by A. Mackay James. Alloway Publishing, Ayrshire, 1993.

Cafarella, L., de Santis, A., and Meloni, A. Secular variation in Italy from historical geomagnetic field measurements. *Physics of the Earth and Planetary Interiors*, 73:206–221, 1992.

Carlowicz, M. J. and Lopez, R. E. *Storms from the Sun.* Joseph Henry Press, 2002.

Carrington, R. C. Description of a singular appearance seen in the Sun on September 1, 1859. *Monthly Notices Royal Astronomical Society*, 20:13–15, 1859.

Cavendish, H. On the height of the luminous arch which was seen on February 23, 1784. *Philosophical Transactions of the Royal Society*, 80:101–105, 1790.

Chapman, S. and Bartels, J. *Geomagnetism.* Clarendon Press, Oxford, 1940. 2 vol.

Clark, S. *The Sun Kings. The Unexpected Tragedy of Richard Carrington and the Tale of How Modern Astronomy Began.* Princeton and Oxford, Princeton University Press, 2007a.

Clark, S. Astronomical fire: Richard Carrington and the solar flare of 1859. *Endeavour*, 31:104–109, 2007b.

Cliver, E. W. The 1859 space weather event: Then and now. *Advances in Space Research*, 38:119–129, 2006.

Cliver, E. W. and Svalgaard, L. The 1859 Solar Terrestrial Disturbance And the Current Limits of Extreme Space Weather Activity. *Solar Physics*, 224:407–422, 2004.

Cortie, A. L. Sun-spots and terrestrial magnetic phenomena, 1898–1911: the cause of the annual variation in magnetic disturbances. *Monthly Notices Royal Astronomical Society*, 73:52–60, 1912.

Courtillot, V. and Le Mouël, J.-L. The study of Earth's magnetism (1269–1950): A foundation by Peregrinus and subsequent development of geomagnetism and paleomagnetism. *Reviews of Geophysics*, 45:RG3008, 2007.

Dall'olmo, U. An additional list of auroras from European sources from 450 to 1466 A.D. *Journal of Geophysical Research*, 84:1525–1535, 1979.

Dalton, J. *Meteorological observations and Essays.* Richardson, Phillips and Pennington, London, 1793.

Davis, T. N. *The Aurora Watcher's Handbook.* University of Alaska Press, Fairbanks, 1992.

Davis, T. N. and Sugiura, M. Auroral Electrojet Activity Index AE and its Universal Time Variations. *Journal of Geophysical Research*, 71:785–801, 1966.

Dennis, B. R. Solar hard X-ray bursts. *Solar Physics*, 100:465–490, 1985.

Dodson, H. W. and Hedeman, E. R. Major Hα flares in centers of activity with very small or no spots. *Solar Physics*, 13:401–419, 1970.

Eather, R. H. *Majestic Lights.* American Geophysical Union, Washington, 1980.

Eddy, J. A. The Maunder Minimum. *Science*, 192:1189–1202, 1976.

Egeland, A. and Brekke, A. The northern light: From mystery to modern space science. *Endeavour New Series*, 8:188–193, 1984.

Farquharson, J. Robert Burns and the aurora. *Journal of the British Astronomical Association*, 116:65, 2006.

Feldstein, Y. I. A quarter of a century with the auroral oval. *EOS Transactions AGU*, 67:761, 765–767, 1986.

Frankel, H. Jan Hosper and the rise of palomagnetism. *EOS Transactions AGU*, 68:577, 579–581, 1987.

Freeman, J. W. *Storm in Space.* Cambridge University Press, 2001.

Frisinger, H. H. Aristotle and his Meteorologica. *Bulletin of the American Meteorological Society*, 53:634–638, 1972.

Frisinger, H. H. Aristole's Legacy in Meteorology. *Bulletin of the American Meteorological Society*, 54:198–204, 1973.

Fritz, H. *Verzeichniss Beobachteter Polarlichter.* C. Gerold's Sohn, Wien, 1873.

Fritz, H. *Das Polar Licht.* Brockhaus, Leipzig, 1881.

Frobesius, J. N. *Nova et Antiqua Luminis Atque Aurorae Borealis Spectacula.* Helmstadt, 1739.

Furuhata, M. Aurora and airglow observations on February 11, 1958. *Report on Ionospheric and Space Research Japan*, 12:40–41, 1958.

Gauss, C. F. Allgemeine theorie des erdmagnetismus. In *Resultate aus den Beobachtungen des magnetischen Verein im Jahre 1838*, pp. 1–52. Göttingen Magn. Ver., Leipzig, 1839.

Gautier, A. Relation entre les taches du Soleil et les phénomènes magnétiques. *Arch. Sci.*, 21:194–195, 1852.

Gemma, C. *De Naturae Divinis Characterismis.* Antwerpiae, 1575.

Gonzalez, W. D., Tsurutani, B. T., and Clúa de Gonzalez, A. L. Interplanetary origin of geomagnetic storms. *Space Science Reviews*, 88:529–562, 1999.

Gonzalez, W. D., Tsurutani, B. T., Lepping, R. P., and Schwenn, R. Interplanetary phenomena associated with very intense geomagnetic storms. *Journal of Atmospheric and Solar-Terrestrial Physics*, 64:173–181, 2002.

Good, G. A. Geomagnetics and scientific institutions in 19th century America. *EOS Transactions*, 66:521, 524–526, 1985.

Good, G. A. The study of geomagnetism in the late 19th century. *EOS Transactions*, 69:218–232, 1988.

Goodman, J. M. *Space Weather and Telecomunications.* The Kluwer International Series in Engineering and Computer Science, 2005.

Graham, G. An account of observations made of the variation of the horizontal needle at London, in the latter part of the year 1722, and beginning of 1723. *Philosophical Transactions of the Royal Society*, 33:96–107, 1724.

Green, R. Sponsored research in geomagnetism 130 years ago. *EOS Transactions AGU*, 53:778–779, 1972.

Greenberg, M. *The Anchor Bible.* Doubleday, New York, *Ezekiel*, pp. 1–20, 1983.

Gubbins, D. and Herrero-Bervera, E. (eds.). *Encyclopedia of Geomagnetism and Paleomagnetism.* Springer-Verlag, Berlin, Heidelberg, New York, 2007.

Guillemin, A. *Les Météores Électriques et Optiques.* Librairie Hachette et Cie, Paris, 1887.

Halley, E. An account of the late surprizing appearance of the lights seen in the air, on the sixth of March last, with an attempt to explain the principal phaenomena thereof. *Philosophical Transactions of the Royal Society*, 29:406–428, 1717.

Halley, E. An account of the Phaenomena of a very extraordinary Aurora Borealis, seen at London on November 10. 1719. both morning and evening. *Philosophical Transactions of the Royal Society*, 30:1099–1100, 1720.

Hamilton, J. C. The Algonquin Manabozho and Hiawatha. *Journal of American Folklore*, 16:231, 1903.

Hardy, R. Theophrastus Observations of Sunspots. *Journal of the British Astronomical Association*, 101:261, 1991.

Harrison, G. Aurora diaries. *Astronomy and Geophysics*, 46:4.31–4.34, 2005.

Hedervari, P. Hungarian auroral observations and the Maunder Minimum. *The Observatory*, 101:21–22, 1981.

Hellman, G. *Die Meteorology in den Deutschen Flugschriften und Flugblattern des XVI Jahrhumderts.* Akademy der Wissenschaffen, Berlin, 1921.

Hess, V. F., Steinmaurer, R., and Demmelmair, A. Cosmic rays and the aurora of January 2526. *Nature*, 141:686–687, 1938.

Hikosaka, T. On the great enhancement of the line [OI] 6300 in the aurora at Niigata on February 11, 1958. *Report on Ionospheric and Space Research Japan*, 12:469–471, 1958.

Hiorter, O. P. Om-magnet-nålens Åtskillige ändringar. *Kongl. Svensk Vetenskaps Akad. Handlingar*, 8:27–43, 1747.

Hodgson, R. On a curious appearance seen in the Sun. *Monthly Notices of the Royal Astronomical Society*, 20:15–16, 1860.

Holzworth, R. H. Folklore and the Aurora. *EOS Transactions AGU*, 56:686–688, 1975.

Hoyt, D. V. and Schatten, K. H. How well was the Sun observed during the Maunder Minimum? *Solar Physics*, 165:181–192, 1996.

Hoyt, D. V. and Schatten, K. H. *The Role of the Sun in Climate Change.* Oxford University Press, New York, 1997.

Hoyt, D. V. and Schatten, K. H. Group Sunspot Numbers: A New Solar Activity Reconstruction. *Solar Physics*, 179:189–219, 1998a.

Hoyt, D. V. and Schatten, K. H. Group Sunspot Numbers: A New Solar Activity Reconstruction. *Solar Physics*, 181:491–512, 1998b.

Jackson, A., Jonkers, A. R. T., and Walker, M. R. Four centuries of geomagnetic secular variation from historical records. *Philosophical Transactions of the Royal Society of London A*, 358:957–990, 2000.

Jago, L. *The Northern Lights.* Hamish Hamilton, London, 2001.

Jamin, J. *Petit traité de Physique.* Gauthier-Villars, Paris, 1870.

Jones, H. S. *Greenwich Observatory Sunspot and Geomagnetic Storm Data.* Her Magesty's Stationery Office, London, 1955.

Jones, W. Notes on the Fox Indians. *Journal of American Folklore*, 24:209–237, 1911.

Jonkers, A. R. T., Jackson, A., and Murray, A. Four centuries of geomagnetic data from historical records. *Reviews of Geophysics*, 41:1006, 2003.

Kakioka Magnetic Observatory. Report of the auroras observed at memambetsu through 1958 and 1960. In *Report of the geomagnetic and geoelectric observations, No. 8*, pp. 109–130. Kakioka Magnetic Observatory, 1969.

Keimatsu, M. A chronology of aurorae and sunspot observed in China, Korea and Japan, Part VII, A brief summary of the records from BC 687 to AD 1600. *Ann. Sci. Kanazawa Univ.*, 13:1–32, 1976.

Krivova, N. A. and Solanki, S. K. The 1.3-year and 156-day periodicities in sunspot data: Wavelet analysis suggests a common origin. *Astronomy and Astrophysics*, 394:701–706, 2002.

Křivský, L. Long-term fluctuations of solar activity during the last thousand years. *Solar Physics*, 93:189–194, 1984.

Křivský, L. and Pejml, K. *Solar Activity, Aurorae and Climate in Central Europe in the last 1000 Years.* Publications of the Astronomical Institute of the Czechoslovak Academy of Sciences Publication No. 75, 1988.

Lamont, J. Über die zehnjährige Periode, welche sich in der Grösse der Täglichen Bewegung der Magnetnadel dargestellt, *Annalen der Physik*, 84 (Series 2): 572–582, 1851.

Lang, A. S. On the aurora borealis of Nov. 17, 1848. *Monthly Notices Royal Astronomical Society*, 9:148, 1849.

Lang, K. R. *Sun, Earth and Sky.* Springer, Berlin, 2006.

Lanzerotti, L. J. Value of historical space weather events. *Space Weather*, 5:6005, 2007.

Lee, E. H., Ahn, Y. S., Yang, H. J., and Chen, K. Y. The sunspot and auroral activity cycle derived from Korean historical records of the 11th 18th century. *Solar Physics*, 224:373–386, 2004.

Letfus, V. Sunspot and auroral activity during Maunder Minimum. *Solar Physics*, 197:203–213, 2000.

Link, F. Observations et catalogue des aurores boréales apparues en occident de -626 a 1600. *Geofysikalni Sbornik*, 173:297–387, 1962.

Link, F. Variations á longues périodes de l'activité solaire avant le 17ème siècle. *Bulletin of the Astronomical Institutes of Czechoslovakia*, 14:226–231, 1963.

Link, F. Observations et catalogue des aurores boréales apparues en occident de 1601 a 1700. *Geofysikalni Sbornik*, 212:501–550, 1964.

Link, F. Solar Activity during the 17th Century. *Astronomy and Astrophysics*, 54:857–861, 1977.

Link, F. Solar Cycles between 1540 and 1700. *Solar Physics*, 59:175–178, 1978.

Liritzis, I. and Vassiliou, H. Were Greek temples oriented towards aurorae? *Astronomy and Geophysics*, 47:14–1, 2006.

Livesey, R. J. Aurora notes January-March 1989. *The Marine Observer*, 60:40–45, 1990.

Livesey, R. J. Robert Burns and the aurora. *Journal of the British Astronomical Association*, 115:295, 2005.

Lovering, J. On the periodicity of the Aurora Borealis. *Memoirs of the American Academy of Arts and Science*, 10:9–351, 1866.

Lowe, E. J. *Natural Phenomea and Chronology of the Seasons.* Bell and Daldy, London, 1870.

Loysha, V. A., Krakovetsky, Yu. K., and Popov, L. N. L. N. *Aurorae: Catalogue from IV to XVIII centuries.* Acad. Sci. USSR Sov. Geophys. Comm., 1989.

Mairan, J. J. *Traité physique et historique de l'Aurora Boréale.* Academie Royale des Sciences, Paris, 1733.

Mairan, J. J. *Traité physique et historique de l'Aurora Boréale.* Second edition. Academie Royale des Sciences, Paris, 1754.

Malin, S. R. C. Historical introduction to geomagnetism. In J. A. Jacobs (ed.), *Geomagnetism*, volume 1, Academic Press, San Diego, CA, pp. 1–49, 1987.

Malin, S. R. C. and Barraclough, D. R. Humboldt and the Earth's Magnetic Field. *Quarterly Journal of Royal Astronomical Society*, 32:279–293, 1991.

Malin, S. R. C. and Bullard, E. The direction of the Earths magnetic field at London, 15701975. *Philosophical Transactions of Royal Society of London*, 299:357–422, 1981.

Mandea, M. and Mayaud, P.-N. Guillaume le Nautonier, un précurseur dans l'histoire du géomagnétisme. *Revue d'histoire des sciences*, 57:161–174, 2004.

Matsushita, S. Ancient Aurorae Seen in Japan. *Journal of Geophysical Research*, 61:297–302, 1956.

Mayaud, P. N. *A Hundred Year Series of Geomagnetic Data, 1868–1967, Indices aa, Storm Sudden Commencements.* IAGA Bull., 33, Int. Union of Geod. and Geophys., Paris, 1973.

Maycock, C. 'Robert Burns and the aurora'. *Journal of the British Astronomical Association*, 116:272, 2006.

Maycock, C. H. *A Passionate Poet: Susanna Blamire 1747–94*. Hypatia Publications, 2003.

McCracken, K. G., Dreschhoff, G. A. M., Smart, D. F., and Shea, M. A. Solar cosmic ray events for the period 1561–1994: 2. The Gleissberg periodicity. *Journal of Geophysical Research*, 106:21599–21610, 2001b.

McCracken, K. G., Dreschhoff, G. A. M., Zeller, E. J., Smart, D. F., and Shea, M. A. Solar cosmic ray events for the period 1561–1994: 1. Identification in polar ice, 1561–1950. *Journal of Geophysical Research*, 106:21585–21598, 2001a.

McCracken, K. G., McDonald, F. B., Beer, J., Raisbeck, G., and Yiou, F. A phenomenological study of the long-term cosmic ray modulation, 850-1958 AD. *Journal of Geophysical Research (Space Physics)*, 109(18):A12103, 2004.

McIntosh, D. H. On the annual variation of magnetic disturbance. *Philosophical Transactions of Royal Society London, Ser. A.*, 251:525–552, 1959.

McIntosh, P. S. August solar activity and its geophysical effects. *Sky and Telescope*, 44:214–217, 1972.

McKinnon, J. A. and Members of the Space Environment Services Center. *August 1972 Solar Activity and Related Geophysical Effects.* NOAA Technical Memorandum ERL SEL-22, Space Environment Laboratory, National Oceanic and Atmospheric Administration, Boulder, Colorado, 1972.

Meadows, A. J. and Kennedy, J. E. The origin of solar-terrestrial studies. *Vistas in Astronomy*, 25:419–426, 1982.

Mendillo, M. and Keady, J. Watching the Aurora from Colonial America. *EOS Transactions AGU*, 57:485–491, 1976.

Merril, R. T., McElhinny, M. W., and McFadden, P. L. *The Magnetic Field of the Earth.* New York, Elsevier, 1996.

Miyaoka, H., Hirasawa, T., Yumoto, K., and Tanaka, Y. Low latitude aurorae on October 21, 1989. *Proceedings of the Japan Academy*, 66:47–51, 1990.

Nakazawa, Y., Okada, T., and Shiokawa, K. Understanding the "SEKKI" phenomena in Japanese historical literatures based on the modern science of low-latitude aurora. *Earth, Planets, and Space*, 56:e41–e44, 2004.

Nevanlinna, H. Founding Father of Finnish Geomagnetic Research. *EOS Transactions AGU*, 86:325–327, 2005.

Nevanlinna, H. and Kataja, E. An extension of the geomagnetic activity index series aa for two solar cycles (1844–1868). *Geophysical Research Letters*, 20:2703–2706, 1993.

Nevanlinna, H. and Ketola, A. *Magnetic Results from Helsinki Magnetic-Meteorological Observatory, Part III, Declination 1854-1880, Geomagnetic Activity 1844–1880.* Geophysical Publications, 33, Finnish Meteorological Institute, 1993.

Nygrén, T. and Silén, J. A. E. Nordenskiöld and the auroral oval. *EOS Transactions AGU*, 63:553,555, 1982.

Ohl, A. I. and Ohl, G. I. A New Method of Very Long-Term Prediction of Solar Activity. volume 2, pp. 258 263, 1979.

Osaki, S. *Kinsei Nihon temmon shiryou (Pre-Modern Japanese Historical Records of Celestial Phenomena) (in Japanese).* Tokyo, Hara Shobou, 1994.

Ovalle, A. *Historica relacion del reino de Chile.* Rome, Francisco Caballo, 1646.

Peirce, B. O. Biographical Memoir of Joseph Lovering 1813–1892. *Biographical Memoirs, National Academy of Sciences*, IV:329–344, 1909.

Pilgram, A. *Wetterkunde.* vols. 1 and 2, Wien, 1788.

Pomerantz, M. A. The Ancestry of Solar-Terrestrial Research. *Eos Transactions AGU* 55:955–957, 1974.

Pope, B. and Jorden, K. *Sentinels of the Sun*. Johnson Books, 2006.

Radelet de Grave, P. and Speiser, D. Le De Magnete de Pierre de Maricourt, traduction et commentaire. *Revue d'histoire des sciences*, 28:193–234, 1975.

Raspopov, O. M., Dergachev, V. A., and Goos'kova, E. G. Ezekiel's Vision: Visual Evidence of Sterno-Etrussia Geomagnetic Excursion? *EOS Transactions AGU*, 84:77–83, 2003.

Rassoul, H. K., Rohrbaugh, R. P., Tinsley, B. A., and Slater, D. W. Spectrometric and photometric observations of low-latitude aurorae. *Journal of Geophysical Research*, 98:7695–7709, 1993.

Rees, M. H. and Roble, R. G. Observations and theory of the formation of stable auroral red arcs. *Reviews of Geophysics and Space Physics*, 13:201–242, 1975.

Rico y Sinobas, M. Noticia sobre las auroras boreales observadas en España durante el siglo XVIII y parte del XIX. *Memorias de la Real Academia de Ciencias Exactas, Físicas y Naturales*, III(3):77–91, 1855.

Rieger, E., Kanbach, G., Reppin, C., Share, G. H., Forrest, D. J., and Chupp, E. L. A 154-day periodicity in the occurrence of hard solar flares? *Nature*, 312:623–625, 1984.

Rishbeth, H. The centenary of solar-terrestrial physics. *Journal of Atmospheric and Solar-Terrestrial Physics*, 63:1883–1890, 2001.

Rosell, M. *Aurora Boreal observada en Valencia en la noche del dia cinco de marzo de este presente año 1764*. Benito Monfort, Valencia, 1764.

Rubenson, R. Catalogue des Aurores Boréales observees en Suede. *Kl. Sven. Veteskapsakad. Handl.*, 15:3–184, 1879.

Rubenson, R. Catalogue des Aurores Boréales observees en Suede. *Kl. Sven. Veteskapsakad. Handl.*, 18:1–300, 1882.

Russell, C. T. and McPherron, R. L. Semiannual variation of geomagnetic activity. *Journal of Geophysical Research*, 78:92–108, 1973.

Rust, D. M. The Great Solar Flares of August, 1972. *Sky and Telescope*, 44:226–230, 1972.

Sabine, E. On periodical Laws discoverable in the mean effects of the larger Magnetic Disturbances.-No. II. *Philosophical Transactions of Royal Society of London*, 142:103–124, 1852.

Sabine, E. On periodical laws discoverable in the mean effects of the larger magnetic disturbances No. III. *Philosophical Transactions of Royal Society of London*, 146:357, 1856.

Sachs, A. J. and Hunger, H. *Astromonical diaries and related texts from Babylonia, Vol. II: 261 B.C. to 165 B. C.* Österreichische Akademie der Wissenschaften, Vienna, 1989.

Sarda, F. *Les Arago: Franois et les autres*. Taillandier, Paris, 2002.

Savage, C. *Aurora: The Mysterious Northern Light*. Firefly Books, 2001.

Schove, D. J. The Sunspot Cycle, 649 B.C. to A.D. 2000. *Journal of Geophysical Research*, 60:127–146, 1955.

Schove, D. J. Auroral Numbers since 500 B.C. *British Astronomical Association Journal*, 72:30–35, 1962.

Schove, D. J. (ed.). *Sunspot Cycles*. Hutchinson Ross Publishing Company, Stroudsburg, 1980.

Schröder, W. Some aspects of the History of auroral research. *EOS Transactions AGU*, 60:1035–1036, 1979.
Schröder, W. *Das Phänomen des Polarlichts.* Wissenschaftliche Buchgesellschaft, Darmstadt, 1984.
Schröder, W. Aurorae during the so-called Spoerer minimum. *Solar Physics*, 151:199–201, 1994a.
Schröder, W. Behavior of auroras during the Spoerer minimum (1450–1550). *Annales Geophysicae*, 12:808–809, 1994b.
Schröder, W. Some aspects of the earlier history of solar-terrestrial physics. *Planetary and Space Sciences*, 45:395–400, 1997.
Schröder, W. Wilhelm Foerster and the development of solar and cosmical physics. *Planetary and Space Sciences*, 47:587–590, 1999.
Schröder, W. Geomagnetic research in the 19th century: a case study of the German contribution. *Journal of Atmospherics and Solar-Terrestrial Physics*, 63: 1649–1660, 2001.
Schröder, W. Missing auroras and sunspots at the end of the 18th century. *Astronomy and Geophysics*, 44:3.06, 2003.
Schröder, W. *The Aurora Borealis. Hermann Fritz and the Foundation of Auroral Research.* Science Edition, 2004.
Schröder, W. The first accurate description of an aurora. *EOS Transactions*, 87:506, 2006.
Schröder, W. and Wiederkehr, K.-H. A history of the early recording of geomagnetic variations. *Journal of Atmospheric and Solar-Terrestrial Physics*, 62:323–334, 2000.
Schröder, W. (ed.). *Catalogue of Aurorae Borealis (502 to 1735).* Science Edition, Bremen, 1996.
Schroeter, J. F. *Catalog der in Norwegen eis Juni, 1878, Beobachteten Nordlichter.* Kristiana, Norway, 1902.
Schwabe, H. Sonnenbeobachtungen im Jahre 1843. *Astronomische Nachrichten*, 21:233–236, 1844.
Senecae, L. A. *Naturales Quaestiones.* Latin text revised and Spanish traslation by Carmen Codoñer Merino, 2 vols., Consejo Superior de Investigaciones Científicas, Madrid, 1979.
Shiokawa, K., Ogawa, T., and Kamide, Y. Low-latitude auroras observed in Japan: 1999-2004. *Journal of Geophysical Research (Space Physics)*, 110(9):A05202, 2005.
Short, T. *A General Chronological History of the Air, Weather, Seasons, Meteors, etc.* London, 1749.
Short, T. *A comparative History of the Increase and Decrease of Mankind, etc.* London, 1767.
Silverman, S. On the "chasms" of Aristotle and Pliny. *Journal of Atmospheric and Solar-Terrestrial Physics*, 24:1108–1109, 1962.
Silverman, S. Ancient writings reveal presence of aurora in 13th-century Canadian Arctic. *EOS Transactions AGU*, 83:367, 2002.
Silverman, S. Comment on "The First Accurate Description of an Aurora". *EOS Transactions AGU*, 88:506–506, 2007.
Silverman, S. M. Joseph Henry and John Henry Lefroy; a common 19th century vision of auroral research. *EOS Transactions AGU*, 70:227, 1989.

Silverman, S. M. The 155-day solar period in the sixteenth century and later. *Nature*, 347:365–367, 1990.

Silverman, S. M. Secular variation of the aurora for the past 500 years. *Reviews of Geophysics*, 30:333–351, 1992.

Silverman, S. M. Low latitude auroras: the storm of 25 September 1909. *Journal of Atmospheric and Solar-Terrestrial Physics*, 57:673–685, 1995.

Silverman, S. M. Early auroral observations. *Journal of Atmospheric and Terrestrial Physics*, 60:997–1006, 1998.

Silverman, S. M. Sporadic auroras. *Journal of Geophysical Research (Space Physics)*, 108:8011, 2003.

Silverman, S. M. The earliest recorded aurora in North America since European colonization. *Journal of Atmospheric and Solar-Terrestrial Physics*, 67:749–752, 2005.

Silverman, S. M. Low latitude auroras prior to 1200 C.E. and Ezekiel's vision. *Advances in Space Research*, 38:200–208, 2006.

Silverman, S. M. and Blanchard, D. C. Wilson Bentley's auroral observations. *Planetary and Space Sciences*, 31:1131–1135, 1983.

Silverman, S. M. and Cliver, E. W. Low-latitude auroras: The magnetic storm of 14-15 May 1921. *Journal of Atmospheric and Solar-Terrestrial Physics*, 63: 523–535, 2001.

Siscoe, G. L. A historical footnote on the origin of aurora borealis. *EOS Transactions AGU*, 26:994, 1978.

Siscoe, G. L. Evidence in the auroral record for secular solar variability. *Reviews of Geophysics and Space Physics*, 18:647–658, 1980.

Siscoe, G. L. and Siebert, K. D. Solar-terrestrial effects possibly stronger in biblical times. *Journal of Atmospheric and Solar-Terrestrial Physics*, 64:1905–1909, 2002.

Siscoe, G. L., Silverman, S. M., and Siebert, K. D. Ezekiel and the northern lights: Biblical aurora seems plausible. *EOS Transactions AGU*, 83:173–174, 2002.

Smith, J. A. Petrus Peregrinus epistola – The beginning of experimental studies of magnetism in Europe. *Earth Science Reviews*, 6:A11–A18, 1970.

Smith, J. A. Precursors to Peregrinus: The early history of magnetism and the mariner's compass in Europe. *EOS Transactions AGU*, 18:21–74, 1992.

Soffel, H. Johann von Lamont: A Pioneer in Geomagnetism. *EOS Transactions AGU*, 87:247, 2006.

Solow, A. R. On celestial events, auroral activity, and the solar cycle in classical antiquity. *Earth and Planetary Science Letters*, 232:67–70, 2005.

Soon, W., and Yaskell, S. H. *The Maunder Minimum and the Variable Sun-Earth Connection*. World Scientific, Singapore, 2003.

Stephenson, F. R., Willis, D. M., and Hallinan, T. J. The earliest datable observation of the aurora borealis. *Astronomy and Geophysics*, 45:15–17, 2004.

Stern, D. P. A Millennium of Geomagnetism. *Reviews of Geophysics*, 40:1007, 2002.

Stothers, R. Ancient aurorae. *Isis. Journal of the History of Science Society*, 70: 85–95, 1979a.

Stothers, R. Solar activity cycle during classical antiquity. *Astronomy and Astrophysics*, 77:121–127, 1979b.

Stuiver, M. Solar variability and climatic change during the current millennium. *Nature*, 286:868–871, 1980.

Gregory of Tours. *Gregorii Turnonousis Episcopi Opera Omnia.* edited by J. P. Migne. Patrologie Cursus Completus, Vol. 71, Garnier, Paris, 1849.

Trigo, R. M. and Vaquero, J. M. An unsung hero. *Astronomy and Geophysics*, 49:2.14–2.16, 2008.

Tsurutani, B., McPherron, R., Gonzalez, W., Lu, G., Sobral, J. H. A., and Gopalswamy, N. (eds.). *Recurrent Magnetic Storms. Corotating Solar Wind Streams.* Geophysical Monograph 167, American Geophysical Union, 2006.

Tsurutani, B. T., Gonzalez, W. D., Lakhina, G. S., and Alex, S. The extreme magnetic storm of 1–2 September 1859. *Journal of Geophysical Research (Space Physics)*, 108:1268, 2003.

Usoskin, I. G., Mursula, K., and Kovaltsov, G. A. Lost sunspot cycle in the beginning of Dalton minimum: New evidence and consequences. *Geophysical Research Letters*, 29:2183, 2002.

Usoskin, I. G., Mursula, K., and Kovaltsov, G. A. The lost sunspot cycle: Reanalysis of sunspot statistics. *Astronomy and Astrophysics*, 403:743–748, 2003.

Usoskin, I. G., Mursula, K., Solanki, S. K., Schüssler, M., and Kovaltsov, G. A. A physical reconstruction of cosmic ray intensity since 1610. *Journal of Geophysical Research (Space Physics)*, 107:1374, 2002.

Vaquero, J. M. Sunspot observations by Theophrastus revisited. *Journal of British Astronomical Association*, 117:346, 2007.

Vaquero, J. M. and Gallego, M. C. Two early observations of aurora at low latitudes. *Annales Geophysicae*, 19:809–811, 2001.

Vaquero, J. M. and Gallego, M. C. Evidence for a sunspot in A.D. 939 in an Arabian Source. *Solar Physics*, 206:209–211, 2002.

Vaquero, J. M., Gallego, M. C., and Garcia, J. A. Auroras observed in the Iberian Peninsula (17001855) from Rico Sinobas catalogue. *Journal of Atmospheric and Solar-Terrestrial Physics*, 65:677–682, 2003.

Vaquero, J. M. and Trigo, R. M. Auroras observed in Portugal in late 18th century obtained from printed and manuscript meteorological observations. *Solar Physics*, 231:157–166, 2005.

Vaquero, J. M., Trigo, R. M., and Gallego, M. C. Sporadic Aurora from Spain Earth. *Earth, Planets and Space*, 59:e49–e51, 2007.

Vaquero, J. M., Valente, M. A., Trigo, R. M., Ribeiro, P., and Gallego, M. C. The 1870 space weather event: Geomagnetic and auroral records. *Journal of Geophysical Research*, 113:10.1029/2007JA012943, 2008.

Vázquez, M., Vaquero, J. M., and Curto, J. J. On the connection Between solar activity and low-latitude aurorae in the period 1715-1860. *Solar Physics*, 238:405–420, 2006.

Veselovsky, I. S., Panasyuk, M. I., Avdyushin, S. I., Bazilevskaya, G. A., Belov, A. V., Bogachev, S. A., Bogod, V. M., Bogomolov, A. V., Bothmer, V., Boyarchuk, K. A., Vashenyuk, E. V., Vlasov, V. I., Gnezdilov, A. A., Gorgutsa, R. V., Grechnev, V. V., Denisov, Y. I., Dmitriev, A. V., Dryer, M., Yermolaev, Y. I., Eroshenko, E. A., Zherebtsov, G. A., Zhitnik, I. A., Zhukov, A. N., Zastenker, G. N., Zelenyi, L. M., Zeldovich, M. A., Ivanov-Kholodnyi, G. S., Ignat'ev, A. P., Ishkov, V. N., Kolomiytsev, O. P., Krasheninnikov, I. A., Kudela, K., Kuzhevsky, B. M., Kuzin, S. V., Kuznetsov, V. D., Kuznetsov, S. N., Kurt, V. G., Lazutin, L. L., Leshchenko, L. N., Litvak, M. L., Logachev, Y. I., Lawrence, G., Markeev, A. K., Makhmutov, V. S., Mitrofanov, A. V., Mitrofanov, I. G., Morozov, O. V., Myagkova, I. N., Nusinov, A. A., Oparin, S. N., Panasenco, O. A., Pertsov,

A. A., Petrukovich, A. A., Podorol'sky, A. N., Romashets, E. P., Svertilov, S. I., Svidsky, P. M., Svirzhevskaya, A. K., Svirzhevsky, N. S., Slemzin, V. A., Smith, Z., Sobel'man, I. I., Sobolev, D. E., Stozhkov, Y. I., Suvorova, A. V., Sukhodrev, N. K., Tindo, I. P., Tokhchukova, S. K., Fomichev, V. V., Chashey, I. V., Chertok, I. M., Shishov, V. I., Yushkov, B. Y., Yakovchouk, O. S., and Yanke, V. G. Solar and heliospheric phenomena in october-november 2003: Causes and effects. *Cosmic Research*, 42:435–488, 2004.

Viera y Clavijo, J. Carta filosofica sobre la aurora boreal, observada en la ciudad de La Laguna de Tenerife la noche del 18 de Enero de 1770. A copy of the original manuscript (transcripted by A. de Ara in 1860) is archived at Museo Canario, Las Palmas de Gran Canaria, 1770.

Weyer, E. M. *The Eskimos, Their Environment and Folkways.* Archon, Hamden, Conn., 1969.

Willis, D. M., Armstrong, G. M., Ault, C. E., and Stephenson, F. R. Identification of possible intense historical geomagnetic storms using combined sunspot and auroral observations from East Asia. *Annales Geophysicae*, 23:945–971, 2005.

Willis, D. M., Stephenson, F. R., and Fang, H. Sporadic aurorae observed in East Asia. *Annales Geophysicae*, 25:417–436, 2007.

Wolf, C. *Gedanken über das ungewöhnliche Phaenomenon.* University of Halle, Halle, 1716.

Wolf, R. Liaison entre les taches du Soleil et les variations en déclinaison de l'aiguille aimantée. *Comptes Rendus H. Acad. Sci.*, 35:364, 1852.

Yau, K. K. C., Stephenson, F. R., and Willis, D. M. *A Catalogue of Auroral Observations from China, Korea and Japan (193 BC–AD 1770).* Rutherford Appleton Laboratory Technical Report RALTR95073, 1995.

Ytter, H. F. *Aurora : The Northern Lights in Mythology, History and Science.* Bell Pond Books, 1999.

Zhang, Z. W. Korean Auroral Records of the Period 1507–1747AD and the SAR Arcs. *Journal of the British Astronomical Association*, 95:205–210, 1985.

7

Reconstruction of Solar Activity During the Telescopic Era

The secular evolution of solar activity and the Sun–Earth relationships have been important research topics of the scientific community throughout the last century. This whole type of investigation needed a variable that was able to provide information on the state of the Sun. For historical reasons, the variables used have depended fundamentally on the sunspots (number of groups of sunspots, Wolf number, sunspot area, etc.). Recently, because the Sun is now monitored practically using almost the whole electromagnetic spectrum and particle flux, the number of indices that give us information on the state of the Sun has increased considerably, although the time series are very short.

The problem of reconstructing the solar activity to obtain long time series of indices related to the behaviour of the Sun has been approached from two points of view: using documentary data and using "proxy" data. The documentary sources provide us with observations of the Sun (or related to it) carried out in remote times, and the "proxy" data, such as the thickness of tree rings or the concentration of certain radioactive isotopes, provide us with time series that are related to the behaviour of the Sun. Both approaches have their advantages and disadvantages. The "proxy" data provide very long and homogeneous series (Usoskin et al., 2003a; Solanki et al., 2004; Usoskin, 2008), but the documentary data have better time resolution during some epochs, and certainly show a direct association with the behaviour of the Sun.

There exists a wide range of possibilities to reconstruct solar activity on the basis of documentary sources (Figure 7.1). On the one hand, one has direct data, i.e. data obtained from observations of the Sun. And on the other, one has indirect data based on terrestrial phenomena linked to the behaviour of the Sun. Nonetheless, the indices that have been most used are those related to the observations of sunspots due to their versatility. In this chapter, we will review the commonest activity indices that have served to carry out reconstructions of long time series related to sunspot observations. This kind of series presents three major drawbacks: (i) the personal bias of the observers, (ii) sensitivity to lost observations, and (iii) they do not take into account the contribution of faculae or the open magnetic field. We will begin necessarily by analysing

J.M. Vaquero, M. Vázquez, *The Sun Recorded Through History*,
Astrophysics and Space Science Library 361, DOI 10.1007/978-0-387-92790-9_7,

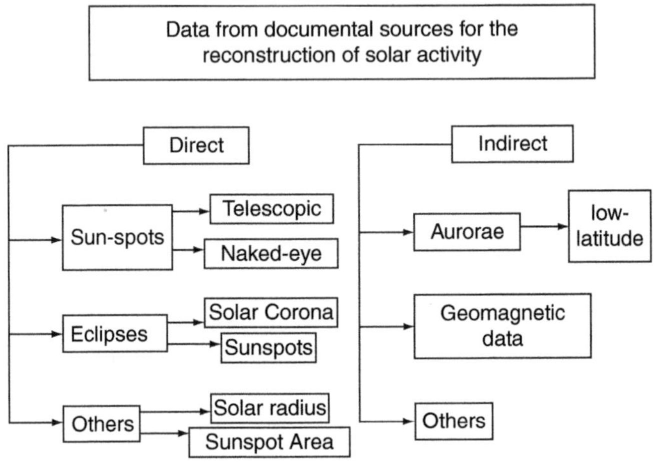

Fig. 7.1. Different types of data from documentary sources for the reconstruction of solar activity.

the Wolf number, and continue with the reconstruction of the solar activity carried out by Hoyt and Schatten (1998a,b) (hereafter HS98) using the number of groups of sunspots.

7.1 Wolf's Reconstruction

7.1.1 Schwabe's Discovery

Sunspots were one of the most observed objects beginning with the first astronomical use of the telescope in 1610. However, the observations of sunspots during the 17th and 18th centuries, although abundant, were dispersed. Thus, nobody found it possible to examine them overall, even ignoring the 11-year cycle so characteristic of the solar activity.

In 1838, Samuel Heinrich Schwabe published his solar observations made with a small telescope over 12 years. In his article, Schwabe (1838) published for every year the number of days that he had observed, the average number of groups of sunspots, and the number of days on which the Sun had no spots. Some years later, Schwabe (1844) recognized an about 10-year apparent rhythm in his observations. Surprisingly, Schwabe's discovery remained ignored by the scientists of the time.

The dissemination of Schwabe's result was really carried out by Alexander von Humboldt who published in the third volume of his monumental treatise *Kosmos* the observations by Schwabe, presenting the hypothesis of the roughly 10-year period, and surmising that this cycle could be present in old sunspot data.

This idea of Humboldt was further developed by Johann Rudolf Wolf who immediately tried to check Schwabe's conjecture looking for old observations of sunspots in several hundred volumes in the libraries in Basel, Berne, and Zürich. Surprisingly, among the documentation that he found, there was the manuscript of Danish scientist Christian Horrebow (1718–1776) that suggested a periodic variation in the sunspots. In 1852, Wolf wrote a long and elaborate paper in which he tried to give a basis from a historical point of view for the periodicity of the sunspots (Wolf, 1852). If the solar activity is truly periodic, then the difference between two dates of maxima of solar activity should be an integer multiple of the duration of the period. And the same would be the case with the dates of the minima. Using the data obtained from historical documentary sources, Wolf determined six peaks of solar activity that would be associated with the maxima, and six dates of very low solar activity that would be associated with the minima (Table 7.1). In this way, he calculated 16 estimates of the period of sunspots (Table 7.2). These estimates have the general form

$$\xi_i \pm \delta_i = c_i \left(p_i \pm u_i\right)$$

where c_i is an estimate of the (integer) number of complete solar cycles in ξ_i years, p_i is the duration (or length) of the cycle and u_i is a "probable error" for p_i. Wolf's estimate of the sunspot period p was a weighted average of the 16 values p_i using the weights $w_i = 1/u_i^2$:

$$p = \left(\sum_{i=1}^{16} w_i p_i\right) \Big/ \left(\sum_{i=1}^{16} w_i\right) = 11.11 \text{ years}$$

Table 7.1. Dates of solar cycle maxima and minima detected by Wolf (1852)

Maxima	Minima
(M1) 1626.0 ± 1.0	(m1) 1645.0 ± 1.0
(M2) 1717.5 ± 1.0	(m2) 1755.5 ± 0.5
(M3) 1816.3 ± 1.0	(m3) 1810.5 ± 1.0
(M4) 1829.5 ± 1.0	(m4) 1823.2 ± 0.5
(M5) 1837.5 ± 0.5	(m5) 1833.6 ± 0.5
(M6) 1848.6 ± 0.5	(m6) 1844.0 ± 0.5

Table 7.2. Wolf's estimates of the solar cycle period (Wolf, 1852)

$M3 - M2 = 9(10.98 \pm 0.16)$	$m3 - m2 = 5(11.00 \pm 0.22)$
$M4 - M2 = 10(11.20 \pm 0.14)$	$m4 - m2 = 6(11.28 \pm 0.12)$
$M5 - M2 = 11(10.91 \pm 0.10)$	$m5 - m2 = 7(11.28 \pm 0.10)$
$M6 - M2 = 12(10.93 \pm 0.09)$	$m6 - m2 = 8(11.06 \pm 0.09)$
$M3 - M1 = 17(11.19 \pm 0.08)$	$m3 - m1 = 15(11.03 \pm 0.09)$
$M4 - M1 = 18(11.31 \pm 0.08)$	$m4 - m1 = 16(11.14 \pm 0.07)$
$M5 - M1 = 19(11.13 \pm 0.06)$	$m5 - m1 = 17(11.09 \pm 0.07)$
$M6 - M1 = 20(11.13 \pm 0.06)$	$m6 - m1 = 18(11.06 \pm 0.06)$

Still today, this value of the period of the solar cycle continues to be commonly used. We should indicate that, although there is no theoretical reason to choose the values p_i close to 11 (one could have chosen 10 which was the initial proposal by Schwabe), Wolf (1852) showed in the second part of his paper an empirical justification for it, based on the historical records of solar activity that he had gathered. Surprisingly, Wolf gave a probable error of 0.038 year (13.87 days) but he never explained how he had arrived at such a result (Izenman, 1985).

7.1.2 The Wolf Sunspot Number

Wolf had begun his observations of sunspots at the end of the year 1847, and he dedicated the following one to familiarizing himself with the observation techniques. Starting in January 1849, Wolf meticulously recorded (1) the total number g of groups of visible sunspots, (2) the total number f of spots, (3) the observation conditions, classified into "clear", "the sunspots can be seen through the clouds" or "the sunspot could not be seen", and (4) the telescope he was using.

The next step Wolf took was to notice that the average number of sunspots was a very poor index with which to represent solar activity. Wolf defined the relative daily (Relativzahlen) sunspot number as $g + (1/10)f$. To obtain monthly values, he calculated the arithmetic mean of the daily values. We should add that Wolf, at the beginning, only used the observations made with his largest telescope (he usually used a refractor f/14, 8 cm diameter, 64×) on clear days. Wolf (1856) redefined his daily relative sunspot number by multiplying by 10 his previous definition, $10g + f$. Wolf intended his relative sunspot number to resemble the area occupied by the sunspots on the solar disk. Two years later, Wolf (1858) carefully explained his formula:

> "Since each group must contain at least one spot, I subtract the number of groups g from the number of spots f, so that the remainder is the number of extra spots $f - g$; this I again break up into groups by multiplying by a fraction q, to whose number I add the earlier number of groups for a number of different groups. I get, them, for the daily state of the spots, the following:
>
> $$\begin{aligned} t = pr &= p\,[g + q\,(f - g)] \\ &= p\,(1 - q)\,g + pqf \\ &= mg + nf, \end{aligned}$$
>
> where I multiply by some factor p, since only relative numbers concern me. The numbers m and n are to a certain extent the relative weights to be applied to the number of groups and the spots for this determination. I believe now that if a new place on the sun is attacked by a spot-building action, then this is more important than if in an

already-present group a new spot arises through a small change, and, therefore, I make m much larger than n. I will not be much mistaken if I make m = 10 and n = 1, because 10 and 1 lie in the neighbourhood of 9 and 1, 11 and 1, etc. if only for the sake of greater comfort. Therefore, I always calculate my daily relative number by the formula t = 10g + f. Naturally, the monthly average of these daily relative numbers gives a truer picture of the month the more numerous the daily numbers are." (translated by Izenman, 1983)

Also, to fill the gaps of the time series, Wolf began to use the data of Schwabe and began to consider the observations that he before discarded (those carried out with clouds that hinder the observation and those carried out with other telescopes). For the incorporation of these observations, Wolf (1861) redefined for the third time his relative sunspot number, incorporating a correction factor whose value depended on the observer, on the telescope used, and on the observation conditions. In this way, the relative sunspot number was transformed into

$$R = k\,(10g + f)\,,$$

where k is a constant for homogenization purposes.

Altas et al. (1998) made a study of the dependence of the constant k with respect to meteorological conditions during the observation. Wolf also defined the smoothed relative sunspot number that continues to be used now:

$$R_{\mathrm{smoothed}} = \left\{ R_{t-6} + 2\sum_{k=-5}^{5} R_{t-k} + R_{t+6} \right\} /24,$$

where R_t is the monthly mean sunspot number series.

These numbers have not changed since the death of Wolf in 1893. From this data, different people have continued the work of Wolf as his successors: A. Wolfer (1854–1931), W. Brunner (1878–1958), and M. Waldmeier (1912–2000). Therefore, one of the time series most widely used in various branches of statistics, physics, and astronomy was configured, and received the name of relative sunspot numbers or R_Z. A description of the work done in the Zürich Observatory in the 1960s can be found in Waldmeier (1968). The daily value of R_Z is calculated using only the data provided by an observer selected as the primary observer for a given period. The primary observers at different times were Staudacher (1749–1787), Flaugergues (1788–1825), Schwabe (1826–1847), Wolf (1848–1893), Wolfer (1893–1928), Brunner (1929–1944), Waldmeier (1945–1980), and A. Koeckelenbergh (from 1980). If the primary observer cannot make the observation on some day (if it is cloudy, for example), then a secondary, tertiary, etc., observer is used to complete the maximum number of days of the series. Waldmeier (1961) presents the hierarchy of the observers. The use of a single observer has as its objective to make the time series as homogeneous as possible. However, in this way all the other available observations are ignored. For dates previous to 1817, the number of lost days was so large that Wolf only tabulated monthly averages. Also, there are no observations during many months from the year 1749 to 1818 and there are missing observations for some months later than the year 1818. Wolf filled in these months by means of interpolations and using data of magnetic needle observations as a parameter related to solar activity. Therefore, it is important to stress that Wolf's numbers are a mixture between sunspot observations and calculated values. Therefore, the quality of the series of numbers of Wolf before the year 1749 is poor. Wolf's series can be used starting from the year 1849 or, with caution, from 1749.

When Waldmeier retired as director of the Zürich Observatory in 1980, the formal responsibility of continuing to upgrade the sunspot series was taken over by the Royal Observatory of Brussels in Belgium. The Solar Influences Data Analysis Centre (SIDC), former Sunspot Index Data Centre, is now responsible for producing the daily, monthly, and annual sunspot values, making the series available through its Web page. To this end, it maintains a network of more than 40 observers distributed all over the world. The series updated by SIDC receives the name of Relative International Sunspot Number[1] (R_I) (Figure 7.2) and it is considered to be the "official" Sunspot Number.

The sunspot number is, therefore, actually formed by a succession of sunspot indices of different intrinsic natures and accuracy (McKinnon, 1986). Clette et al. (2007) distinguished four main eras in the sunspot number series (Table 7.3). The historical era is entirely based on reconstruction from sparse data, and some details of the Wolf era have been presented above. The Zürich era began in 1882, when A. Wolfer introduced an important change in the counting method (Hossfield, 2002) including all small sunspots and

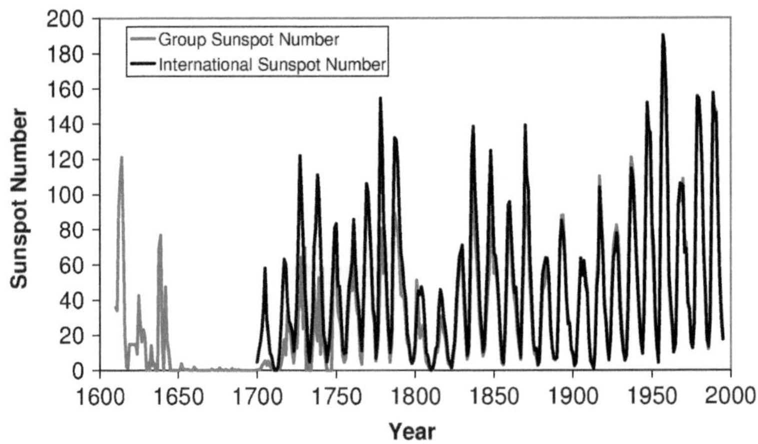

Fig. 7.2. Sunspot time series: Group Sunspot Number and International Sunspot Number.

[1] Also, Wolf or Zürich Sunspot Number (R_Z).

Table 7.3. Main characteristics of the sunspot number for the four periods indicated in the text (Clette et al., 2007)

Epoch	Historical (1610–1848)	Wolf (1848–1882)	Zürich (1882–1980)	SIDC (1981–now)
Observations	Sparse	Systematic	Systematic	Systematic
Obs. technique	Variable	Eyepiece	Drawing (proj.)	Drawing (proj.)
Stations (daily)	Base: 1	Base: 1	Base: 1 (triple)	20–50 (all)
		Aux: a few	Aux: ≈ 30	
Geog. distrib.	W-Europe	W-Europe	Europe + Asia	Worldwide
Processing	Manual	Manual	Manual	Computer
Reference	Wolf (Zürich)	Wolf (Zürich)	Zürich + Lorcano	Lorcano
Small spots	No	No	Yes	Yes
k coefficients	Estimates	Regular	Systematic (yearly)	Systematic (monthly)
Est. yearly error	10–45%	10%	< 5%	< 1%
Est. accuracy	20–50%	20%	5–10%	≈ 5%
Drift control	Loose, indirect	Loose, indirect	Direct (RGO, $F_{10.7}$)	Direct (many)

multiple umbrae in the counting, leading to a much more robust definition of the sunspot number. Moreover, Wolfer determined the scaling ratio between the new count and the Wolf sunspot series. In 1981, the SIDC introduced a global multi-station treatment using computerized statistical processing.

7.1.3 The Wolf Number in Depth

The Wolf Sunspot Number was quickly transformed into a useful index in solar physics and geophysics. However, its definition may seem somewhat arbitrary and with little physical meaning since it is simply a weighted mean of the number of individual sunspots and the number of groups of sunspots. Wolf chose a weighting factor equal to 10 for the groups and equal to 1 for the individual spots. The need to combine old and modern observations was the main reason for this choice. We should point out that the main problem in comparing solar observations is the computation of the individual sunspots. Thus, Wolf decided to give greater weight to the number of groups. As we will see, his choice was very fortunate.

In the definition provided by Wolf, g accounts for the sunspot clusters belonging to a single emerging magnetic flux rope. It is evident that g cannot offer information on the wide range of sizes and morphological classes that different sunspot groups have. With respect to f, this factor takes into account all the observed sunspots. However, it must be emphasized that multiple spots inside a large group are elements of the same magnetic structure, and that single isolated sunspots are markers of an entire magnetic concentration.

The definition of Wolf therefore allows the underlying magnetic flux strength and spatial density to be reflected. A good index should balance both contributions. Wolf probably chose the factor 10 in an arbitrary way. However, recent modern statistical studies indicate that in mean values $f = 10g$. Therefore, the definition made by Wolf gives the same value to the two contributions, providing a balance between them. The high linear correlation of the Wolf Number with other solar indices confirms the utility of the sunspot definition. Figure 7.3 shows the close relationship between sunspot number and the solar radio flux at 10.7 cm.

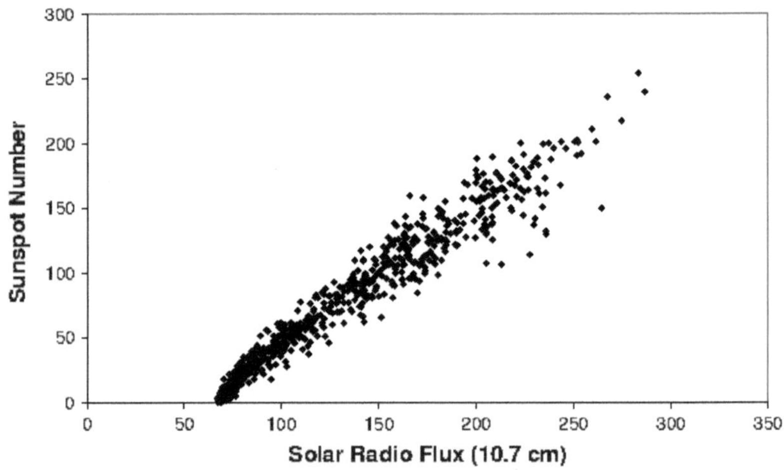

Fig. 7.3. Relationship between monthly Wolf Sunspot Number and solar radio flux (10.7 cm) during the period 1947–2006.

7.1.4 Other Sunspot Numbers

There are other indices in the scientific literature based on Wolf's ideas but calculated with different observer networks. One of the most important indices is the Boulder Sunspot Number. This index is calculated and published by the Space Environment Center, National Oceanic and Atmospheric Administration (NOAA). Another index is the American relative number of sunspots or R_A. It was born during the Second World War, under the auspices of the Department of Terrestrial Magnetism of the Carnegie Institute and the American Association of Variable Star Observers (AAVSO). R_A has an interesting history due to the constant interest in guaranteeing that the series stays consistent and precise (Shapley, 1949; Schaefer, 1997a,b).

Although it is not strictly an index based on sunspot numbers, we must also cite the reconstruction made by Schove (1983) using different documentary

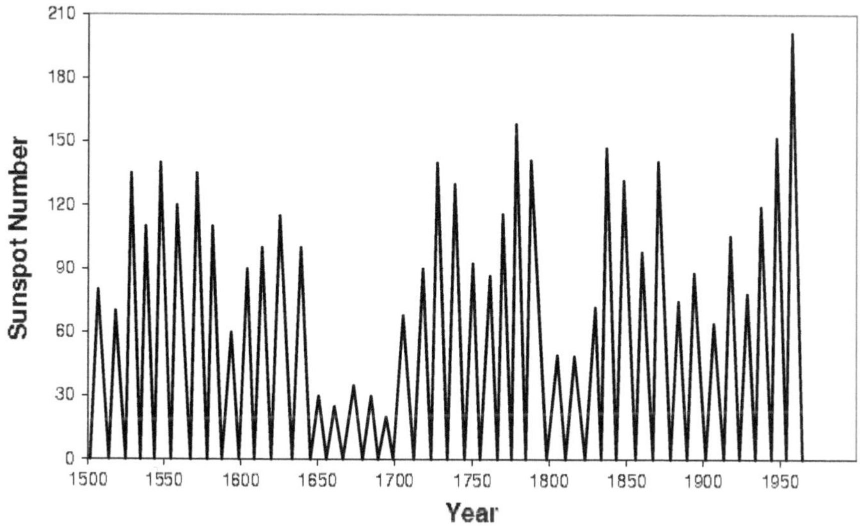

Fig. 7.4. Reconstructed solar activity according to the amplitudes and dates of solar cycles during the last 500 years. Data: Schove (1983).

proxies of solar activity. Schove estimated the amplitudes and dates of maxima and minima for solar cycles from numbers −22 to 21. These estimates of the sunspot numbers during the last 500 years are shown in Figure 7.4 where the sunspot number for the minima of solar cycles have been approximated as zero, and for the maxima as the amplitudes provided by Schove (1983).

7.1.5 Other Solar Indices

During the 20th century, some new global solar activity indices have been developed. Of course, these indices have a much shorter temporal extent that

Table 7.4. Main characteristics of the global long-term solar indices (modified from Clette et al., 2007). Duration is expressed in solar cycles

Index	Duration	Since	Observation
Sunspot area A_s	12	1874	Problems with definition of boundaries
CaII-K index	8	1915	Several uncalibrated series
$F_{10.7cm}$	6	1940	Undersampling
Total Solar Irradiance	3	1976	Mixed contribution from spots and faculae
MgII, HeII index	3	1976	Space-based
Total/polar magnetic flux	3	1970	Inaccurate near-limb measurement

the indices based on historical sunspot observations. Some interesting characteristics of a selection of these indices are summarized in Table 7.4.

It is necessary to point out that there are some indices closely related to the chromosphere ($F_{10.7cm}$, CaII, MgII, HeII) or containing a chromospheric component, such as the total solar irradiance (TSI). However, the sunspot number or total sunspot area are purely photospheric indices. In this context, we can explain the nonlinear (or even non-univocal) relationship between chromospheric and photospheric indices. In fact, weak open plage and network magnetic fields play a significant (and sometimes even dominant) role compared to strong closed magnetic fields (sunspots). Figure 7.5 shows an example of a nonlinear relationship between modern satellite measurements of solar irradiance (Fröhlich, 2000) and contemporaneous sunspot area values.

The data of the solar radio flux at 10.7 cm $F_{10.7cm}$ (Baranyi et al., 2001; Balmaceda et al., 2005), space-based MgII core-to-wing ratio (Viereck and Puga, 1999; Snow et al., 2005), and HeII flux (Floyd et al., 2002) are produced by a single station or instrument. Therefore, it is difficult to guarantee continuity over very long time periods, and they prevent the possibility of independent control of calibration and shift.

From the point of view of historical documents, the case of total sunspot area is particularly interesting. The total area of all sunspots visible on the solar hemisphere was measured systematically from 1874 until 1976 by the Royal Greenwich Observatory (RGO). It is the longest and most complete record of sunspot areas measured by a small network of observatories. However, several astronomical observatories measured this parameter in different periods

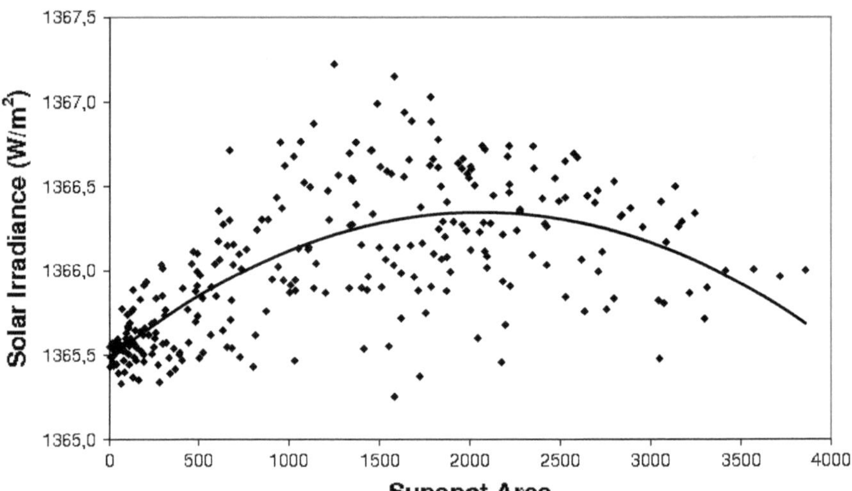

Fig. 7.5. Relationship between the monthly average sunspot area and the corresponding measured total solar irradiance. The fit of a polynomial of degree 2 to the data is included as a solid line.

(pre- and post-RGO era). Balmaceda et al. (2005) compiled a complete and cross-calibrated time series comparing the data from different observatories and finding the corresponding correction factors. The RGO data set was used until 1976. The data of observatories from the former USSR was used between 1977 and 1985 and the Mt. Wilson data since 1986. Other data sets provided by the Rome, Yunnan, and Catania observatories were used to fill in the gaps. Thus the utilization of various sets of data can improve the analysis of these long series. Some data sets have still been no more than very lightly studied. An example of a set of information that has yet to be studied is represented by the observations of sunspots by the Astronomical Observatory of the University of Valencia (Spain) during the first third of the 20th century. Figure 7.6 shows the measurements of the areas of sunspots during the years 1926 and 1927 obtained by the Observatory of Valencia (Almer, 1930, 1931) and the RGO network. The relationship between the sunspot area measured by the Valencia Observatory and the RGO network is

$$\mathrm{A_{RGO}} = (0.917 \pm 0.015)\mathrm{A_{Valencia}} + (73 \pm 21)$$

where $\mathrm{A_{RGO}}$ is the daily sunspot area measured by the RGO network and $\mathrm{A_{Valencia}}$ is the daily sunspot area measured by the Valencia Observatory. The correlation coefficient is r = 0.939 with 570 pairs of data.

Also, before the measurement programme of the RGO, De la Rue et al. (1869, 1870) made a great effort in observing the positions and areas with the Kew photoheliograph during the years 1862–1866. They also compiled

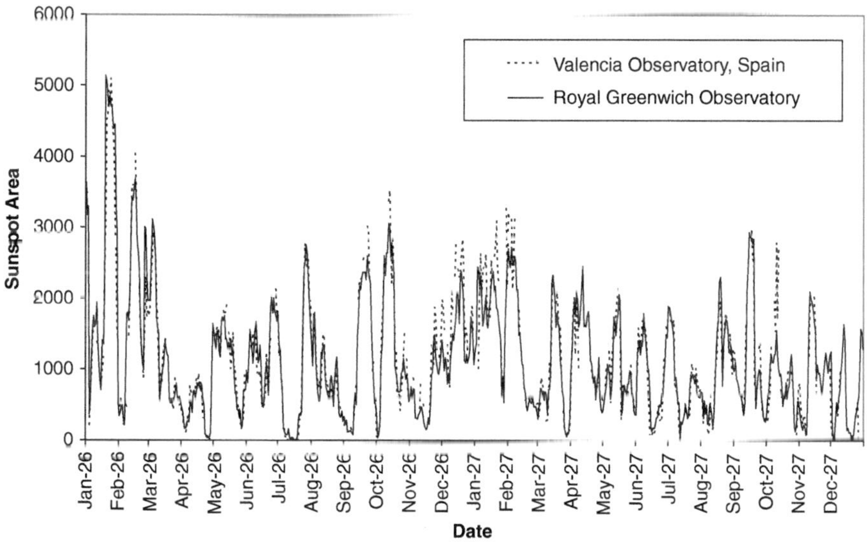

Fig. 7.6. Measurements of area of sunspots during the years 1926 and 1927 obtained by the Observatory of Valencia and the RGO network. The unit is one millionth of the solar disk area.

drawings and photographs of sunspots in the period 1832–1868. From these drawings and photographs, they determined fortnightly values of the sunspot areas. The measurement method and the reliability of this series have been studied by Vaquero et al. (2002). Some reconstructions of sunspot areas are available (Nagovitsyn et al., 2004; Vaquero et al., 2004; Nagovitsyn, 2005).

7.2 The Reconstruction by Hoyt and Schatten

At the end of the 20th century, some researchers began to demonstrate that there were some problems with Wolf's reconstruction, especially in some historical dates (Sonett, 1983; Hoyt et al., 1994; Wilson, 1998). Sonett (1983) suspected that there was an error in the R_Z series in 1780–1800, and later Usoskin et al. (2001) suggested that one solar cycle was lost in that epoch (the beginning of the Dalton minimum) because of sparse and partly unreliable sunspot observations. Hoyt et al. (1994) concluded that R_Z is probably overestimated during the 18th century. Another problem was indicated by Wilson (1998) who compared the annual number of sunspots reconstructed by Wolf with the annual number of "clusters of spots" recorded by Schwabe. According to this comparison, Wolf would have misplaced and underestimated the maximum amplitude of solar cycle 7. These works suggested that researchers in long-term solar activity should use great caution in applying the sunspot record.

HS98 published a new reconstruction of solar activity based on telescopic sunspot observations after some previous work (Hoyt and Schatten, 1992a, b, c, 1995a, b, c, d, 1996).

7.2.1 The Dataset

The first step to make this reconstruction was the storage and subsequent mechanization of sunspot observations from 1610. HS98 only mechanized the number of groups. They began digitizing the observations published by Wolf and his successors in a journal printed in Zürich called *Mitteilungen über der Sonenflecken* (which would later change its name to *Astronomische Mitteilungen*) that was published from 1858 to 1947. This source of information provided 224 503 observations made by 306 different observers.

The following step was the search for modern (starting from 1947) and old observations. HS98 looked for these in scientific journals and magazines. As Wolf had indicated which journals he had consulted, HS98 tried to locate observations in Italian, Dutch, and English scientific journals that Wolf had not consulted. Another of their greatest sources of new data consisted of the unpublished manuscripts of astronomers. The search of HS98 was very fruitful. If we consider active observers before 1874 (when the series of RGO begins), HS98 were able to gather 147 462 observations made by 330 different observers. Wolf, on the other hand, only had 81 521 observations made by 213 different observers.

The final database produced by HS98 contains 455 242 observations made by 463 different observers. There are 140 986 days with at least one observation from 1610 to 1995; that is to say, the average number of observations per day is 3. Unfortunately, the observations are not distributed evenly in time. Even so, the database allows estimation of the solar activity during 111 358 days (79% of the days during the period 1610–1995).

It is also interesting to point out that HS98 differentiated 12 types of observers. This classification can be re-elaborated in accordance with various criteria. Tables 7.5, 7.6, and 7.7 list the types of sunspot observers according to the Zürich original record, the HS98 database, and some characteristic of the record, respectively.

7.2.2 The Group Sunspot Number

HS98 defined the Group Sunspot Number, R_G, as

$$R_G = \frac{12.08}{N} \sum k_i' G_i$$

where G_i is the number of sunspot groups recorded by the ith observer, k_i' is the ith observer's correction factor, N is the number of observers used to form the daily value, and 12.08 is a normalization factor chosen to make the mean R_G's identical with the mean R_Z's for 1874 to 1976 when the Royal Greenwich Observatory (RGO) actively made sunspot observations. HS98 determined this equation for deriving the sunspot number because 90% of the sunspot

Table 7.5. Types of sunspot observers according to the Zürich record

Name	Description
Zürich recorded observers	Observations tabulated in the *Astronomische Mitteilungen*. They cover the period from 1610 to 1947 (306 observers with 224 503 observations). HS98 corrected occasional typographical errors. These observations plus the unpublished observations for 1948 to the present form the raw database for the Wolf or Zürich Sunspot Number time series.
"New non-Zürich" observers	These are the observations that HS98 collected from journals and unpublished archives (163 new observers with 230 739 observations).
"Effectively new" observers	Wolf received tabulations of the observations by Thaddeus Derfflinger (for 1802 to 1824) and Schwarzenbrunner (for 1825 to 1830) in 1893, just before he died. These observations were never incorporated in the R_Z. HS98 indicate that they may be labeled as effectively new.

Table 7.6. Types of sunspot observers according to the HS98 database

Name	Description
"Enhanced" observers	Wolf did not acquire all the observations from a particular observer in some cases. Observers where HS98 obtained more observations than Wolf included G.B. Riccioli, J. Hevelius, J. Picard, P. La Hire, V.F. Stancarius, J. Flamsteed, Rost, J.L. Alischer (called Alishez by Wolf), C. Horrebow, W. Herschel, J. Schmidt, and G. Spoerer. Obviously, HS98 suspect that their database probably has the same deficiency.
"Partially recorded" observers	For some observers, not all their observations were published. One example is the San Miguel Observatory in Argentina. Its record is not complete because HS98 could not locate a complete publication. This is one of the potential improvements that can still be made to the Hoyt–Schatten database.
"Corrected" observers	In a couple of cases the tabulations sent to Wolf appear to have been erroneous. Hoyt and Schatten (1995) revised the cases of J.W. Pastorff (from 1819 to 1833) and C. Horrebow.

number variance is caused by changes in the number of groups. Moreover, many observers during the historical epoch specify only the number of groups rather than both the number of groups and the number of individual spots (Hoyt et al., 1994). Therefore, R_G is designed to be more internally self-consistent and less noisy than R_Z.

This series has complete or nearly complete coverage from about 1800 to 1995 and from 1645 to 1727 (Figure 7.7). From the first half of the 17th century (Figure 7.8) and from some periods of the 18th century (Figure 7.8), there are many years with only sparse observations. For six years (1636–1637, 1641, 1744–1745, and 1747) there exist no reports of sunspot observations. The main periods with poor coverage are (1) 1610–1644, (2) 1722–1753, (3) 1758–1765, (4) 1777–1795, and (5) 1805–1806. From 1848, there is almost one sunspot observation per day. Surprisingly, the worst covered period is the 18th century, even though one might have expected that the worst period would be the 17th century because older documents should be more difficult to locate.

The number of sunspot records during the period that spans from 1777 to 1795 is extremely low. The pronounced decrease in the number of available observations during those two decades corresponds to a significant drop of in-

Table 7.7. Types of sunspot observers according to some characteristic of the record

Name	Description
"Vague" observers	Some observers are "vague" in one way or another so that their observations could not be used. These observers generally comment on whether spots are present or not, but do not estimate the number of groups. Vague observers include J.H. Schroter, Hahn, Sturmer, and many others.
"Summary" observers	Some observers do not supply details of their daily observations. This is particularly true among modern observers who publish only monthly means. Another type of summary observer are those who comment that they have seen no sunspots from one date to another, despite actively observing the Sun. There are about 20 of these observers in the HS98 record, mostly before 1700.
"Misplaced" observers	Another type of observer are those whose observations HS98 know to exist, but repeated efforts to locate the observations failed. Prominent observers in this category include J.G. Fink (active 1788–1816), S.T. Soemmering (active 1826–1829), and T. Chevallier (active 1847–1849).
"Lost" observers	HS98 know of some observers who were active and whose observations were either definitively lost such as those of Horrocks (active 1638) whose manuscripts were burned. For some observers, such as Scheiner, who observed sunspots on a nearly daily basis from 1611 to 1633, only a small portion of his observations survive in *Rosa Ursina* and his other publications. Another observer in this category is J.L. Alischer who kept a sunspot diary called *Diaria macularum solarium* that may have observations from 1727 to 1746 when hardly any observations were made. Lost manuscripts also include observations by J. Picard (before 1665), M. Fogel (1662–1670), E. Weigel (1662–1664), C. Weickmann (1666–1667), and H. Siverus (1675–1690).
"Unknown" observers	Despite considerable searching, there undoubtedly remain completely unknown observers.
"Poor" observers	Some observers may be classified as poor and can be dropped entirely for analysis. Most of these observers miss too many sunspot groups.

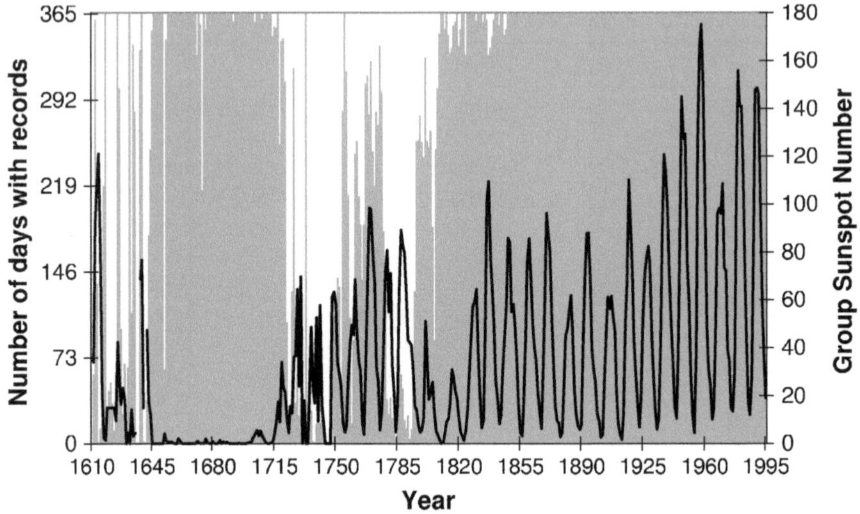

Fig. 7.7. Number of days with records (grey bars) and values of R_G (lines) from HS98 during the entire period. All the days have at least one record from 1848.

terest in solar activity monitoring by major astronomical observatories (Hoyt and Schatten, 1997) in spite of important contributions to solar studies such as that of the paper of Wilson (1774). Consequently, the number of people engaged in regular solar observations was extremely low.

According to Lalande (1771), the regular surveying of sunspots was placed at a very low order of priority on the observatories list of observations (14th

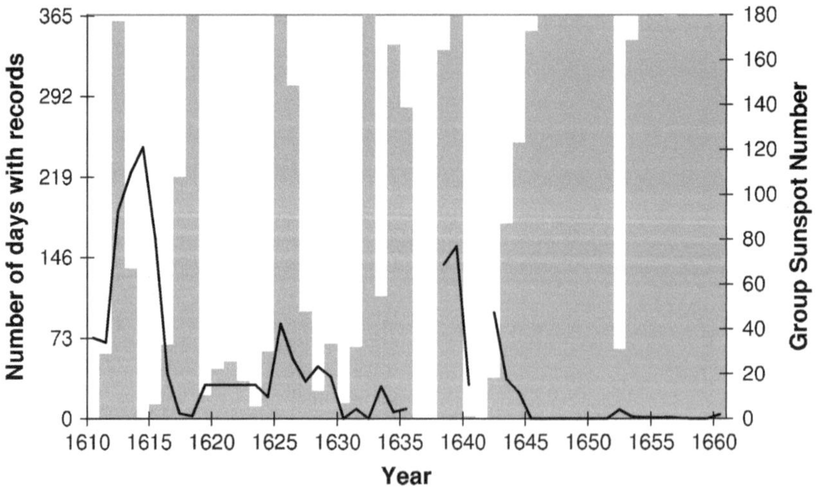

Fig. 7.8. Number of days with records (grey bars) and values of R_G (lines) from HS98 during the first half of the 17th century.

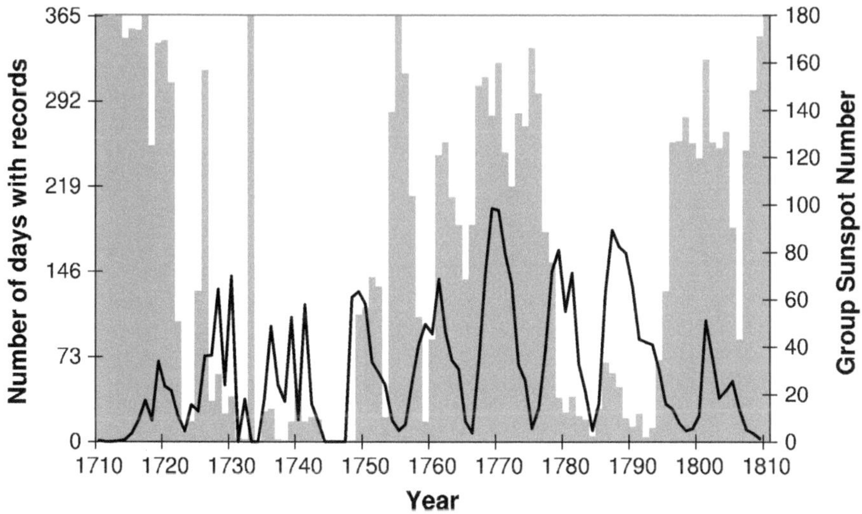

Fig. 7.9. Number of days with records (grey bars) and values of R_G (lines) from HS98 during the 18th century.

out of 18 duties). This dependence on a small number of observers can be appreciated if we split the number of observations for the last 25 years of the 18th century by country (Figure 7.9), where it is immediately striking that more than 90% of all sunspot observations were performed by observers located in just three countries, namely Denmark, France, and Germany. Observers from the rest of the world (including England) were responsible for a tiny number of observations. Germany kept a relatively constant number of observations throughout the period, obtained mostly by just one observer (J.C. Staudacher). On the contrary, during the early 1770s the vast majority of observations were performed by two Danish observers (C. Horrebow and E. Lievog), while the 1790s were dominated by French observations, particularly by H. Flaugergues. This pattern highlights the enormous dependence of total sunspot observations on a handful of dedicated observers. Moreover, Vaquero and Trigo (2006) point out two additional factors for this significant decline in observations, namely the outstanding eruption of Laki, Iceland (1783–1784), and the French revolution (1789).

The unusually large eruption of Laki started in June 1783 and lasted for eight months and corresponded to the largest outpouring of lava in recorded history (roughly 15 km^3), a value larger than Tambora or Krakatau (Boer and Sanders, 2002). The predominant westerly and north-westerly winds transported the intense plume of smoke and dry sulphurous fog over the British Isles and Northern Europe (where most astronomical observations were performed), therefore limiting the capacity to observe astronomical events, as depicted by the extremely low number of observations obtained for 1784 (first vertical line in Figure 7.10). After an apparently small recovery in the follow-

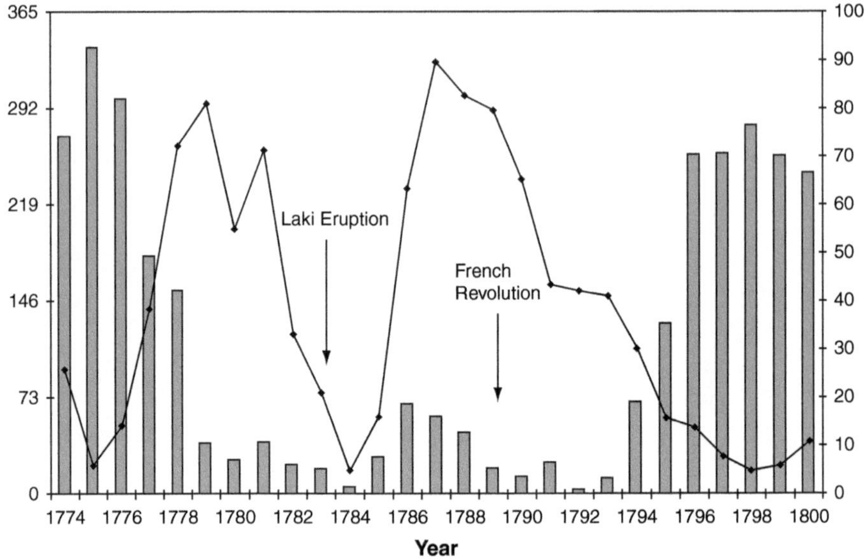

Fig. 7.10. Number of days with records in the solar reconstruction of HS98 and yearly values of Group Sunspot Numbers (solid curve). Vertical arrows correspond to the two major events that had a direct impact on the number of observations performed.

ing years (1785–1788) there is another major "hole" afterwards. This decline could be human in nature, probably associated with the socio-political upheavals that spread over Europe shortly after the French revolution in 1789 (second vertical line in Figure 7.10), including the enforced closure of the French Academy of Sciences in 1793 (Serres, 1989). This elimination of a key scientific institution in Europe was bound to deter many scholars (or upper class citizens) fron continuing their regular scientific activities, including astronomical (and meteorological) observations. Despite the obvious impact of these two factors, we must acknowledge that additional causes are required to explain the earlier drop in the number of observations (1777–1782).

7.2.3 Uncertainties

It is very difficult to estimate the uncertainties for the historical sunspot number derived by Wolf. HS98 provided systematic errors of the values of his reconstruction. The systematic errors in R_G can be separated into four components:

1. Errors arising from missing observations;
2. Errors arising from uncertainties in the values of k';
3. Errors arising from random errors in the daily values; and
4. Errors arising from drifts in the k' values.

The dominant error term corresponds to errors arising from missing observations. HS98 proposed a method to evaluate this error source. Thus, for 20 or more days of observations D, the error E follows a linear relationship:

$$E = 0.217 - 0.00059D$$

Obviously, this systematic error approaches zero when D approaches 365 or 366. However, erratic results were found for D less than 20. Hence, the annual mean values of R_G calculated with less than 20 days of observation per year are unreliable.

The preliminary analysis that HS98 made from its solar activity reconstruction provided very attractive results. Firstly, the solar activity prior to 1882 was lower than had generally been assumed. Therefore, the solar activity of the last decades has been more intense than in previous centuries. Secondly, there was a maximum of the 11-year cycle in 1801 (not in 1805), thus disappearing an anomalous solar cycle with 17 years of duration (from maximum to maximum) that the series of Wolf presented. Lastly, the Wolf number presents inhomogeneities due to the noise of the observers, affecting the daily, monthly, and annual values. The number R_G is also affected by the noise of the observers, but in less markedly. Also, the index R_G presents differences with respect to R_Z, especially in the daily values.

7.2.4 Group Versus Wolf Number

The Group and Wolf Sunspot Numbers are not different alternative proxies of solar activity. R_G should be considered an upgrade of R_Z numbers. Moreover, the Group Sunspot Number is strongly recommended for the study of solar activity before 1850 (Usoskin and Kovaltsov, 2004; Hathaway et al., 2002; Kane, 2002). The two time series are plotted in Figure 7.2.

Figure 7.11 shows the difference between the annual values of R_G and R_Z. It is striking that there are three values that surpass the value −80. They correspond to the years 1727, 1738, and 1778. These three values correspond to maximum years of the undecennial solar cycle in the Wolf numbers that do not correspond to maximum years (one year lag) in the HS98 reconstruction. The R_G and R_Z numbers during the period 1720–1790 are plotted in Figure 7.12, an enlarged version for this epoch of Figure 7.2.

From the long-term variability point of view, we can compare a 23-year moving average of the two indices as shown in Figure 7.13. In the period 1700–1750, the values of the curve of R_Z are practically double the values of the curve of R_G. During the period 1750–1900, the curve of R_Z is always above that of R_G, but in the last period that covers the 20th century, the values of the two curves become almost identical although with small variations.

Some authors have used the series of R_G to try to verify some of the laws and correlations implied by R_Z. Hathaway et al. (2002) have presented some of these results. They compare smoothed monthly R_G to R_Z, the 10.7-cm

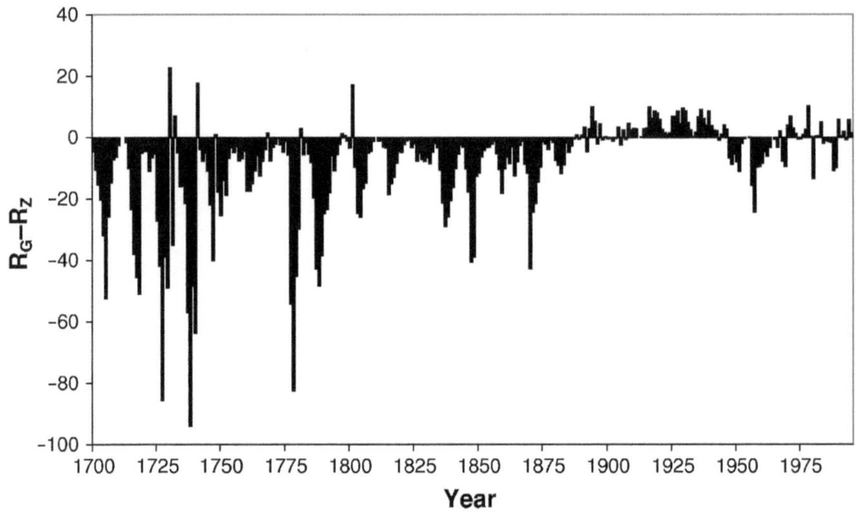

Fig. 7.11. Annual values of Group Sunspot Number minus Wolf Sunspot Number.

radio flux, and total sunspot area, and find that R_Z follows the 10.7-cm radio flux and total sunspot area measurements only slightly better than the R_Z. With respect to significant characteristics of the sunspot cycle, they found the following results using R_G and R_Z:

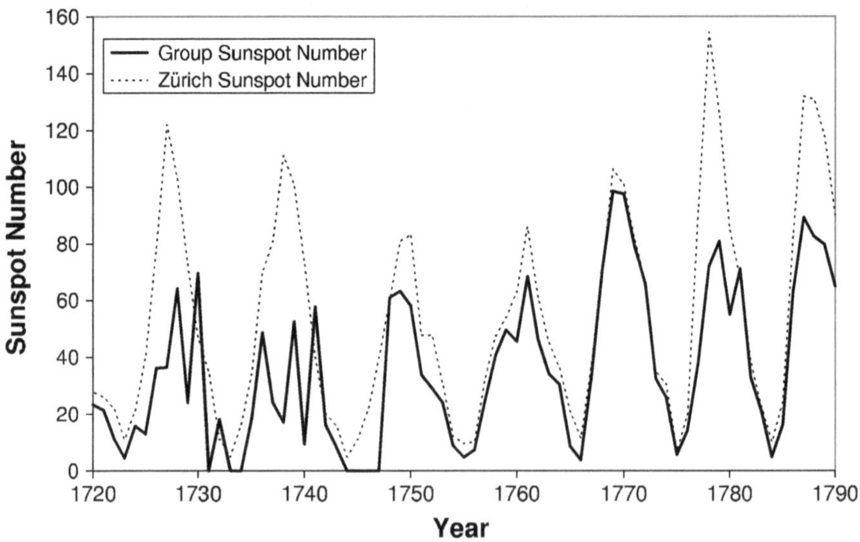

Fig. 7.12. A comparison between the annual Group and Wolf Sunspot Numbers for the period 1720–1790.

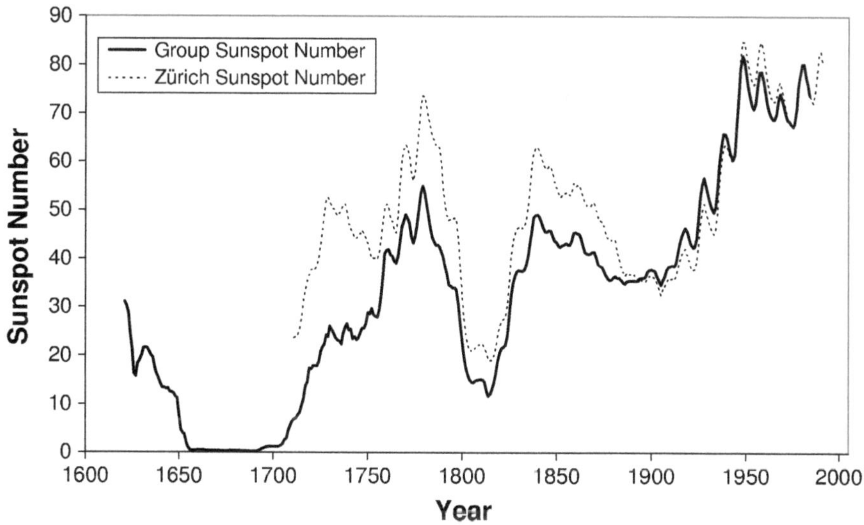

Fig. 7.13. 23-year moving averages of the annual Group and Wolf Sunspot Numbers.

1. The "Waldmeier effect" (anticorrelation between cycle amplitude and the elapsed time between minimum and maximum of a cycle) is much more apparent in R_Z.
2. The "amplitude–period effect" (anticorrelation between cycle amplitude and the length of the previous cycle from minimum to minimum) is also much more apparent in the Zürich numbers.
3. The "amplitude minimum effect" (the correlation between cycle amplitude and the activity level at the previous minimum) is equally apparent in the R_Z and the R_G.
4. The "even–odd effect" (odd-numbered cycles are larger than their even-numbered precursors) is somewhat stronger in the R_G but with a tighter relationship in the R_Z.
5. The "secular trend" (increase in cycle amplitudes since the Maunder minimum) is much stronger in R_G.

Hathaway et al. (2002) concluded that the R_G numbers are more useful for extending the sunspot cycle data further back in time, but the R_Z are slightly more useful for characterizing the on-going levels of solar activity.

A comparison of the spectral features of Group Sunspot Numbers and the traditional Wolf Sunspot Numbers for the period 1700–1995 was made by Faria et al. (2004) using the multi-taper method (MTM) and wavelet analysis. The MTM analysis identified the well-known Gleissberg and Schwabe cycles (98.6 and 10–11 years, respectively) in both time series. However, a MTM analysis of two subsets (1700–1850 and 1851–1995) indicated that the main differences occurred in the first subset, i.e. during the historical period due to uncertainties in the early observations. The wavelet analysis showed a strong and persistent 11-year periodicity during the whole period. Therefore, the main spectral characteristics of the two series are very similar, and the long-term variability has the same behaviour except in the secular trend.

7.2.5 Some Problems

HS98 performed a great work in their reconstruction of solar activity from documentary sources because their reconstruction is much better than that of Wolf. The detailed compilation work done both by Wolf and by HS98 is tedious and unrewarding, and we must be grateful to both for their laborious work. However, the reconstruction does contain some minor errors and inexact statements.

Some mistakes in the reconstruction came from calendar problems, the differences between spots or group, changes of date, and partial information in solar eclipse observations. Moreover, we can point out one methodological inconsistency and the presence of observations made using a camera obscura (Vaquero, 2007).

All observations in the database of reconstructed solar activity must be referred to the same calendar (the Gregorian Calendar). However, some countries still used the Julian (or Old Style) Calendar well into the telescope era. Examples are Great Britain (including the British colonies) up to 1752 and Russia up to 1918. Thus, it is very easy to confuse the Julian and Gregorian dates.

Vaquero (2007) gives the example of the sunspot observations by Jeremiah Horrocks or Horrox (1619–1641), an English astronomer who made the first observation of a transit of Venus (see Chapter 5). The dates of his sunspot observations are erroneously interpreted in HS98. HS98 found six observations by Horrocks during the year 1638 in Liverpool (England), using the *Opera posthuma* of Horrocks (1673). The observation dates are 22–24 May (two sunspot groups) and 20–22 October (two sunspot groups). However, these dates refer to the Julian calendar. E.g., the Venus transit of 4 December 1639 was observed by Horrocks who, however, wrote in his book (Horroccii, 1673, p. 393) the date of 24 November. Horrocks also observed the solar eclipse of 1 June 1639 but wrote in his book (Horroccii, 1673, pp. 387–389) the date 22 May 1639. It is evident that he used the Julian Calendar (10 days of difference with respect to the Gregorian Calendar in that epoch).

HS98 corrected some of Wolf's estimates (e.g. Horrebow) since Wolf had not differentiated between groups and spots. Equally, some records of HS98 may suffer from this problem. As an example, Vaquero (2007) puts as an example the observation of the solar eclipse on 15 July 1730. According to HS98 (and Wolf), the observer is Hallerstein from Beijing. However, we know that although Hallerstein's observations are also compiled in Hell (1768), he was not the observer of this eclipse. The observers were the Jesuits Koegler (or Kegler) and Pereira (1733, 1799). In the description of the contacts of the Moon with the sunspots, the observers describe seven spots on the solar disk.

HS98 took this to be seven groups of sunspots. Nevertheless, a close reading of the observation shows that these seven spots can easily be grouped into three groups. The first group consists of just the largest sunspot observed near the solar limb. The second group consists of four spots in the solar northwest, and the third consists of two spots in the southwest. Thus, the number of groups should be reduced from seven (HS98 and Wolf) to three. This might partly explain the anomalously high number of groups during 1730, because in that year there were very few other observations.

Another problem in the reconstructions is the change of dates (year or month). As an example, Vaquero (2007) put the observation of J. Huxham (included by HS98) published in the *Philosophical Transactions* describing the Mercury transit in 1736 (Huxham, 1744). However, the year of observation given in HS98 is erroneously 1739 (because it was an errata in the paper) creating a spurious sunspot observation in that year.

A considerable part of the information in the work of HS98 about sunspots during the 18th century comes from observations of solar eclipses. A usual task undertaken by astronomers of the 18th century was to time the moments of apparent contact between the Moon and the sunspots during the progress of the eclipse. The Moon covers all the spots on some occasions, and on others not. Therefore, to use the sunspots that appear in a list of times of contact during eclipses can be dangerous because it can underestimate the number of sunspots and groups.

Vaquero (2007) put as an example of this situation the solar eclipse observed in Lyon by French Jesuits on 25 September 1726. HS98 lists this observation among the observations of Souciet (1729), although Souciet was only a compiler of the work. Figure 7.14 shows the table of contacts of the eclipse (page 228 in Souciet, 1729). In this case, we can read four occulted sunspots in the table of contacts of the eclipse, and, indeed, HS98 included four groups of sunspots (one group for every occulted sunspot).

Nevertheless, one can see from the drawing that the observers provided in their report (Figure 7.15) that six spots were observed on the solar disk. Spots C and D (occulted by the Moon) are in the penumbra itself and have to be considered as one group of sunspots. Therefore, one should instead include five groups of sunspots from this eclipse observation.

Commencement de l'éclipſe	4h.50'.quelques ".
La tâche A diſparoît	4. 58. 30".
La tâche B diſparoît	4. 59. 57.
La tâche C diſparoît	5 2. 9.
La tâche D diſparoît	5. 2. 14.
Le Soleil étoit éclipſé de 3. doigts à	5h.15'.34".
de 4. doigts à	5. 16'. 1.
de 5. doigts à	5. 24. 15.

Fig. 7.14. Table of contacts of the eclipse observed on 25 September 1726 by French Jesuits in Lyon (Souciet, 1729, p. 228).

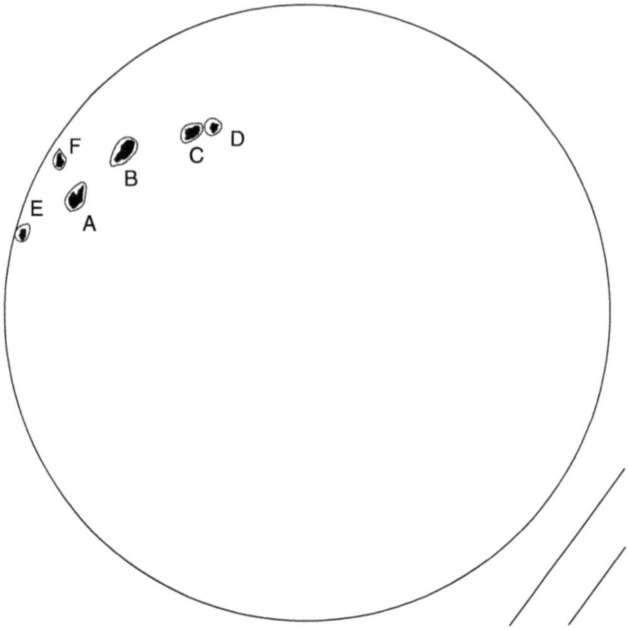

Fig. 7.15. Drawing of the sunspots on 25 September 1726 by French Jesuits in Lyon (Souciet, 1729).

HS98 deviated from their reconstruction methodology on one occasion. They found in the Cambridge University Library (Flamsteed paper collection) a letter by William Crabtree (1610–1644) with information on sunspots. They wrote in their Bibliography: "According to a letter by Crabtree the average number of spot groups seen in 1638 and 1639 were 45 per day. The database has Greenwich fill values to give 45 groups per day. This substitution technique was used to simplify the analysis. This is the only place in the entire database where we do this type of substitution."

Figure 7.16 shows a comparison between the observed (by J. Horrocks and P. Gassendi) and the estimated (from Crabtree's comment) values from HS98 for the sunspot groups during the years 1638–1640. One can appreciate that, from a general point of view, the number of observed groups is smaller than the number of estimated groups.

Another problem in the HS98 reconstruction is the use of observations made by astronomers who used a camera obscura. The great majority of sunspot observations compiled by HS98 obviously were made with telescopes. But we must keep these camera obscura observations in mind. The camera obscura was already very useful for astronomers in the 16th and early 17th centuries. Small camera obscuras were used from very early times (Sigismondi and Fraschetti, 2001), but during the 16th century until the 1750s meridian lines were used in major cathedrals to determine the Sun's position and

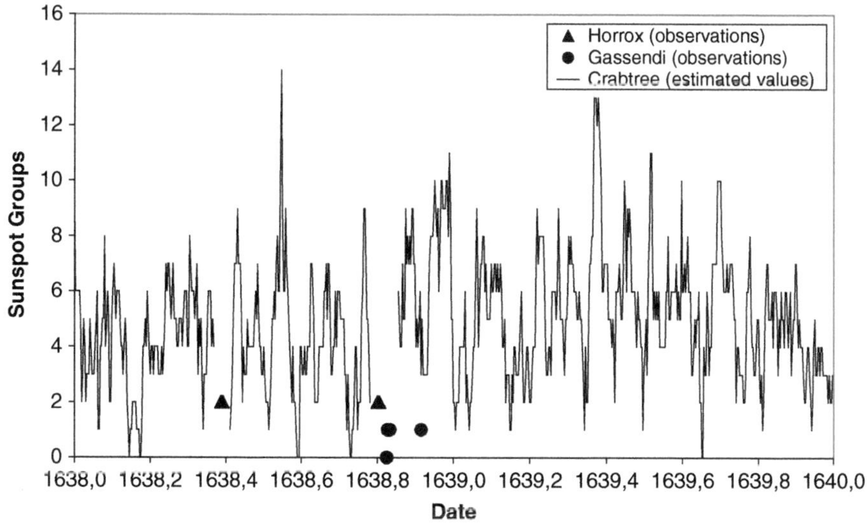

Fig. 7.16. A comparison between the sunspot group values from observations (by Horrocks and Gassendi) and the estimate made by HS98 from the Crabtree letter.

the Earth's orbit (Heilbron, 1999). 2218 solar observations made with the San Petronio camera obscura were used in HS98 to reconstruct solar activity (taken as zero spot days) during the Maunder minimum (Table 7.8) from the book published by E. Manfredi in 1736.

Sunspots were also observed with small camera obscuras. There is a famous text of Galileo where he invites persons who do not want to use the telescope to use a broken window of a temple as a camera obscura to observe sunspots.

Table 7.8. Solar observations during the Maunder minimum from Manfredi (1736) taken as zero spot days by HS98

Observer	Period	Number of records
J.C. Calcina	1674	20
G.D. Cassini	1656–1670	13
A. Fabrius	1668–1675	90
J.F. Gulielmini	1675–1696	406
D. Manzius	1673	1
F. Maraldi	1673	273
P. Mengoli	1663–1670	66
C. Mezzavacca	1663–1695	6
G. Montanari	1671–1676	107
G. Riccioli	1655–1661	92
V.F. Stancarius	1696–1702	1103
I. Uccelli	1695–1696	41

However, the first known observation of a sunspot using a camera obscura was done by Kepler who observed the Sun with a camera obscura on 28 May 1607 and detected a small spot on the solar disk. He was convinced that he was observing a transit of Mercury (see Chapter 5). He published his discovery in a short treatise on the comet of 1607 and in more detail in his *Phaenomenon singulare* (Kepler, 1609).

7.2.6 Some Unfinished Tasks

There are some pending tasks in the comments on the observers in the Bibliography of HS98. One could classify these tasks into six categories:

1. Lost original observations. Many manuscripts are lost. In some cases, we know the observations because they were copied by other persons. Discovery of the original observations would allow one to verify the information. In other cases, the observations have been lost, and locating them would help to improve notably our knowledge of past solar activity.
2. Lost daily data. In some cases, we only know monthly or annual averages from some observers. Locating the daily information would allow us to improve the database of observations.
3. Observations not included in the database. For some reason, some published information was not incorporated into the database compiled by HS98.
4. Observations too vague, ambiguous or narrative style. On occasions, observational reports are written in a narrative form that makes it very difficult to construct trustworthy tables with that information.
5. Incomplete observations. On other occasions, HS98 obtained information of an observer knowing that there existed other volumes of observations that were not accessible to them. Retrieval of the remaining observations would be a simple task for those persons who do have access to the complete collection.
6. References need to be re-checked. On some occasions, HS98 themselves note a failure to annotate the correct reference of books or manuscripts.

7.3 Improving and Finding Lost Observations

Undoubtedly, a great problem in reconstructions from historical solar observations is the influence that the days with no available information have on the monthly and annual averages. Some authors have treated this problem by trying to fill the gaps in the series (Letfus, 1999) or by developing statistical methods that estimate the margin of error of the average even if the number of observations per year is very small (Usoskin et al., 2003b).

Another possibility is to find new observations to fill the gaps in the series. For example, Vaquero (2003) describes observations of Andrea Argoli (1570–1657). He recorded in his book *Pandosion sphaericum* (Argoli, 1653) that he

Table 7.9. Annual Group Sunspot Number during the 18th century derived by HS98. Values in italics are values revised in some papers (Vaquero, 2004; Vaquero et al., 2005, 2007b, c). We use n.a. as not available

Year	R_G	Year	R_G	Year	R_G	Year	R_G	Year	R_G
1700	0.4	1720	23.4	1740	9.3	1760	45.5	1780	55.0
1701	0.5	1721	21.4	1741	57.7	1761	68.5	1781	71.1
1702	0.6	1722	11.0	1742	16.1	1762	46.2	1782	32.9
1703	2.7	1723	*09.0*	1743	8.3	1763	34.2	1783	21.1
1704	4.1	1724	15.6	1744	n.a.	1764	30.5	1784	*0.3*
1705	5.5	1725	12.8	1745	n.a.	1765	8.4	1785	*16.4*
1706	3.2	1726	36.2	1746	*n.a.*	1766	3.7	1786	*63.7*
1707	5.3	1727	36.5	1747	n.a.	1767	33.9	1787	89.2
1708	2.8	1728	64.2	1748	61.0	1768	71.3	1788	82.5
1709	1.6	1729	24.0	1749	63.2	1769	98.5	1789	79.7
1710	0.4	1730	*67.2*	1750	58.0	1770	97.6	1790	65.1
1711	0.0	1731	*n.a.*	1751	*33.5*	1771	79.4	1791	43.2
1712	0.0	1732	18.0	1752	29.0	1772	66.2	1792	42.0
1713	0.3	1733	0.0	1753	23.9	1773	32.4	1793	41.0
1714	0.9	1734	0.0	1754	8.8	1774	25.8	1794	30.2
1715	3.6	1735	18.3	1755	4.7	1775	5.6	1795	15.7
1716	9.1	1736	*53.6*	1756	7.3	1776	14.1	1796	13.7
1717	17.5	1737	*32.3*	1757	24.8	1777	38.3	1797	7.7
1718	9.0	1738	*48.0*	1758	40.7	1778	72.0	1798	4.7
1719	33.9	1739	*42.2*	1759	49.5	1779	80.8	1799	5.6

had not seen any sunspot in 1634 from 19 July until mid-September. During this period P. Gassendi also observed the Sun with the same result. Therefore, this observation by Argoli does not fill any gap in the series, but only confirms the information that we already have. In the last year, some attempts to improve the series derived by HS98 (especially during the 18th century when the sunspot observations are scarce) have been made. Table 7.9 lists a revised version of R_G values for the 18th century. In the following subsections, these attempts to improve the series will be briefly reviewed in spite of the fact that the advances made in improving the HS98 series have been very small.

7.3.1 Records in Spanish and Portuguese

Most of the records of the database of HS98 are written in Latin, English, or German. Those in other languages, such as Spanish and Portuguese, are practically non-existent. However some works that we shall review briefly have presented some historical observations of sunspots in the Iberian-American environment during the 18th century. Some information on historical astronomical observations made in Latin America can be consulted (Keenan, 1991;

Moreno-Corral, 1986; Gonzalez, 1955). Also, Vaquero and Moreno-Corral (2008) have reviewed historical observations of sunspots made from Mexico.

José A. Alzate (1737–1799) was a well-known Mexican scientist of the 18th century (Saladino, 1990). Alzate's published work is dispersed over several Mexican journals (Alzate, 1831; Aureliano et al., 1996). The following is a compilation of Alzate's records on sunspots (Vaquero, 2004; Vaquero et al., 2007a) in chronological order:

1. One observation (with drawing) during the transit of Venus on 3 June 1769.
2. Comments by Alzate on sunspots during August 1769.
3. One observation (with drawing) during the transit of Mercury on 9 November 1769.
4. Comments by Alzate on sunspots during July 1786.
5. Spotless period from 25 August to 29 October 1784.

The spotless period reported by Alzate is very valuable (see Figure 7.17) for the reconstruction of solar activity because HS98 only found five observations during that year all performed by J.C. Staudacher leading to an HS98 value of 4.8 in 1784. However, using conjointly the data provided by Alzate and Staudacher for 1784, one can determine a value of R_G equal to 0.3 ± 0.1 with eighty records for that year. It would be very interesting to find other observations of sunspots by J.A. Alzate for a better knowledge of solar activity during the 18th century.

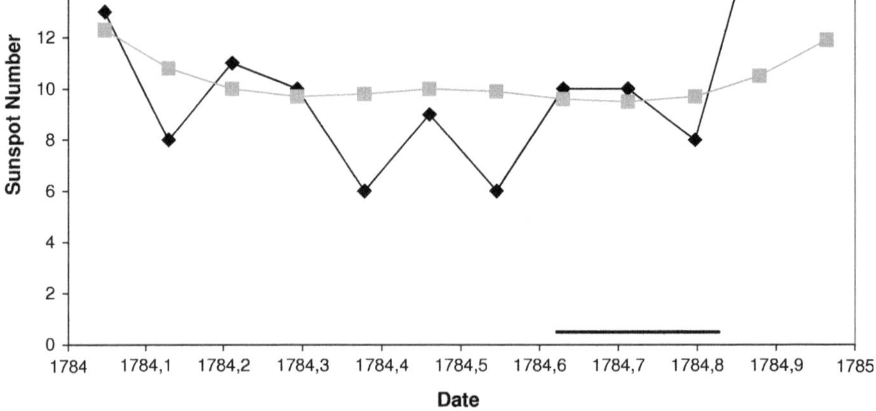

Fig. 7.17. International Sunspot Number during the year 1784 and spotless period according to J.A. Alzate.

Vaquero et al. (2005) presented a "lost" sunspot observation made by a Portuguese scientist Sanches Dorta during his observation of the solar eclipse of 9 February 1785 from Rio de Janeiro (Brazil). This record was not included in the database compiled by HS98. A value of R_G equal to 16.4 ± 3.4 for the year 1785 is obtained using this observation. The lack of observations in the late 18th century is not a superfluous issue as it roughly corresponds to a secular minimum of solar activity (Dalton minimum) but also because of the unusual length of cycle number 4 between 1784 and 1798. This can be either regarded as a 14 year long cycle (peak-to-peak) or two consecutive unusually short cycles (Usoskin et al., 2002, 2003c), implying the existence of a so-called "lost" solar cycle (see also Krivova et al., 2002). Whatever the case, the fact remains that this period was badly covered in terms of observations and also that the cycle of solar activity exhibited a highly unusual pattern of behaviour between 1784 and 1798.

Vaquero et al. (2007a) recovered some descriptions of sunspot observations in Portugal during the 18th century but these are poorly dated. In particular, the ideas of Teodoro de Almeida (1722–1804) about sunspots were analysed using his didactic and popular science works (in particular, *Recreação Filosofica*). A sunspot observation published in this last work was made by Almeida on either 10 April 1760 or 10 April 1761. If this observation was in fact made in 1760, then one could speculate that the maximum value of cycle number 1 took place in 1760 and not in 1761 (as has been proposed by HS98), resembling other solar cycles with similar maximum values in two consecutive years. In the second case (observation corresponding to 1761), the R_G value obtained by HS98 for April 1761 would be slightly lower.

7.3.2 The Great Gap in the 1740s

We have pointed out in previous sections that records of solar activity through the second quarter of the 18th century are very sparse. One of the sources used by Wolf (and later by HS98) for this period is the compilation by Maximiliano Hell (1768) of the astronomical observations of Jesuits in China from 1717 to 1752. Vaquero et al. (2007b) reviewed the sunspot observations included in this compilation that are based exclusively on the solar eclipse observations. Note that the solar (and lunar) eclipse observations were very important at that time for the accurate determination of geographical coordinates. Moreover, a very usual task undertaken by astronomers of the 18th century was to time the moments of apparent contact between the Moon and the sunspots during the progress of the eclipse. Table 7.10 shows all solar eclipses observations recorded by Hell including the date of the observation, the record of HS98, and the revised record by Vaquero et al. (2007b).

The most interesting result is that HS98 includes only one record in the years 1731 and 1746 (both compiled by Hell). However, as we indicated in Table 7.10, no sunspots were recorded in those dates. Therefore, both years must be included in the list of years without sunspot records (1636, 1637,

Table 7.10. Dates of the eclipse observations compiled by M. Hell. The numbers of groups included in the reconstruction of Hoyt and Schatten (HS98) are presented, as are the revised values by Vaquero et al. (2007b)

Date	HS98	Vaquero et al. (2007b)
24 Sep 1718	0	No sunspot information
19 Feb 1719	–	No sunspot information (Carolo Slaviczek)
04 Aug 1720	–	No sunspot information
24 Jul 1721	–	No sunspot information
03 Jun 1723	0	No sunspot information (Slaviczek, Nanchang, Kiang-Si and anonymous, Sinoae, Conchinchina)
15 Jul 1730	7	3 (Koegler and Pereira)
29 Dec 1731	0	No sunspot information
16 Oct 1735	1	1 (Jac. Phil. Simonelli, Sin-Fun-Hien, Kiang-Si)
03 Jun 1742	1	1
22 Mar 1746	0	No sunspot information
25 May 1751	–	1 (Antonio Gogaisl, Beijing and anonymous)

Table 7.11. Dates of the eclipse observations compiled by Hell. The numbers of groups included in the reconstruction of Hoyt and Schatten (HS98) are presented, as are the revised values by Vaquero et al. (2007b)

	Daily values		Monthly values		Annual values	
Date	HS98	Revised	HS98	Revised	HS98	Revised
03 Jun 1723	0	n.a.	0	n.a.	4.5	9.0
15 Jul 1730	106	45	61	30	69.7	67.2
29 Dec 1731	0	n.a.	0	n.a.	0	n.a.
22 Mar 1746	0	n.a.	0	n.a.	0	n.a.
25 May 1751	n.a.	15	49.0	47.5	33.7	33.5

1639, 1641, 1731, 1744, 1745, 1746, and 1747). Note that there is currently a major gap of information in the period 1744–1747 for which there are no sunspot observations. The revised values of R_G are listed in Table 7.11.

7.3.3 The Sunspot Numbers During 1736–1739

Some periods in the HS98 sunspot number reconstruction composed are based on very few records. The period 1736–1739 is an illustrative example. Vaquero et al. (2007c) attempted to improve the reliability of the sunspot number reconstruction in HS98 for this four-year period based on information about solar activity published in three journals of that epoch: *Philosophical Transactions*, *Histoire de l'Académie Royale des Sciences*, and *Nova Acta Eruditorum*. They were able to identify 42 papers with solar observations, including 30 with relevant information on sunspots. Based upon this new information, a reconstruction of the monthly solar activity for these years was proposed.

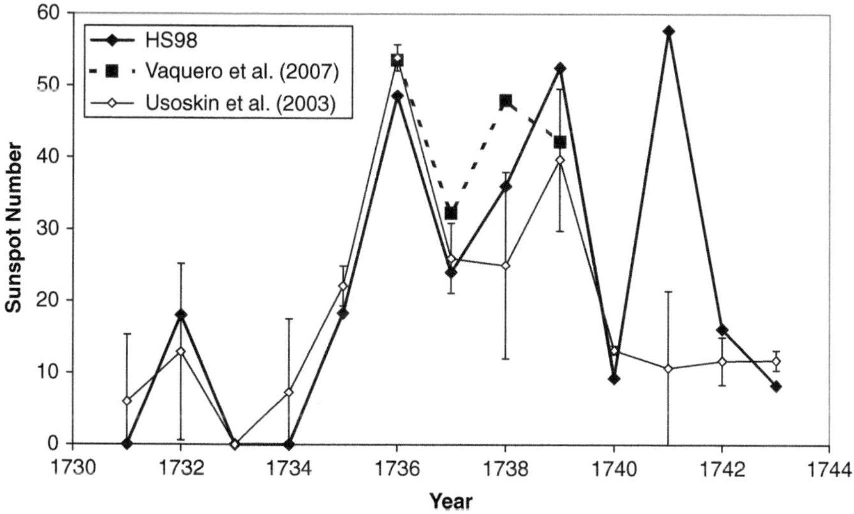

Fig. 7.18. Yearly Group Sunspot Number around the solar cycle number −1.

Figure 7.18 shows the yearly sunspot number according to the HS98 values and the values obtained by Vaquero et al. (2007c). They corrected the annual value for 1738 given by HS98 as it clearly corresponds to a typographical error. Also plotted are statistical estimates given by Usoskin et al. (2003b) because annual values from a few daily observations are quite uncertain. The new shape of solar cycle number −1 is more uniform than that in the HS98 reconstruction, with a less pronounced abrupt decline within the expected peak years. Nevertheless, they believe that it may be possible to improve this shape based on further historical sunspot observations that have yet to be analysed.

7.3.4 The Hemispheric Numbers

The hemispheric asymmetry of solar activity is one of the key parameters to incorporate into dynamo models (see, for example, Goel and Choudhuri, 2009). If the short-term variations are averaged out, the solar activity is reasonably symmetric in the two solar hemispheres. Some solar cycles appear to have been stronger in one hemisphere. However, Carbonell et al. (2007) reached the conclusion that the quantitative results concerning the statistical significance of the N-S asymmetry of solar activity must be considered with care because they depend on the values chosen for different parameters and on the units considered.

Vaquero (2007) pointed out that an effort must be made to incorporate the historical information on solar activity into studies that require data about the position of sunspots, such as differential rotation, active longitudes, north–south asymmetry, etc. It is evident that the simplest information involving the sunspot position is the hemispheric number of sunspot groups. However, there is little information about hemispheric asymmetry during historical times. Ribes and Nesme-Ribes (1993) demonstrated that during the Maunder minimum there was no sunspot activity for approximately 60 years, and solar activity was concentrated near the solar equator. Zolotova and Ponyavin (2007) have suggested that the puzzling length of solar cycle number 4 could be explained by outstanding phase asynchrony between hemispheric activities with a delay of up to 4.5 years and a strong north-south asymmetry during the course of the ascending phase of the solar cycle.

From the definition of R_G, the Hemispheric Group Sunspot Numbers (R_Gn and R_Gs) can be defined as

$$R_{Gn} = \frac{12.08}{N} \sum k_i' G_{in}$$

and

$$R_{Gs} = \frac{12.08}{N} \sum k_i' G_{is}$$

where G_{in} (G_{is}) is the number of sunspot groups recorded in the north (south) hemisphere by the ith observer. Newton and Milsom (1955) proposed a normalized measure of sunspot asymmetry (NA) using hemispheric sunspot areas. We can modify this definition using the Hemispheric Group Sunspot Number:

$$NA = \frac{R_{Gn} - R_{Gs}}{R_{Gn} + R_{Gs}}$$

An evaluation of Hemispheric Group Sunspot Numbers during the period 1860–1870 can seen in Figure 7.19 using the sunspot position provided by

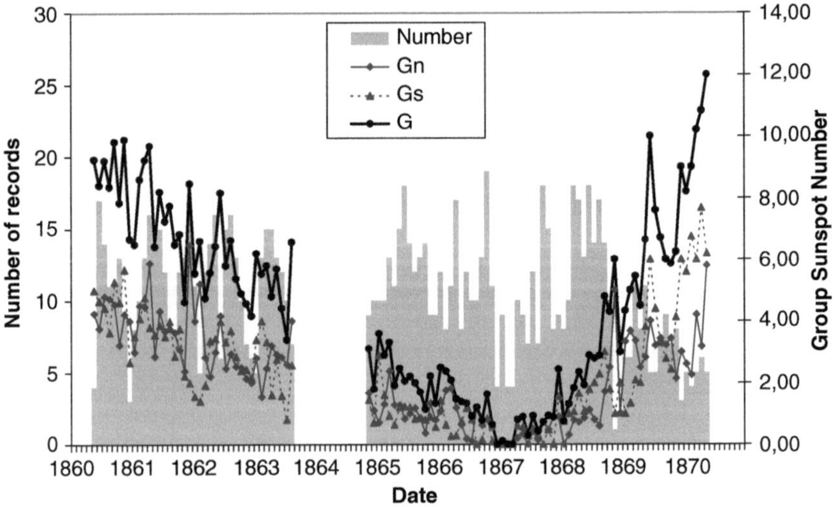

Fig. 7.19. Monthly Hemispheric Group Sunspot Number from the solar observations of Peters (1907).

C.H.C. Peters (1907). According to this record, the descent phase of solar cycle solar number 10 was characterized by a preponderance of the north hemisphere. However, the south hemisphere was dominant during the ascent phase of solar cycle number 11.

7.4 Final Comments

In this book we have shown different types of information related to solar activity that can be obtained from historical documents. The observation of aurorae and naked-eye sunspots provides us with continuous information through the last few centuries that can be used to improve our knowledge of the long-term solar activity. Other historical records, such as the eclipse and planetary transit observations, are valuable to improve our knowledge of the temporal evolution of the terrestrial rotation and solar diameter over time spans of centuries.

Observations of sunspots with telescopes recorded in historical documents that are conserved in libraries and archives are a source of data of exceptional interest to reconstruct the history of the Sun during the last four centuries. Since the 19th century, the improvement and the constant upgrading of the sunspot number has been a task of enormous interest for astronomy in particular, and for Earth and space sciences in general. Thanks to the collaborational effort of many people (especially we would mention Wolf and Hoyt and Schatten), the scientific international community now has a sunspot number series that is used in a multitude of studies of diverse subject matter.

The documentary data can improve our knowledge of the long-term solar activity and historical short-term solar activity. Figure 7.20 shows the solar activity during the last millennium (AD 1000–1900) using several reconstruction methods. The dark dashed line represents the smoothed Group Sunspot Number (Hoyt and Schatten, 1998a,b) using a 25-year moving-average window. The thin dashed line represents the smoothed number of aurora in the Křivský and Pejml (1988) catalogue using a 25-year moving-average window and re-scaled from a logarithmic scale. The grey line represents the smoothed number of naked-eye sunspot observations recorded in the catalogues (from Vaquero et al., 2002) using a 25-year moving-average window and re-scaled. Moreover, we have included in the figure the reconstruction of solar activity made by Solanki et al. (2004) based on dendrochronologically dated radiocarbon concentrations.

We would stress that the Group Sunspot Number is only available from early 1600, and the behaviour of solar activity on longer time scales is studied using indirect proxies. In a general form, all the indices show similar behaviour throughout the period considered. One can also find some remarkable discrepancies within specific periods. The main characteristic disagreements are located basically in the last few centuries.

On the one hand, naked-eye sunspot observations from 1620 approximately were recorded in local chronicles instead of in imperial chronicles. This involved a major loss of quality of the observations and a drastic reduction in their number. This is especially severe throughout the 18th century. The advantage of the series resides in that the data correspond to direct observations of the Sun. However, some authors have demonstrated the importance of other important non-solar factors (especially meteorological or socio-historical) in these observations.

And on the other hand, the auroral observations show a minimum around AD 1750 (identified by Silverman, 1992) that have no correspondence in the other indices. The Silverman auroral minimum is therefore an interesting anomaly that deserves careful attention. Also, we should emphasize that the auroral series represented in Figure 7.20 has been re-scaled from a logarithmic scale due to the great variability in the number of observed aurorae. We should be cautious in the use of aurorae for the reconstruction of the activity because an aurora, in spite of its clear solar origin, is the product of a wide variety of heliospheric and magnetospheric factors.

Our main conclusion with respect to long-term solar activity reconstruction is that aurorae occurrences or naked-eye sunspot observations are qualitative indicators of solar activity that can be quantitatively interpreted with caution. We would stress that the cosmogenic isotope records provide a better basis for quantitative estimation of past solar activity. Moreover, it is evident that while the sunspot number is the most suitable index, it is only available for the last four centuries.

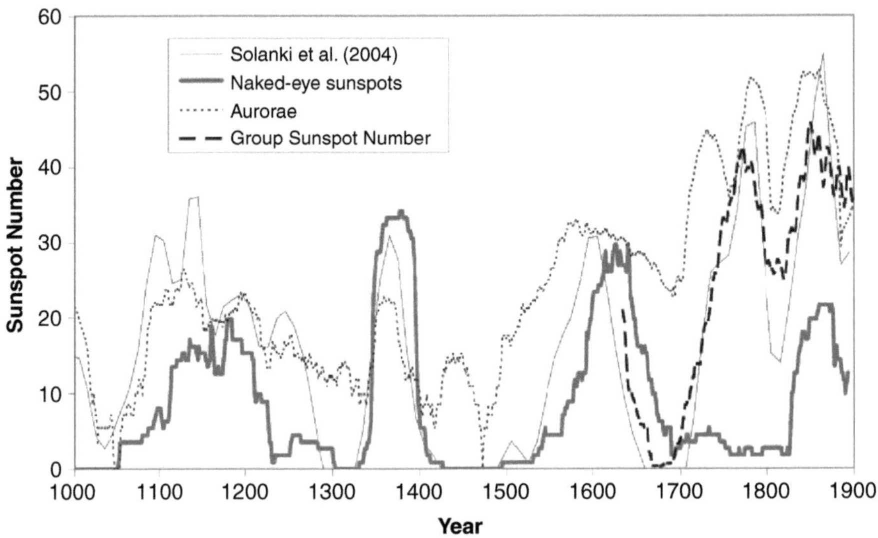

Fig. 7.20. Solar activity reconstruction during the period AD 1000–1900 from different sources and methods.

We should also emphasize here that the Sun has been very active during the last few decades. Probably this has been the episode of greatest solar activity that has occurred during the history of civilization (Solanki et al., 2004). Many instruments, on Earth and in space, have been witnesses of this major episode of solar activity. However, solar activity has had several episodes of minima (such as the Maunder minimum) during the last centuries. Certainly, a greater solar activity would have generated greater interest among astronomers, and there would have been more documents and studies on the Sun in historical times.

Historical short-term solar activity is a branch of solar–terrestrial studies that has only begun to be developed during recent years. The detection of 155-day periodicity in auroral occurrences using historical auroral records (Silverman, 1990) or recurrent geomagnetic storms around AD 1128 using combined historical records of aurorae and sunspots (Willis and Stephenson, 2001) are examples of the great interest of this type of study. Moreover, the increasing importance of space weather for our technological systems has led to growing interest in the major historical solar-geomagnetic storms, such as the Carrington storm (Clark, 2007; Cliver, 2006).

The Sun controls our life and all the processes that occur in our Solar System. Its study is of great importance for mankind. We would like to finish by emphasizing that any information on the state of the Sun in past times is interesting because the Sun is not a laboratory experiment that can be controlled. This also implies the importance of the appropriate storage of the vast amount of information that solar scientists have been obtaining in the last few decades. This will be historical material (of course of quite another type to the material reviewed here) for future generations.

References

Almer Arnau, T. *Estadística de manchas solares del año 1926.* Publicaciones del Observatorio Astronómico, Cuaderno número 6, Valencia, 1930.

Almer Arnau, T. *Estadística de manchas solares del año 1927.* Publicaciones del Observatorio Astronómico, Cuaderno número 7, Valencia, 1931.

Altas, L., Duzgelen, A., and Barlas, O. The determination of the reduction factor for sunspot observations and its variation with meteorological parameters. *Solar Physics*, 178:267–270, 1998.

Alzate, J. A. *Gaceta de literatura de México.* Oficina del Hospital de San Pedro, Puebla, México, 1831. 4 vols.

Argoli, A. *Pandosion sphaericum in quo singula in elementaribus regionibus, atque aethera, matematicé pertractantur.* Typis Pauli Frambotti, Patavii, 1653.

Aureliano, R., Buriano, A., and López, S. *Índice de las Gacetas de Literatura de México de José Antonio Alzate y Ramírez.* Instituto Mora, México, 1996.

Balmaceda, L., Solanki, S. K., and Krivova, N. A cross-calibrated sunspot areas time series since 1874 . *Memorie della Societa Astronomica Italiana*, 76: 929–932, 2005.

Baranyi, T., Gyori, L., Ludmány, A., and Coffey, H. E. Comparison of sunspot area data bases. *Monthly Notices Royal Astronomical Society*, 323:223–230, 2001.

Boer, J. Z. and Sanders, D. T. *Volcanoes in Human History.* Princeton, Oxford, Princeton University Press, 2002.

Carbonell, M., Terradas, J., Oliver, R., and Ballester, J. L. The statistical significance of the North-South asymmetry of solar activity revisited. *Astronomy and Astrophysics*, 476:951–957, 2007.

Clark, S. *The Sun Kings. The Unexpected Tragedy of Richard Carrington and the Tale of How Modern Astronomy Began.* Princeton and Oxford, Princeton University Press, 2007.

Clette, F., Berghmans, D., Vanlommel, P., van der Linden, R. A. M., Koeckelenbergh, A., and Wauters, L. From the Wolf number to the International Sunspot Index: 25 years of SIDC. *Advances in Space Research*, 40:919–928, 2007.

Cliver, E. W. The 1859 space weather event: Then and now. *Advances in Space Research*, 38:119–129, 2006.

de la Rue, W., Stewart, B., and Loewy, B. Researches on solar physics. Heliographical positions and areas of sun-spots observed with the Kew Photoheliograph during the years 1862 and 1863. *Philosophical Transactions*, 159:1–110, 1869.

de la Rue, W., Stewart, B., and Loewy, B. Researches on solar physics. No. II. The positions and areas of the spots observed at Kew during the years 1864, 1865, 1866 also the spotted area of the Sun's visible disk from the commencement of 1832 up to May 1868. *Philosophical Transactions*, 160:389–496, 1870.

Faria, H. H., Echer, E., Rigozo, N. R., Vieira, L. E. A., Nordemann, D. J. R., and Prestes, A. A comparison of the spectral characteristics of the Wolf Sunspot Number (R_Z) and Group Sunspot Number (R_G). *Solar Physics*, 223:305–318, 2004.

Floyd, L., McMullin, D., and Herring, L. Intercomparison of SEM irradiances and solar indices. In A. Wilson (ed.), *From Solar Min to Max: Half a Solar Cycle with SOHO*, volume 508 of *ESA Special Publication*, pp. 197–200, 2002.

Fröhlich, C. Observations of irradiance variations. *Space Science Reviews*, 94:15–24, 2000.

Goel, A., and Choudhuri, A. R. The hemispheric asymmetry of solar activity during the last century and the solar dynamo, *Research in Astronomy and Astrophysics* 9:115–126, 2009.

Gonzalez, A. D. (ed.). *El primer observatorio de Montevideo.* Instituto Histórico y Geográfico del Uruguay, Montevideo, Uruguay, 1955.

Hathaway, D. H., Wilson, R. M., and Reichmann, E. J., Group sunspot numbers: Sunspot cycle characteristics. *Solar Physics*, 211:357–370, 2002.

Heilbron, J. L. *The Sun in the Church.* Harvard University Press, 1999.

Hell, M. *Observationes Astronomicae ab Anno 1717 ad Annum 1752 Pekini Sinarum Factae.* Vindobonae, 1768. 2 vols.

Horroccii, J. *Opera Posthuma.* London, 1673. 2 vols.

Hossfield, C. H. A history of the Zurich and American relative sunspot number indices. *Journal of the American Association of Variable Star Observers*, 31: 48–53, 2002.

Hoyt, D. V. and Schatten, K. H. A new look at Wolf sunspot numbers in the late 1700's. *Solar Physics*, 138:387–397, 1992a.

Hoyt, D. V. and Schatten, K. H. New information on solar activity, 1779–1818, from Sir William Herschel's unpublished notebooks. *Astrophysical Journal*, 384: 361–384, 1992b.

Hoyt, D. V. and Schatten, K. H. Sir William Herschel's notebooks – Abstracts of solar observations. *The Astrophysical Journal Supplement Series*, 78:301–340, 1992c.

Hoyt, D. V. and Schatten, K. H. A new interpretation of Christian Horrebow's sunspot observations from 1761 to 1777. *Solar Physics*, 160:387–392, 1995a.

Hoyt, D. V. and Schatten, K. H. A revised listing of the number of sunspot groups made by Pastorff, 1819 to 1833. *Solar Physics*, 160:393–399, 1995b.

Hoyt, D. V. and Schatten, K. H. Observations of sunspots by Flamsteed during the Maunder Minimum. *Solar Physics*, 160:379–385, 1995c.

Hoyt, D. V. and Schatten, K. H. Overlooked sunspot observations by Hevelius in the early Maunder Minimum, 1653 1684. *Solar Physics*, 160:371–378, 1995d.

Hoyt, D. V. and Schatten, K. H. How well was the Sun observed during the Maunder Minimum? *Solar Physics*, 165:181–192, 1996.

Hoyt, D. V. and Schatten, K. H. *The Role of the Sun in Climate Change*. New York, Oxford University Press, 1997.

Hoyt, D. V. and Schatten, K. H. Group sunspot numbers: A new solar activity reconstruction. *Solar Physics*, 179:189–219, 1998a.

Hoyt, D. V. and Schatten, K. H. Group sunspot numbers: A new solar activity reconstruction. *Solar Physics*, 181:491–512, 1998b.

Hoyt, D. V., Schatten, K. H., and Nesmes-Ribes, E. The one hundredth year of Rudolf Wolf's death: Do we have the correct reconstruction of solar activity? *Geophysical Research Letters*, 21:2067–2070, 1994.

Huxham, J. A Letter from John Huxham, M.D.F.R.S. to C. Mortimer, M.D. Secr. R.S. concerning an Extraordinary Hernia Inguinalis; and Observation of the Passage of Mercury over the Sun, Oct. 31. 1738 [sic]. *Philosophical Transactions*, XLI:640–697, 1744.

Izenman, A. J., Wolf, J. R., and Wolfer, H. A. An historical note on the Zurich Sunspot Relative Numbers. *Journal of the Royal Statistical Society A*, 146: 311–318, 1983.

Izenman, A. J., Wolf, J. R., and the Zürich Sunspot Relative Numbers. *The Mathematical Intelligencer*, 7:27–33, 1985.

Kane, R. P. Some implications using the Group Sunspot Number reconstruction. *Solar Physics*, 205:383–401, 2002.

Keenan, P. C. The earliest national observatories in Latin-America. *Journal for the History of Astronomy*, 22:21–30, 1991.

Kepler, J. *Phaenomenon singulare seu Mercurius in Sole*. Lipsiae, 1609.

Koegler, I. and Pereira, A. Observatio Eclipsis Solis die 15 Julii 1730. *Philosophical Transactions*, XXXVII:179 181, 1733.

Koegler, I. and Pereira, A. Accedit Observatio Eclipsis Solis die 15 Julii 1730. *Memorias de Mathematica e Phisica da Academia R. das Sciencias de Lisboa*, 2:36–38, 1799.

Krivova, N. A., Solanki, S. K., and Beer, J. Was one sunspot cycle in the 18th century really lost? *Astronomy and Astrophysics*, 396:235–242, 2002.

Křivský, L. and Pejml, K. *Solar Activity, Aurorae and Climate in Central Europe in the last 1000 Years*. Publications of the Astronomical Institute of the Czechoslovak Academy of Sciences Publication No. 75, 1988.

Lalande, J. F. *Astronomie.* Paris, 1771. 4 Vols.

Letfus, V. Daily relative sunspot numbers 1749-1848: reconstruction of missing observations. *Solar Physics*, 184:201–211, 1999.

Manfredi, E. *De Gnomone Meridiano Bononiensi ad Divi Petronii.* Laeli a Vulpa, Bononiae, 1736.

Mc Kinnon, J. A. *Sunspot Numbers 1610-1985.* Report UAG-95, WDC-A, Boulder, Co, 1986.

Moreno-Corral, M. A. (ed.). *Historia de la astronomía en México.* SEP–FCE, México, 1986.

Nagovitsyn, Y. A. To the description of long-term variations in the solar magnetic flux: The sunspot area index. *Astronomy Letters*, 31:557–562, 2005.

Nagovitsyn, Y. A., Ivanov, V. G., Miletsky, E. V., and Volobuev, D. M. ESAI database and some properties of solar activity in the past. *Solar Physics*, 224:103–112, 2004.

Newton, H. W. and Milsom, A. S. Note on the observed differences in spottedness of the Sun's northern and southern hemispheres. *Monthly Notices Royal Astronomical Society*, 115:398–404, 1955.

Peters, C. H. F. *Heliographic Positions of Sun Spots Observed at Hamilton College from 1860 to 1870.* Carnegie Institution of Washington, Washington, DC, 1907. Edited by E. B. Frost.

Ribes, J. C. and Nesme-Ribes, E. The solar sunspot cycle in the Maunder minimum AD1645 to AD1715. *Astronomy and Astrophysics*, 276:549–563, 1993.

Saladino, A. *Dos Científicos de la Ilustración Hispanoamericana: J. A. Alzate y F. J. de Caldas.* Universidad Nacional Autónoma de México, México, 1990.

Schaefer, B. E. Automaticc Inflation in the AAVSO Sunspot Number. *Journal of the American Association of Variable Star Observers (JAAVSO)*, 26:40–46, 1997a.

Schaefer, B. E. Improvements for the AAVSO Sunspot Number. *Journal of the American Association of Variable Star Observers (JAAVSO)*, 26:47–49, 1997b.

Schove, J. D. *Sunspot Cycles.* Stroudsburg, Pennsylvania, Hutchinson Ross Publishing Company, 1983.

Schwabe, H. Ueber die Flecken der Sonne. *Astronomische Nachrichten*, 15:223–248, 1838.

Schwabe, H. Sonnen-Boebachtungen in Jahre 1843. *Astronomische Nachrichten*, 21:233–236, 1844.

Serres, M. *Éléments Pour une Histoire des Sciences.* Bordas, Paris, 1989.

Shapley, A. H. Reduction of sunspot-number observations. *Publications of the Astronomical Society of the Pacific*, 61:13–21, 1949.

Sigismondi, C. and Fraschetti, F. Measurements of the solar diameter in Kepler's time. *The Observatory*, 121:380–385, 2001.

Silverman, S. M. The 155-day solar period in the sixteenth century and later. *Nature*, 347:365–367, 1990.

Silverman, S. M. Secular variation of the aurora for the past 500 years. *Reviews of Geophysics*, 30:333–351, 1992.

Snow, M., McClintock, W. E., Woods, T. N., White, O. R., Harder, J. W., and Rottman, G. The Mg II Index from SORCE. *Solar Physics*, 230:325–344, 2005.

Solanki, S. K., Usoskin, I. G., Kromer, B., Schüssler, M., and Beer, J. Unusual activity of the Sun during recent decades compared to the previous 11,000 years. *Nature*, 431:1084–1087, 2004.

Sonett, C. P. The great solar anomaly ca. 1780–1800 - an error in compiling the record. *Journal of Geophysical Research*, 88:3225–3228, 1983.

Souciet, E. (ed.). *Observations mathematiques, astronomiques, geographiques, chronologiques et physiques, tires des anciens livres chinois*. Chez Rollin, 3 vols., Paris, 1729.

Usoskin, I. G. A history of solar activity over Millennia. *Living Reviews in Solar Physics*, 5(3), 2008.

Usoskin, I. G. and Kovaltsov, G. A. Long-term solar activity: Direct and indirect study. *Solar Physics*, 224:37–47, 2004.

Usoskin, I. G., Mursula, K., and Kovaltsov, G. A. Was one sunspot cycle lost in late XVIII century? *Astronomy and Astrophysics*, 370:L31–L34, 2001.

Usoskin, I. G., Mursula, K., and Kovaltsov, G. A. Lost sunspot cycle in the beginning of Dalton minimum: New evidence and consequences. *Geophysical Research Letters*, 29(24):2183, 2002.

Usoskin, I. G., Mursula, K., and Kovaltsov, G. A. Reconstruction of monthly and yearly group sunspot numbers from sparse daily observations. *Solar Physics*, 218:295–305, 2003b.

Usoskin, I. G., Mursula, K., and Kovaltsov, G. A. The lost sunspot cycle: Reanalysis of sunspot statistics. *Astronomy and Astrophysics*, 403:743–748, 2003c.

Usoskin, I. G., Solanki, S. K., Schüssler, M., Mursula, K., and Alanko, K. Millennium-scale sunspot number reconstruction: Evidence for an unusually active Sun since the 1940s. *Physical Review Letters*, 91(21):211101, 2003a.

Vaquero, J. M. New data for the reconstruction of solar activity: The case of Andrea Argoli. *The Observatory*, 123:150–151, 2003.

Vaquero, J. M. On the solar activity during the year 1784. *Solar Physics*, 219: 379–384, 2004.

Vaquero, J. M. Historical sunspot observations: A review. *Advances in Space Research*, 40:929–941, 2007.

Vaquero, J. M., Gallego, M. C., and Sánchez-Bajo, F. Reconstruction of a monthly homogeneous sunspot area series since 1832. *Solar Physics*, 221:179–189, 2004.

Vaquero, J. M., Gallego, M. C., and Sanchez-Bajo, F. Improving sunspot records: the observations by M. Hell revisited. *The Observatory*, 127:221–224, 2007b.

Vaquero, J. M., Gallego, M. C., and Trigo, R. M. Sunspot numbers during 1736 1739 revisited. *Advances in Space Research*, 40:1895–1903, 2007c.

Vaquero, J. M. and Moreno-Corral, M. A. Historical sunspot observations from Mexico. *Geofsica Internacional*, 47:189–192, 2008.

Vaquero, J. M., Sánchez-bajo, F., and Gallego, M. C., On the reliability of the de la rue Sunspot Area Measurements. *Solar Physics*, 209:311–319, 2002.

Vaquero, J. M. and Trigo, R. M. Identification of possible intense historical solar storms during the years 1781 1788 inferred from aurorae and geomagnetic observations in Rio De Janeiro. *Solar Physics*, 235:419–432, 2006.

Vaquero, J. M., Trigo, R. M., and Gallego, M. C. A "lost" sunspot observation in 1785. *Astronomische Nachrichten*, 326:112–114, 2005.

Vaquero, J. M., Trigo, R. M., Gallego, M. C., and Moreno-Corral, M. A. Two early sunspots observers: Teodoro de Almeida and José Antonio Alzate. *Solar Physics*, 240:165–175, 2007a.

Viereck, R. A. and Puga, L. C. The NOAA Mg II core-to-wing solar index: Construction of a 20-year time series of chromospheric variability from multiple satellites. *Journal of Geophysical Research*, 104:9995–10006, 1999.

Waldmeier, M. *The sunspot Activity in the years 1610-1960.* Zurich Schulthess and Company AG, 1961.

Waldmeier, M. The Swiss Federal Observatory, Zürich. *Solar Physics*, 5:423–426, 1968.

Willis, D. M. and Stephenson, F. R. Solar and auroral evidence for an intense recurrent geomagnetic storm during December in AD 1128. *Annales Geophysicae*, 19:289–302, 2001.

Wilson, A. Observations of the solar spots. *Philosophical Transactions*, 64:1–30, 1774.

Wilson, R. M. A Comparison of Wolf's Reconstructed Record of Annual Sunspot Number with Schwabe's Observed Record of 'Clusters of Spots' for the Interval of 1826 1868. *Solar Physics*, 182:217–230, 1998.

Wolf, J. R. Neue Untersuchungen über die Periode der Sonnenflecken und ihre Bedeutung. *Mittheilungen der Naturforschenden Gessellschaft in Bern*, 255: 249–270, 1852.

Wolf, J. R. Mittheilungen über die Sonnenflecken. *Vierteljahrsschrift der Naturforschenden Gessellschaft in Zürich*, 1:151–161, 1856.

Wolf, J. R. Mittheilungen über die Sonnenflecken. *Vierteljahrsschrift der Naturforschenden Gessellschaft in Zürich*, 3:124–139, 1858.

Wolf, J. R. Mittheilungen über die Sonnenflecken. *Vierteljahrsschrift der Naturforschenden Gessellschaft in Zürich*, 6:157–198, 1861.

Zolotova, N. V. and Ponyavin, D. I. Was the unusual solar cycle at the end of the XVIII century a result of phase asynchronization? *Astronomy and Astrophysics*, 470:L17–L20, 2007.

Index

Recently Published in the ASSL series

Volume 361: *The Sun Recorded Through History*,
by J. M. Vaquero and M. Vázquez. 978-0-387-92790-9, April 2009

Volume 360: *ThePrinciplesofAstronomicalTelescopeDesign*,
by Jingquan Cheng. 978-0-387-88790-6, February 2009

Volume 354: *Sirius Matters*, by Noah Brosch 978-1-4020-8318-1, March 2008

Volume 353: *Hydromagnetic Waves in the Magnetosphere and the Ionosphere*,
by Leonid S. Alperovich, Evgeny N. Fedorov.
Hardbound 978-1-4020-6636-8

Volume 352: *Short-Period Binary Stars Observations, Analyses, and Results*,
edited by Eugene F. Milone, Denis A. Leahy, David W. Hobill.
Hardbound ISBN: 978-1-4020-6543-9, September 2007

Volume 351: *High Time Resolution Astrophysics*, edited by Don Phelan, Oliver Ryan,
Andrew Shearer. Hardbound ISBN: 978-1-4020-6517-0, September 2007

Volume 350: *Hipparcos, the New Reduction of the Raw Data*, by Floor van Leeuwen.
Hardbound ISBN: 978-1-4020-6341-1

Volume 349: *Lasers, Clocks and Drag-Free Control, Exploration of Relativistic Gravity in Space*, edited by Dittus, Hansjörg; Lámmerzahl, Claus; Turyshev, Slava.
Hardbound ISBN: 978-3-540-34376-9

Volume 348: *The Paraboloidal Reflector Antenna in Radio Astronomy and Communication, Theory and Practice*, edited by Baars, Jacob W.M. Hardbound ISBN: 978-0-387-69733-8

Volume 347: *The Sun and Space Weather*, edited by Hanslmeier, Arnold.
Hardbound ISBN: 978-1-4020-5603-1

Volume 346: *Exploring the Secrets of the Aurora*, edited by Akasofu, Syun-Ichi.
Softcover ISBN: 978-0-387-45094-0

Volume 345: *Canonical Perturbation Theories, Degenerate Systems and Resonance*,
edited by Ferraz-Mello, Sylvio. Hardbound ISBN: 978-0-387-38900-4

Volume 344: *Space Weather: Research Toward Applications in Europe*,
edited by Jean Lilensten. Hardbound 1-4020-5445-9

Volume 343: *Organizations and Strategies in Astronomy: volume 7*,
edited by A. Heck. Hardbound 1-4020-5300-2

Volume 342: *The Astrophysics of Emission Line Stars*, edited by Tomokazu Kogure,
Kam-Ching Leung. Hardbound ISBN: 0-387-34500-0

Volume 341: *Plasma Astrophysics, Part II: Reconnection and Flares*,
edited by Boris V. Somov. Hardbound ISBN: 0-387-34948-0

Volume 340: *Plasma Astrophysics, Part I: Fundamentals and Practice*, by Boris V. Somov.
Hardbound ISBN 0-387-34916-9, September 2006

Volume 339: *Cosmic Ray Interactions, Propagation, and Acceleration in Space Plasmas*,
by Lev Dorman. Hardbound ISBN 1-4020-5100-X, August 2006

For other titles see www.springer.com/astronomy

MIX
Papier aus verantwortungsvollen Quellen
Paper from responsible sources
FSC® C105338

If you have any concerns about our products,
you can contact us on
ProductSafety@springernature.com

In case Publisher is established outside the EU,
the EU authorized representative is:
Springer Nature Customer Service Center GmbH
Europaplatz 3, 69115 Heidelberg, Germany

Printed by Libri Plureos GmbH
in Hamburg, Germany